INTELLIGENT INTEGRATED MEDIA
COMMUNICATION TECHNIQUES

Intelligent Integrated Media Communication Techniques

COST 254 & COST 276

Edited by

Jurij F. Tasič
University of Ljubljana, Slovenia

Mohamed Najim
University of Bordeaux I, France

and

Michael Ansorge
University of Neuchâtel, Switzerland

SPRINGER-SCIENCE+BUSINESS MEDIA, B.V.

A C.I.P. Catalogue record for this book is available from the Library of Congress.

ISBN 978-1-4757-7771-0   ISBN 978-0-306-48718-7 (eBook)
DOI 10.1007/978-0-306-48718-7

Printed on acid-free paper

Contents

List of Figures

List of Tables

Preface

We can understand the multimedia technology as a digital integration of texts, graphics, audio, animation, still images and motion video in a way that provides a high level of personal interaction and control to individual users. The evolution of multimedia is a convergence of the mentioned media types to a unique system. Before the term of multimedia became popular in the early 90's, the field of lively presentation and communication had been dedicated to a limited number of professionals. The first high-quality programmes ran on specialised, expensive machineries such as laser disc players. From 1994, multimedia materials have been distributed on CD-ROMs, available to anyone with a compatible personal computer. But, while distribution was becoming easier, producing of programmes still remained expensive and time-consuming. At the same time, the debut of the World Wide Web was there. Now, shortly ten years thereafter, anyone with access to the Web has the power to share his or her ideas with millions of others.

The today's digital technology is offering many advantages and is promising a lot of amazing services for the near future in the area of fixed and mobile personal terminals. The digital transmission and reception technology guarantees a better picture and sound quality even in bad transmission conditions. The digital technology will be providing us with exciting new features, such as shopping through mobile screens, requesting teleconferencing services, remote learning, watching movies of our choice, participating in interactive TV shows, and many other exclusive services.

Judging from the present state-of-the-art, such situation is expected to develop within the next few years, when the mobility supporting technology will be entirely based on the Internet. The new technologies will push the Internet-based services into the pockets of millions of users. Today we can already witness the beginnings of these developments, which will further influence the evolution of future networks, system architectures, backbones, applications and mobile accesses. The immediate challenge for the industry is to develop terminal architectures, networks and services able to support multiple mutimedia services in real time across communication networks and at a guaranteed end-to-end level of quality.

Such personal and user oriented communications will change the behaviour of the nowadays society. For the third generation of mobile phones it is expected that they will become an everyday accessory or a personal portable assistant for millions of users. We can consequently believe that in the next few years the development of the third generation of mobile phones will be shaped by the convergence of computing, communication and broadcasting technologies. While mobile communications are presently offering speech communication anywhere and at anytime, the modern mobile terminals and services are rapidly growing into new personal interactive mobile multimedia services and terminals. To put it short, the end user will have a retail outlet in his or her pocket, with a set of abilities, like ticket reservation, banking transactions, parking fee payments, personal health care and so on. Mobile multimedia services will simplify the implementation of virtual enterprises. The solution will be similar at home where the appliance-to-appliance, appliance-to-people and edutainment communication applications will grow in importance, vastly improving our security and efficiency.

As understood from its title, *Intelligent Integrated Media Communication Techniques*, the intention of this book is to discuss methods and techniques required for the third generation of mobile multimedia terminals and trends for the third generation services, such as have been addressed by the specific activities of COST 254 and COST 276 actions.

Origin and structure of the book

The book originates from the European scientific cooperation networks COST Action 276 *Information and knowledge management for integrated media communication*, and COST Action 254 *Intelligent processing and facilities for communication terminals* and provides an insight into selected research activities deployed within this cooperation framework.

The chapters of the book are thematically grouped according to the topics covered. Part I of the book is devoted to Hypermedia data management, dealing with multimedia information organization methods for classification, indexation, storage, and retrieval purposes, in particular for the construction and management of complex multi-modal databases. Part II is handling the robust data hiding methods, where it is noticed that connected techniques can be used for video data authentication purposes. The advent of wireless wideband data communication stimulates a renewed interest in video and still image coding, which is covered in Part III. Next, voice-driven interfaces, speech coding, and speech recognition are addressed in Part IV. Dedicated hardware interfaces relying on miniature image sensors for the design of embedded image/video processing components and devices are then presented in Part V, whereas the closing Part VI discusses the potentialities offered by distributed media technologies, in the light of the illustrative application example of a real-time virtual-reality sailing regatta webcasting.

Part I: Hypermedia Data Management

In the first chapter of this section the authors U. Burnik and M. Pogačnik present their ideas based on adaptive personalised services. In their paper **Content and Presentation Adaptation in Hypermedia Systems** they introduce adaptive strategies that can be applied to provide an optimal level of service quality. The idea is to understand the web content as a process evolving from a simple keyword based model into a machine semantic approach.

The personalisation procedure given in this chapter can be presented by individual filtering instead of a collaborative one. Sometimes the two can be combined in a common filtering process named social filtering. The reason why personalisation as one of the major techniques is focused on in this chapter is because there are not many examples of such retrieval procedures of audio/video and images found in literature. The given examples are based on the domain of television broadcasting where both authors are very active (MyTV). In the proposed approach each user profile contains two types of information, i.e. domain and programme preferences. Based on the system descriptors, adaptive presentation and quality measurement approach, the authors introduce a model assuring balance between quality, retrieval and latency of the filtered content presentation. The approach they propose for the personalised content presentation is of a modular type. It enables certain components of the system to adapt themselves to restrictions imposed by the current status of environment resources, which is not the case with the popular scalable approach.

The second chapter entitled **Annotation, Storage, Retrieval and Analysis of Digital Video** is written by H. Li, Y. Zhao, J.-L. Sancho-Gómez and S. C. Ahalt. Its focus is on the bunch of new multimedia content producers, individual owners of a digital camcorder and professional producers. They also discuss the future impact of the foreseeable multimedia technologies. Currently, all professional producers and some amateurs are using relatively mature methods, such as Sequential Querry Language, enabling access to traditional data. Therefore, the need for efficient and flexible acces methodologies is becoming increasingly important. Users would like to see a video system that provides for easy browsing, retrieving and analysing. In the light of these problems, the authors propose a typical video system that consists of three major blocks, i.e. *Information Extraction Block* responsible for extracting information from the raw video data, *Information Organisation and Storage Block* responsible for organisation and storage of the raw video data and video meta-data, and *Information Processing Block* responsible for video access, retrieval and analysis of the video meta-data. The authors present functions of each of the three blocks. Considering the block organisation, they try to find answers to a number of issues important for effective extraction of video meta-data and organisation of video features. They propose, according to the Multimedia Content Description Interface (MPEG 7), a systematic approach to structurising the various components of the video analysis, which they demonstrate on examples of the proposed components.

As explained in the previous chapter, the raw video data should first undergo a video segmentation process. This process is described in the chapter entitled **Segmentation Techniques for Video Sequences in the Domain of MPEG-Compressed Data** written by I. Koprinska and S. Carrato. The authors provide a comprehensive taxonomy and critically survey a great variety of techniques for temporal video segmentation operating on a MPEG compressed video stream. As most of them are in a close analogy with the approaches operating on uncompressed video, this chapter starts with a brief overview of video segmentation in uncompresed domain. The next to follow are a review of the relevant parts of MPEG compression standard and a survey of segmentation approaches on compressed video. This chapter addresses also tailoring of both the compressed and uncompressed methods.

Part II: Robust Video Data Protection and Watermarking

The authors of the first chapter of this section bearing the title **Digital Watermarking for the Copyright Protection of Compressed Video** are D. Simitopolous, S. A. Tsaftaris, N. V. Boulgouris and M. G. Strintzis. They analyse new techniques for digital content protection, i.e. watermarking of the MPEG-1 and MPEG-2 compressed audio-video multiplexed streams. Each multiplexed stream contains minimally two elementary streams, namely audio and video. The authors developed a new watermarking scheme that operates with a multiplexed stream as input. The data to be watermarked are extracted from the stream as an intraframe (I frame). There the quantised AC coefficients of each discrete cosine transformed block of the luminance component are watermarked and then placed back into the stream. In this chapter the basic requirements for video watermarking are examined and the process of watermark embedding and the verification procedure are presented.

The chapter **Robust Watermarking of Video for Copyright Protection,** written by M. Barni, F. Bartolini, R. Caldelli, V. Cappellini, A. De Rosa and A. Piva, is an extension of watermarking processes in video data protection in the field of MPEG-4 objects. The development of network services is conditioned by the achievement of efficient methods to safeguard data owners against nonauthorised copying and redistribution of the multimedia content. In the framework of video watermarking for the intellectual property management and protection (IPMP) of multimedia data, a specific watermarking procedure is discussed in depth, carefully considering also its robustness aspects. Three different approaches for code embedding are described in details; two of them are foreseen for the non-coded video domain and the third for the coded video domain. Special attention is paid to watermarking MPEG-4 video streams. The first and the third approach deal with raw video and allow for a direct interaction with video frames for watermarking purposes. The focus of the third approach is also on video objects viewed from the perspective of the MPEG-4 specifications.

Part III: Video and Image Coding

As the transmission of multimedia materials nowadays runs over more or less unreliable networks, protection of transmitted data against possible channel failures has become an imperative.

The goal of this section is to review source and channel coding methods, providing the basis for the development of new approaches. The chapter **Error-Resilient Coding for Multimedia Communications,** written by N. V. Boulgouris, N. Thomos and M. G. Strintzis, describes methods for a reliable transmission of images and video over unreliable channels. The authors advertise three robust transmission methods based on error resilient coding, forward error correction coding and fragment retransmission using Automatic Repeat Request (ARQ) protocols. The general framework for coding and organisation of visual information given in subchapters deals with information streams and endowing errors resilient capabilities. Also presented are investigation results of image and video transmission over noisy channels using the forward correction method as well as the method for error removal from the corrupted stream. The resulting coders produce scalable bitstreams able to support the very good quality of the transmitted stream over different bandwidth channels. The proposed error resilient structures of the data stream allow the system to localize and discard the corrupted portion of the transmitted bitstream and to attain a superior reconstruction quality.

The chapter **Wavelet Centric Video Coding and Coding Artifact Concealment**, written by B. Marušič, P. Skočir and J. F. Tasič, is an introduction into the video coding technique inspired by multidimensional extensibility of the wavelet transform. A full video coding system including a post-processing stage for video enhancement is described and discussed. The video coding approach is built on a three dimensional motion compensated wavelet transform and context based coding of bitplanes. An approach based on *MC3DWT* and context based arithmetic coding is described. Experiments reveal the effectiveness of the proposed 3D transform based approach compared to hybrid paradigm based coders and the more advanced *3D SPIHT coder*. The authors also analyse processing methods for which they offer two low complexity artifact removal algorithms. The proposed coding algorithm in conjunction with the shown enhancement stages provides a viable alternative to hybrid video coding procedures at a higher bit rate.

The JPEG international industry standard for compression of continuous-tone still images is presented in the next chapter entitled **Bit-Rate Control for the JPEG Algorithm**. It is written by J. Bracamonte, M. Ansorge and F. Pellandini. The standard, which was proposed by the Joint Photographic Expert Group, is one of the most popular image compression standards. The chapter reports on a study in the characteristics of two algorithms of compression ratio control. The first algorithm is based on defining a single slope model improved by a piecewise linear model requiring less iteration to produce the same results. Also reported is a more generalised control method, showing effectiveness of the bit-rate control techniques. The chapter ends by showing that application of these algorithms does not affect the quality of decompressed images.

Part IV: Voice Interfaces and Processing

Speech communications with electronic devices and in the user's own language are one of the most natural and flexible communications. For many users spoken language interfaces are fast and flexible.

The first chapter addresses content-based retrieval of the voice information from multimedia databases. The focus of the second chapter is on voice man-machine interfacing. The title of this chapter written by A. Drygajlo is **Man-Machine Voice Enabled Interfaces**. A methodology is proposed for voice interfaces adapted to autonomous mobile tour-guide robots. The development of the methodology is described throughout all its phases from the beginning to end when a complete voice interfacing system is obtained.

The second chapter entitled **Speech Coding and Recognition in Noisy Environments for Communication Terminals** is written by the same author. It is an extension to the previous chapter in the direction of robust speech processing methods, paying due regard also to limitations of auditory perception. Two methods are presented. They enable a combination of some speech enhancement algorithms with speaker/speech recognition and speech coding procedures. The author offers a new solution to the problem of robust speaker verification and speech recognition in noisy environments in the light of the missing data theory. The proposed technique, based on merging the speech enhancement method with coding in the area of auditory modelling, extends the unreliable feature detection and compensation to application in a noisy environment.

Part V: Dedicated Hardware Interfaces

Electronic imaging has become very important for multimedia communication devices, such as voice systems for manipulation and data or transmission management. The deep sub-micron technologies enable implementation of very complex imaging systems also in portable *Intelligent Integrated Media Communication Systems*. Methods and techniques needed for the third generation of mobile multimedia terminals with integrated video, possible MPEG players, vision based identification and recognition systems, cost-effective solutions and low-power consumption terminal architectures are important for new services in the interactive multimedia mobile systems.

The hardware approach is highlighted in the chapter on **Low-Power Micro-Cameras for Mobile Multimedia Devices**, written by S. Tanner, M. Ansorge, F. Pellandini and N. Blanc. The authors present various concepts for the design of low-power micro cameras. They first describe the photo-detection principle and basic CMOS sensor architectures and then give an insight into the design of low-power analog to digital converters adapted for video applications. They also discuss a chip control aproach and pre-processing functionalities and describe two applications of micro-camera realisations together with their specifications, architectures and performances. In the years to come, the recent advances in the deep sub-micron technologies will enable the high resolution camera with on chip MPEG real time compression to consume only 100mW and to occupy less than 50 mm^2 of silicon. Technically, such camera will satisfy expectations of a vast range of applications imposed on mobile multimedia devices.

Part VI: Distributed Media

In the final section, an example of intelligent integrated media communication application is shown in the chapter **Retrospective of the Distributed**

Media Server Technique. The chapter is written by K. Skala, Z. Zelenika, I. Nikolić and T. Skala. It presents the early generation of Virtual Reality Webcasting and defines new media models. It discusses ideas of the Virtual Reality Modelling Language (VMRL) technology, which is mostly focused on integration of VRML with other technologies applied in Internet services. The example given in this chapter is a VRML of a university sailing regatta that is the first attempt of the authors of a real time VR web-cast of such an event.

Readership

The intended readership is large, ranging from specialized scientists and engineers in the field of intelligent media communication technologies to a more general scientific audience interested in collecting conceptual or practical information in the presented domain.

The goal of this book is to highlight methods for graduate content management and presentation. We hope that academics, students and researchers in industrial labs will find the book a useful aid for their work.

Acknowledgements

The editors wish to express their gratitude to all authoring teams for their important contribution to this book and to thank Dr. Urban Burnik and Mr. Tomaž Finkšt, both from the University of Ljubljana, Slovenia, for the efficient technical support they deployed all along the book preparation. Also, the appreciated guidance received from Mr. Mark de Jongh, Senior Publishing Editor at Kluwer Academic Publishers, and his team is thankfully acknowledged.

Special thanks are addressed to COST TIST / Brussels for the international scientific cooperation framework offered within the scope of COST Actions 254 and 276 from which this book has emerged.

JURIJ F. TASIČ

MOHAMED NAJIM

MICHAEL ANSORGE

I

HYPERMEDIA DATA MANAGEMENT

Chapter 1

CONTENT AND PRESENTATION ADAPTATION IN HYPERMEDIA SYSTEMS

Urban Burnik and Matevž Pogačnik
University of Ljubljana, Faculty of Electrical Engineering,
Tržaška 25, 1000 Ljubljana, Slovenia
{urban.burnik,matevz.pogacnik}@fe.uni-lj.si

Abstract   The present chapter is dedicated to personalised real–time multimedia communication services. The challenge of modern communication services is a combination of personalised information gathering and customised data retrieval technology. A concept of personalised communication service is presented, followed
by an overview of information search, retrieval and adaptive presentation technologies. Both generalised strategies and supporting technologies are presented
throughout the chapter.

Keywords: Personalized services, agent technologies, image compression, real-time imaging.

1.   Personalized Communication Services

New trends in telecommunications are revolutionary as they migrate from
traditional broadcasting services to active information explorations. The service
user becomes an active participant, freely selecting the range and presentation
form of information. The ubiquity of communication services means not only
generalised service availability, it points toward distributed information sources
as well.

The key issue in modern communications is therefore to establish and to
manage the knowledge structure in a heterogeneous system of distributed multimedia data. With a dominant role of graphics and video elements that provide
the information in a clear, evident form, careful resource exploitation is also
required. It is necessary to ensure a reliable and responsive real–time service.

The lack of a universal service approach equally satisfying each potential
service user is evident. An average person is not expected to control and to

·3

J.F. Tasič et al. (eds.), Intelligent Integrated Media Communication Techniques, 3-42.
© 2003 *Kluwer Academic Publishers.*

manage the entire process of information retrieval on his/her own. Not only such a process requires a certain level of expertise not common for a majority of casual users, with a growing quantity of accessible information manual seek and retrieval mechanisms are becoming unmanageable even for the experts. Each user is looking to have an individual guide who assists him/her in data exploration. The guidance is to be provided automatically and should combine current user requests with additional information describing his general preferences, location and accessibility.

The two main issues in personalised communication services are addressed by the present work, namely *which* is the information actually requested and *how* to present the (hopefully positive) seek results in the retrieval stage of a real-time service.

The software assistants that seamlessly provide adequate level of the personalised service are commonly referred to as software agents. They are most commonly used to assist in the search phase of the information gathering process. The same approach, however, may be used to prepare the most adequate instance of the data to be presented on the user terminal. The agent technology is to be backed up by additional information stored in user, service and equipment profiles. These special mechanisms seamlessly provide the required information and quality management.

We may speak of a virtual environment, from within an individual may seek, browse and display his or her own networked resources, shared data and public available materials. The heterogeneous range of equipment, all from personal computer workstations, handheld computers, video and television as well as mobile phones with related services converge to a common information management concept. The centre of a modern communication system is a user with his or her own personal needs and expectations, who wants to fulfil his or her expectations as easily as possible. Rather than reading user manuals, coping with equipment and service handling, he/she wants to use the service.

2.   Personalised Information Search

Nowadays, searching for information is cumbersome. The number of documents grows with a tremendous rate and so does the number of users. Some estimates suggest that there is a couple of billion Web pages available to over 500 million users[1], but Web is only one of the information sources. Others include on-line digital libraries (document, image, video and audio), virtual museums, research publications, news (Usenet), product and service catalogues and many more. With the ascent of peer-to-peer systems, user devices are to become

---

[1] According to Nua Internet Survey (Feb. 2002):
http://www.nua.com/surveys/how_many_online/world.html

sources of information, which will increase the number and heterogeneity of sources even further. It is already today, that searching for a particular information (document, image...) usually results in a vast number of hits, with a high amount of irrelevant ones. It is unlikely that 500 million users are so similar in their interests that one approach to information search fits all needs. The main problem is that there is too much information available, and that keywords, being the most common query mechanism, are rarely an appropriate means of locating the information in which a user is interested. Information retrieval can be more effective if individual users' interests and preferences are taken into account.

There are a number of personalisation approaches, which differ depending on the type and source of information. Personalisation can be focused on an individual user or can be based on recommendations of "similar" users. Some of the techniques require explicit feedback and some don't. In order to better understand their functioning and efficiency, we should first take a look at the most common information retrieval and indexing techniques.

As the content of the vast majority of information is unstructured[2], selection and classification of textual documents is therefore not a trivial task. Retrieval of textual documents is mostly based on keyword extraction. Mechanisms for content descriptions of Web documents are available in the form of the Dublin Core metadata structure [1], but they are not used as much as expected. Image and video items are even worse off. Research in this area is very intensive and is mostly based on the analysis of low-level features (colour histograms, textures, object shapes...), but from here to automatic content recognition is, admittedly, still a very long way. There are some cases of on-line image libraries with keyword descriptions. These are mostly generated through an (automatic) textual analysis of Web pages (Google[3], Yahoo[4], Ditto[5]...), which contain images and are therefore not entirely reliable.

In general, there are two[6] ways of getting the information users want. They can *search* for it or *browse*. In the Web, search engines index documents and allow users to enter keywords to retrieve documents. On the other hand, browsing is normally done by clicking through a hierarchy of subjects until the area of interest has been found. Search can be based on broader or narrower terms. Broad search terms retrieve a lot of useful information items together with a significant number of irrelevant ones, while narrow search terms retrieve fewer documents and may miss some relevant ones.

---

[2]HTML tags cannot be taken for content structure.

[3]http://www.google.com

[4]http://www.yahoo.com

[5]http://www.ditto.com

[6]There is a third alternative, namely *push* of the documents. In its essence it is a kind of automatic search, based on pre-stored user profiles and is usually performed by a recommendation service.

Two evaluation measures are used to describe retrieval effectiveness, *recall* and *precision*:

$$recall = \frac{NRR}{NRC} \qquad (1.1)$$

$$precision = \frac{NRR}{TNR} \qquad (1.2)$$

where NRR is Number of Relevant items Retrieved i.e. number of retrieved items, recognised as relevant by the user, NRC is Number of all Relevant items existing in the Collection and TNR is Total Number of Retrieved items (relevant and irelevant ones). Ideally, we tend to achieve both high recall and precision. In reality we must strike a compromise. Broader search (and indexing) terms result in higher recall at the cost of precision and the other way around. Therefore an information retrieval system's effectiveness is measured by the precision parameter at various recall levels [2].

How users search for information, depends on the type of features (metadata) the information items provide. Indexing of information (extraction of metadata) can be performed either automatically or manually. There are a number of techniques for automatic feature extraction for text and image, not (yet) so many for video. Automatic indexing, apart from being much faster, does not require human effort and is therefore very beneficial, but despite many years of study it remains at a primitive level of development. Manual indexing on the other hand is more precise, but it would be impossible to manually index all the available information.

In the following sections we'll take a look at different approaches to indexing and retrieval of different content types (texts, images, video), which should represent a basis for better understanding of personalisation techniques.

2.1   Content Types and Indexing Strategies

2.1.1   Textual documents.   There are a number of ways to index (assign context terms to) textual documents.

Automatic methods can be based either on single term or on multiple terms indexing [2, 3, 4]. With single term indexing the documents are most often represented with a term set, named word vector. The simplest model is the *Boolean vector space model*, where the entries of the vector are either 1 or 0, depending on whether the keyword associated with this entry occurs in the document or not. Queries are Boolean expressions, using logical operators such as AND, OR and NOT. The problem with this model is that no ranking exists since all the retrieved documents have the same rank (e.g. 1). With this model it may as well happen that a query consisting of 10 terms, linked by logical AND, does not retrieve a document that has 9 of these terms. Obviously, this model has major drawbacks since it does not take into account word frequency and is therefore too coarse. The rest of automatic single term frequency methods can be grouped into statistical methods, information theoretic methods, probabilistic methods and hybrid methods [2].

Statistical methods assume that the importance of a term is directly proportional to the frequency, with which it appears in a document: $tf_{ij} = \frac{T_j}{D_i}$, where tf_{ij} denotes the term frequency of term T_j in document D_i. This fulfils one indexing aim, namely, recall. However, terms, which appear in a smaller number of documents, are better discriminators between documents in a collection and should therefore be assigned higher weight. Let df_j denote the document frequency of the term T_j in a collection of N documents. Then, by combining the term frequency and inverse document frequency components we obtain the weight formula for the term T_j in document D_i: $w_{ij} = tf_{ij} \cdot \log(\frac{N}{df_j})$. This measure is well known as the TF-IDF (Term Frequency, Inverse Document Frequency) measure [5]. A related measure is the term relevance measure [2, 4], which gives preference to terms that help differentiation between classes of documents e.g. increase the average distance between them. A combined weighting scheme is also used (TF-IDF & term relevance). Retrieval is based on the premise that documents in collections are represented by n-dimensional vectors in a space, spanned by n-normalised term vectors. User queries are similarly represented by vectors and document relevance to the query is calculated as a scalar product of the query and document vectors. The model is simple and allows for the relevance feedback, however the expressiveness of the query specification in the Boolean model is lost [2, 4].

Information theoretic methods [2] are based on the fact that the least predictable terms carry the greatest information value, because they occur with the lowest probability. They favour terms that are concentrated in fewer documents and are as such similar to inverse document frequency methods.

Probabilistic methods take into consideration the difference in the distributional behaviour of words over all documents in a collection. Assigning weights to terms requires training sets of documents, obtained by users through the relevance feedback. Users are asked to evaluate the retrieved documents, as either relevant or irrelevant. Then the training set is used to compute term weights, by estimating conditional probabilities that a term occurs, given that a document is relevant (or irrelevant). Two conditional probabilities are estimated for each term: the probability that the term appears in the document, assuming that the document is relevant, and the probability that the term appears in an irrelevant document. The term weights are derived using the Bayes' theorem. Retrieval of documents is again based on the probabilities of relevance and nonrelevance of a document to a user query. The model also uses cost parameters to represent the loss associated with the retrieval of an irrelevant document and nonretrieval of a relevant document [2]. Among probabilistic methods are the Bayes' classifier, neural network approach, decision trees and some others [6].

Hybrid models are also being used, such as the *extended Boolean model* [2]. This model uses the generalised scalar product between the corresponding vectors in a document space. The generalisation uses the L_p norm, defined for an n-dimensional vector. The interpretation of a query can be altered, when computing the query, e.g. document similarity, and can consequently vary from the vector space model to the strict Boolean and fuzzy model.

Complementarily to the single term indexing methods, multiple term indexing methods are considered as well. They strive to generate descriptions of documents in forms of term phrases and/or thesaurus groups. It is reasonable to believe that these approaches can be more efficient than single term indexing, since single term meanings, when taken out of the context, can often be ambiguous. According to [2] they can similarly be grouped into statistical, probabilistic and linguistic methods. They include Latent Semantic Indexing (LSI) and weighted n-grams [6].

Opposed to the automatic indexing methods are manual methods. As already mentioned, they are extremely time consuming, when compared to automatic methods. On the other hand, manually generated indexes are much more exact, and some conceptual terms can be used to describe content even though they do not appear in the content. The first attempts of manual content descriptions date back to the middle ages, when the first library classification schemes were made. In the era of the Web, the DublinCore initiative [1] has made one of the first efforts towards organizing descriptions of Web documents. The idea was to enable meta-tags in the HTML documents to provide content and other information about the document. The proposed metadata model is simple and therefore very general. It provides basic information about the document content (title, subject, description), intellectual property (creator, publisher, rights, etc.) and instantiation (date, format, language, etc.).

The idea of describing the Web content has continued to grow and is evolving from a simple keyword based model into a machine semantical approach. The current situation in the Web is such that data is generally "hard coded" in HTML files. The concept terms used are semantically ambiguous, so there is no way of telling which of the possible meanings is the right one. For example, a search query "north pole" may provide thousands of result pages with a content about the famous geographical location, a company with the same name or even a pub. In order to resolve the increasing "messiness" of the Web, an initiative called the Semantic web has been started [7, 8]. To put it simply, the idea is to describe specific term meanings and their relationships with the help of schemata and ontologies. The descriptions are made using RDF (Resource Description Format). With the development and use of inference logic, the Semantic web may become a very powerful tool for information processing and retrieval. The idea of the authors [7, 8] is for the Semantic web to become a huge machine-processable and accessible database. However, Semantic Web technologies are still very much in their infancies, and although the future of the project in general appears to be bright, there seems to be little consensus about the likely direction and characteristics of the early Semantic Web [7].

2.1.2   Image search.   Digital images (and video) are becoming an integral part of human communications. The ease, with which nowadays digital imagery is created and captured, has resulted in widespread on-line collections of images, which are growing larger and more common. Therefore, tools are needed to efficiently manage, organize, and navigate through image repositories. Images (including video) differ significantly from conventional data items (textual documents), because the extractable features are not high-level concepts like word lists, which enable relatively direct matching of search queries against data item representations. At first glance, content-based querying may appear deceptively simple, because we humans seem to be so good at it. Perceptual organization - the process of grouping image features into meaningful objects and attaching semantic descriptions to scenes through model matching - is an unsolved problem in image understanding [9]. Humans are much better than computers at attaching semantic meaning to images. What computers can do is measuring and processing of image properties.

Basic image features are the so called low-level features, which contain information about colours, textures, shapes, etc. We should stress that, at this moment, no metadata standard seems to be widely used for image descriptions. An effort in this direction has been made by the MPEG working group [10]. Namely, the MPEG-7 standard aims to describe all types of the multimedia content (audio and speech, moving video, still pictures, graphics and 3D models) including information on how objects are combined in scenes. The standard has powerful mechanisms for content description, but is on the other hand very complex, so a widespread use is still questionable.

The image indexing methods can basically be grouped in three major approaches. The first one is characterized by simple keyword based descriptions, manually or automatically assigned to images. Today, text-based search techniques are the most direct, accurate, and efficient methods for finding images (and video) [11]. Text annotations can be obtained either by manual effort, embedded text, or from hyperlinked documents containing images (Google, Yahoo Ditto). In these systems, keyword and full text searching may be enhanced by language processing techniques using thesauri for example, that improve in this way categorization and retrieval of images. The problem is that, for truly accurate textual descriptions, manual indexing of images is usually required; other (semi)automatic methods provide more or less satisfactory, but not entirely accurate results.

Very often, textual indexing is not sufficient. Imagine for example a fashion designer looking for a particular pattern or an archaeologist looking for a texture or shape, similar to the one that has just been found. In such cases, visual low-level features of the images (and video) are the best means for generation of suitable queries.

There are a number of approaches to low-level indexing of images. Two of the most common approaches, according to [12], are histogram and texture based representations [12, 9, 13, 14]. Some authors [13] have even combined texture and colour features into low-level objects (blobworld objects), thus providing information about texture, colour and location of low-level image regions.

On the part of low-level query generation, there has been a substantial progress in developing powerful tools, which allow users to specify image queries by giving examples, drawing sketches, selecting visual features (e.g., colour and texture), and arranging spatial structure of features. The QBIC system [9], for example, allows a user to query an image collection using features of image content colours, textures, shapes, locations, and layout of images and image objects. Such queries use visual properties of images, so the user can match colours, textures and their positions without describing them in words. An example query would be: "Find images that have 25 % of red colour and a blue textured object".

To break the barrier of decoding semantic content in images, user-interaction and domain knowledge are needed. Some of the research [15, 11, 16, 17] has focused on systems, where low-level image queries are combined with text and keyword predicates. In this way, we get powerful retrieval methods for image and multimedia databases. In [16] the system learns from the user input as to how the low-level visual features are to be used in the matching of images at the semantic level. Learning and other techniques in artificial intelligence provide a great potential for these systems. One example of these approaches is the VisualHarness system [15], an extension of the InfoHarness [15] system. Users can arbitrarily combine keyword, attribute and low-level feature based search, using relative weights, directly adjustable through the user interface. Iterative refinements of search results are also possible.

Similar systems are proposed by [16] and [17]. The first one combines semantic keyword search with low-level visual features. In the proposed system, keyword image annotations are organized as a semantic network. Keywords are not only assigned to images, a certain weight is also associated with them. Weighted keywords are first provided manually, but are gradually adjusted through the search process. Weight adjustments are based on the image relevance feedback provided by the user. Authors claim that such semiautomatic method of image annotations is efficient. Furthermore, hierarchical two-level image annotations are used in order to avoid the influence of synonymy and polysemy[7], in natural language. Twenty-four first level categories are defined and each image is certainly classified into one (or more) of them (like animal, business, sports, etc.). At the second level, more keywords are used to present the image content (horse, racing). Visual low-level features used in the proposed system are colour (hue, saturation, histogram, etc.) and texture (tamura, MRSAR, etc.) based. In [17] a database of visual representation of semantic concepts is used. Concepts are assigned different visual features such as colours (histogram & coherence), textures (wavelet & tamura), specific images, audio and video. Since different media representations can be more or less relevant, relevance factors are assigned to concept representations. Search queries can either be visual or textual. Since there are no textual image annotations in the database, the textual queries are translated into visual data and the search in the database is performed based on visual features. In addition, WordNet [18] taxonomy was used to generate synonyms/antonyms[8], hyponyms/hypernyms and meronyms/holonyms[9], of each concept and store them in the MediaNet knowledge base. Authors claim to have improved the retrieval effectiveness, compared to the standard content-based retrieval systems.

2.1.3   Video search.   The amount of the available audio-video material has increased significantly in the past decades. Database systems hold vast quantities of audio-visual data, which comprise TV programmes (movies, news, etc.), surveillance monitoring, geo-spatial footages, home recordings, etc. In order to be able to easily retrieve these data, they have to be adequately indexed and stored in databases. Users may wish to browse through databases or perform queries based on the content of a video.

Since video is a sequence of still images, which usually comes together with audio recording, it is not surprising that indexing and retrieval of video is in many ways similar to indexing and retrieval of images. Speaking very generally, video indexing is, again, based on low-level and high level features. Low-level features include shape, colour, texture, position, direction and speed of motion [19, 20, 21, 10]. Objects can be indexed using these feature values and retrieved based on the similarity to the feature values of other objects. High-level features most often used are keyword annotation, title, speech transcript, headline, synopsis, genre classification etc. [22, 19, 20, 10, 23]. While low-level features are mostly obtained automatically, high-level features are usually a result of manual annotations. Retrieval is most often based on the combination of low- and high-level features.

---

[7] Single word having multiple meanings-opposite of synonymy

[8] Antonyms are words with the opposite meaning (ex. big/small)

[9] Meronymy and holonymy are a part/whole relations (ex. *wings* and *beak* are meronyms of the *bird*, and *bird* is the holonym of the *wings* and the *beak*)

In [20] a system is presented, where the video database is decomposed in hierarchically organised semantic entities: video clusters, videos, chapters, scenes, shots, etc. Every entity is characterised by low-level features. Sometimes more features are available, like extracted text or camera motion information. Shots or sequences are decomposed into visual and sound objects, which are, again, annotated by low- or high-level features. Every entity can also be described with an annotation about sky presence or buildings, for example. As a consequence, queries are based on low- and high-level features, including temporal predicates and interestingly, mood dependent profiles. Similarly, a video data model for content based search is presented in [19]. The model named VIDAM uses two distinct classes of objects to describe audio-visual and spatio-temporal content of videos. Audio-visual aspects are described by high-level semantic objects, while spatio-temporal concepts are described by low-level structural objects. Semantic objects include catalogue descriptions (title, director, producer, etc.), segment descriptions (a news item within a news video), speech transcripts and shot lists (type of camera motion in a scene). It is worth noticing that most of these descriptions are textual and are annotated manually. On the other hand, the structural objects describe camera motion, frames and time-stamps. While frames include information about colour, texture and shape, camera motion tells us about pan and zoom. This system enables queries like: "A wide shot of a rain forest" or "A close-up of Paul Keating in front of the blue university building" and is a good example of coupling the structural and semantic aspects of the video.

On the other hand, retrieval and indexing of the video in some systems is based only on textual annotations [23, 22]. In [22] a four-step hybrid approach for retrieval and composition of a video newscast is proposed. It is based on different metadata sets: annotated (content and structural) and unstructured metadata. Content metadata contain information about an entity (persons, vehicles, etc.), location, event and news category, while structural metadata include a headline, introduction, body and enclosure of the news item. As their name already suggests, these metadata are manually generated. On the other hand, unstructured metadata are automatically obtained and are derived from audio transcripts of the news items. Indexing of unstructured metadata is organised in a form of keyword vectors, similar to indexing of textual documents.

The retrieval mechanism is a four-step procedure in which a keyword query is matched with unstructured metadata. A threshold is used to cut-off less appropriate items. Query results are then clustered using a graph theoretic method. In the third step new queries are generated based on unstructured metadata of each object in every query and finally structural metadata are used to further improve recall and obtain additional (similar) items. By using the four-step approach, the recall of the system is improved significantly, but at the same time some of the precision is lost, especially in the third step. The technique also achieves clustering, which is necessary for creation of cohesive video stories.

In [23] a video retrieval model is presented, based on TVAnytime [24] metadata. The TVAnytime metadata are a result of work of the TVAnytime forum, which seeks to develop specifications to enable audio visual and other services based on mass market, high volume digital storage in consumer platforms. As such, the TVAnytime metadata enable content referencing, content description and rights management. Video items used in the system are TV programmes, which are manually annotated by broadcasters and accompanying third party services.

Metadata fields, used for content description, include title, genre, synopsis, language and others. Some of the fields are mandatory (title) some are recommended (genre, synopsis, language), while other fields are optional. It is even possible to use arbitrary metadata fields, to be used by arbitrary applications. There are many possible retrieval scenarios, which involve search by genre categories (ex. sports, drama, news & documentaries), search by a keyword, performed on all existing metadata fields, or even enhanced search using thesauri.

When talking about video indexing and retrieval, we should mention the MPEG-7 standard [10]. MPEG-7 is an upcoming standard, designed to be generic for a broad range of applications (personalised video services, biomedical or surveillance applications, etc.). It concerns not only video but also all types of multimedia: audio and speech, moving video, still pictures, graphics, 3D models, and information on how objects are combined in scenes. As such, it tries to select descriptors and description schemes that are common to a big number of applications. The consequence is a rather complex standard, which might prevent some authors from using it. Nevertheless, the first MPEG-7 based applications have already been implemented, namely in the frame of the EU funded Sambits project [25]. The Sambits broadcast application demonstrates the relevance of MPEG-7 compliant metadata for offering new services.

2.2   Personalisation (user modelling)

In previous sections we took a look at a number of indexing and retrieval techniques, for different media types, available today. But even the most advanced retrieval techniques cannot prevent avalanches of query results, coming down to the user. There are simply too many data items available today, corresponding to a typical user query, regardless of the required information type (video, image, document, etc.). The user would like to have the most "relevant" query results displayed first. To be able to do that, the system must have a mechanism to model users. We should stress that the possibilities for personalisation are not limited to personalisation of the content. Other dimensions for personalisation, such as the user attitude towards the service, service time scale and different parameters regarding the actual delivery of information, can also be identified [26]. Nevertheless, we will focus only on personalisation of content retrieval.

Almost everybody has experienced user modelling, even if they were not aware of it. Any application or service, which behaves differently for different users, employs user modelling. User models can be big or small, complex or simple and often have different names: personal profiles, user profiles and even consumer databases, etc. An incomplete list of personalisation domains includes selection of hyper documents, news, video programmes, music, e-mail filtering, electronic commerce (shopping, brokerage etc.), education, job search and many more. According to [27], the majority of research papers devoted to Web, published since 1996, are about adaptive hypermedia systems.

The personalisation techniques and structure of the user model depend on the type of the metadata or other features describing data items. We could say that true personalisation can only be employed on high-level features, e.g. word-based descriptions of multimedia items. To the best of our knowledge, there are no known mechanisms that would enable personalisation based on low-level image features. It is difficult to describe user preferences in terms of colours and textures and be able to retrieve images based on their content. There are a number of image retrieval systems [9, 16, 17], which refine queries and/or filter search results based on explicit user feedback, but they are used only for refinement of each separate query and not for adaptation of the user profile.

Categorisation of personalisation approaches can be based on many different criteria, like the type of the media retrieved, feedback techniques (implicit, explicit), profile structure and adaptation, individual vs. collaborative, etc.

There has been a lot of discussions about benefits and drawbacks of implicit and explicit feedback techniques. In general it is clear that explicit feedback requires more effort from the user and can as such be annoying for him or her, but the information obtained from explicit feedback bring better results, since the system is not guessing about user preferences or suitability of recommended data items. On the other hand, implicit feedback is less disturbing for the user, but the conclusions about users' likings usually are more imprecise.

Individual and collaborative filtering differs in the number of user profiles involved in the personalisation process. Individual filtering uses a single profile and is based on the estimations of similarity of one individual profile with the corpus of the available content. The collaborative approach, on the other hand, measures the similarity between multitudes of user profiles. Recommendations are made based on the data items, which similar users have liked. The advantage of collaborative filtering is lack of need for sophisticated content-profile similarity metric and rich content representations. On the other hand, the drawbacks are latency problems, since new items will not be recommended until they have found a way into a sufficient number of user profiles and the fact that it requires a minimum number of profiles. Anyhow, both individual and collaborative approaches are complementary and are often used together.

In order to stay consistent with the structure of the chapter, we are going to take a look at different personalisation approaches, used for retrieval of different media types. Even though the majority of personalisation approaches belong to the domain of textual documents, there have been some efforts in retrieval of multimedia data items, especially video.

2.2.1   Personalisation in retrieval of textual documents.

One of the consequences of the overwhealming development of Internet have undoubtedly been increased research efforts of the adaptive hypermedia systems. Apart from personalisation in (hypermedia) documents [3, 28, 29, 30, 31, 32, 33], personalisation approaches have also been widely used in retrieval and filtering of news (newsgroups and newspapers) [31, 34, 35] and e-mail filtering [36].

Representation of the user profile depends on the application type and document representation, but systems differ also in the profile learning technique. User profile structures vary from (un)weighted keyword vectors [29, 30, 28], Boolean feature vectors, and more advanced approaches, where keywords are organised in a semantic network [3, 32]. Most of the user profiles are adaptive, which means they are trying to converge to the best possible representation of user interests. There are a number of learning techniques from the Bayes classifier [29, 31] to neural networks and genetic algorithms [34, 33]. Some of the systems require explicit user feedback in order to evaluate the results, whereas others are using only implicit feedback. As mentioned above, outcome shows, that explicit feedback leads to better results, although users may find it annoying and obtrusive.

Browsing assistants, such as Syskill and Webert [28], make suggestions about interesting pages. Suggestions are based on the observation of the user. Initially, the user supplies the system with the name of the topic and a URL of an index of pages for that topic. Syskill and Webert learn one profile per topic, which consists of a list of weighted keywords (probability ratings). A set of user profiles for different topics enables the system to become more specialised. While browsing the Web, the user rates suggested topics as good or bad. The system uses these examples as a training set and employs the Bayes classifier to revise the user profile accordingly. Related, keyword vector based, approaches can be found in [29, 30, 31].

A different approach to the browsing assistant has been developed within the SiteIF project [32]. In contrast to the majority of retrieval systems, whose user profiles are based on simpler keyword lists, SiteIF models the user profile as a semantic net, where every node is a word or a concept, and the arcs between nodes are the co-occurrence relations between words. Every node and arc has a weight, which represents a different level of interest to the user. With the introduction of the word co-occurrence, the problem of word synonymy and polysemy is reduced, which improves the retrieval precision. The system employs a number of agents (modules), whose tasks are divided into control of the user interface (Interface agent), generation of personalised pages (SiteIF agent) and retrieval and selection of documents, interesting to the user (WUP Agent). The user accesses the suggested results through the browser, as on-the-fly generated pages are presented to him or her. User feedback is based on the selection of suggested results and is gathered implicitly.

Similarly, the semantic networks are used in the OBIWAN system [3]. The system uses a publicly accessible hierarchy of concepts Magellan, which is comprised of approximately 4400 nodes. Figure 1.1 shows an excerpt of this hierarchy. Each website can be characterised with respect to this ontology. After the TF-IDF keyword vectors for the page are calculated, they are compared with the ontology node vectors to locate the top matching node(s) or categories. Browsing is done by clicking on the nodes of the standard ontology, i.e. the subject hierarchy, which results in displaying the pages of the sites corresponding to that subject. The user profile is represented as a weighted subject hierarchy, organised in a semantic network. The learning algorithm is based on the analysis of surfed pages, using implicit feedback. Adjustments to the interests in the profile are made based on the surfed page category, the time spent on the page and the length of the page. The experiments have shown that the length of the page has less influence than the time spent on it, and is therefore not strongly represented in equations. In future, the authors intend to use explicit feedback in order to improve the categorisation of the pages. The convergence of the profiles (adaptation rate) has also been analysed and the experiments have shown that on average it takes 17 days and a little over 300 pages for a profile to converge (become more or less static). Since the number of profile categories is quite high (50-150), the convergence results seem reasonably good.

An alternative approach to adaptation of user profiles is presented in [34]. The Newt system is used for selection and filtering of Usenet News. The representation of documents in this system is keyword based. A document consists of many fields and text (content) is just one of them. Other fields can include the author, location (the geographic origin of the news article), date, (number of) lines, etc. The newsgroup field needs special mentioning, since it is perhaps the only domain-specific field used in this representation. Each field is assigned terms used for identifying purposes. All the terms are not equally important, and are therefore assigned weights. Since a document consists of many fields, it is represented by a set of field-vectors. The representation of a profile is similar to that of a document. A profile also consists of a number of fields, like newsgroup, author, location, keyword, etc. Each field is a vector of terms, each of

```
4062  Sports
      4248  Recreation
      ...
      4314  Water-Sports
      4315  Boating
            4316  Boat Manufactures
            4317  Boat Shows
            ...
            4326  Yacht-Clubs
      4327  Canoeing
      4328  Fishing
      ...
      4339  Kayaking
```

Figure 1.1.   Excerpt from the Magellan ontology.

which is weighted in proportion to its importance for identification purposes. A user is represented with a number of user profiles. A fitness score is associated to every profile.

Retrieval of documents is as follows. The documents, which are presented to the user, are selected from among the documents, which score well with respect to the different profiles. Profile adaptation is based on explicit user feedback, which has an effect at two levels. One is at the population level, where the fitness of members of the population increases or decreases depending on the feedback. This is required for genetic learning. The other effect is on the profile, which is modified based on the features of the article the feedback is given for. The genetic operators, namely crossover and mutation, refresh the population every generation by introducing new members into the population and flushing out the unfit ones.

A similar approach in retrieval, using genetic algorithms, is used in [33].

2.2.2   Nontextual data items.   While user modelling techniques are mostly focused on retrieval of textual data items, there are not that many examples of personalisation in retrieval of audio/video and images. While a lot of image/video retrieval systems rely on extraction of low-level features, efficient personalisation models require high-level, keyword-based representations of user models. Nevertheless, some personalisation efforts have been made, especially in the domain of television broadcasting. The evolution of digital TV systems has resulted into an unprecedented number of available programs. Appearance of digital and interconnected devices, such as digital TVs and set-top boxes, and current generation of electronic programme guides go some way to assist users in selection of interesting programmes, but that is hardly enough. Users still need to manually search programme listings and with the ever-increasing number of programmes, this task is getting more and more cumbersome. A solution to this problem is an automatic recommendation system, aware of user preferences, which searches and filters programme listings.

The PTV – Personal Television Listings system [37] is an example of a personalised video retrieval system. Its users record, browse and watch their television programmes on-line, using their Web browser or WAP (Wireless Application Protocol) phone. The users can select and record programmes, which are listed by channel, theme and personal recommendation, provided by the ClixSmart personalisation engine.

In this system each user profile contains two types of information: domain and programme preferences. The former describe general user preferences such as a list of available TV channels, preferred viewing times, subject keywords, genre preferences, and guide format preferences. Programme preferences are represented as two lists of programme titles, a positive list containing programmes that the user has liked in the past, and a negative list containing programmes the user has disliked. Obviously not all of the profile information is used for content selection, but genre preferences, keyword lists and positive/negative programme lists are essential for programme selections. On the other hand, TV programmes are described by so called "programme cases", which include information about title, genre, director (creator), cast, language etc.

The personalization process is based on two recommendation strategies: individual (content based) and collaborative, also known as social filtering. For the purpose of individual filtering the user profile is converted into a profile schema, which represents a summary of the programme preferences. Profile schema is structured in such a way, that it is comparable to programme cases. The similarity between a profile and a programme can then be computed using the standard weighted sum similarity metric, as shown in Equation 1.3 [37], where Schema(u) is the profile schema of the user u, $f_i^{Schema(u)}$ and f_i^p are the ith features of the schema and the programme case (p) respectively (ex. genre, origin, etc.). w_i is the weight, representing importance of the ith feature in terms of similarity.

$$\Pr gSim(Schema(u), p) = \sum w_i \times sim(f_i^{Schema(u)}, f_i^p) \qquad (1.3)$$

The collaborative filtering is based on computation of user similarity by using a simple graded difference metric shown in Equation 1.4; where $p(u)$ and $p(u')$ are the ranked programmes in each user profile, and $r(p_i^u)$ is the rank of programme p_i in profile u.

$$\Pr fSim(u, u') = \frac{\sum\limits_{p(u) \cup p(u')} \left| r(p_i^u) - r(p_i^{u'}) \right|}{4 \times |p(u) \cup p(u')|} \qquad (1.4)$$

$$\Pr gRank(p, u) = \sum_{u \in U} \Pr fSim(u, u') \qquad (1.5)$$

Once the similar profiles have been identified, a list of programmes from these profiles is generated, ranked and suggested to the user. The ranking metric is presented in Equation 1.5. During the evaluation only 3% of the testing population found the recommendation system of poor quality, 6% found it satisfactory, and 61% good.

The previously described MyTV system includes a very similar recommendation service [23]. Since content descriptions are based on the TVAnytime metadata fields (title, genre, synopsis, language, etc.), the difference with the PTV system is mostly in the fact that there is no need for conversion of the user profile to a content-description like form. The user profile has been designed in such a way that the content-profile similarity is not difficult to evaluate. It consists of approximately 100 fields, which describe the affinity of the user towards every single genre[10]. Additionally the profile includes a list of recorded programmes and information about them (title, genre) and some user information needed for identification purposes. Because of the language differences in content descriptions, a small thesaurus functionality has been implemented. It enables identification of a similar content, described with different synonyms (ex. movie = film). The system also includes mechanisms for collaborative filtering, while the design permits further extensions and integration of additional, more sophisticated approaches (extended thesauri, use of domain specific terminology, etc.).

Finally, we should mention that there are a few more systems in the domain of personalised programme selection. Among them are TiVo [38] and ReplayTV [39], both designed as a personalised digital video recorder.

It is apparent that personalised retrieval of the mass-market consumer video (TV programmes) is a domain on which a lot of attention is focused. The description schemes, allowing for personalised navigation, browsing, automatic filtering based on genre and actors, etc. have also been included in the MPEG-7 standard. Currently, much of the content descriptions must be obtained through a manual process although several advances have been made in summarization technology like audio transcript text extraction, key-frame extraction, etc.

---

[10]According to the TVAnytime classification there are approximately 100 different programme genres.

3. Adaptive Information Retrieval

In modern communication services, it is desired also to optimally use the currently available resources. The information in a multimedia environment takes a wide range of presentation forms, such as text, voice, image and video. The generic data quantity therefore often exceeds the limits imposed by the current terminal equipment and available communication channel. The multimedia objects themselves represent only fragments in a distributed data warehouse, which are given their final meaning in a relation with other media objects. Additional descriptive elements as well as the mentioned relations can be given in a separate form, known as *metadata*. There exist two main issues related to this problem, namely, how to refer only to the relevant pieces of information, and how to select an optimal retrieval and presentation form of selected objects. We shall now focus on the second mentioned issue.

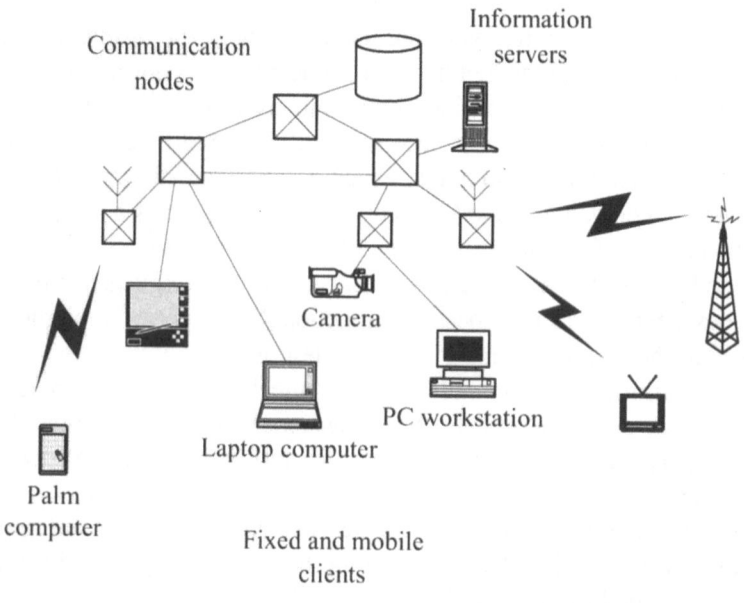

Figure 1.2.   Communication service environment.

The hardware structure of a typical heterogeneous communication service environment is shown in Figure 1.2. The key elements of such a system are information servers holding multimedia data and related descriptors, communication infrastructure, access terminals and adaptive procedures for optimal service execution under specific conditions. Processes of information management, such as seeking for a most relevant media object, can in this aspect share the same importance with adaptive information retrieval procedures. The later methods are based on lossy information encoding and take into account system properties such as bandwidth restrictions, perceptual properties, as well as specifics of the terminal equipment.

Such methods assure an operational level of quality of interactive communication services in a restrictive environment. Assuming true mobility of service users, *usability* of a specific service should not depend on the current location and available resources. An integrative point is required at which service-specific requirements, technical resources and encoding mechanisms are fine-tuned to allow for a balanced service operation. The descriptors of a specific user, service requirements, represented data and operation environment may well take the same structure as descriptors known from fundamental information management, referred to as *metadata*. Such metadata structures may provide additional information, not specifically required by typical "search and retrieval" procedures, but valuable in the process of service quality evaluation and adaptation.

Following the entire path from the information source to a service user, the main factors that contribute to a subjective judgement of the *overall* multimedia telecommunication service quality may include:

- data acquisition procedures,

- multimedia data encoding methods,

- telecommunication channel,

- display equipment properties, and

- user expectations.

One should be aware of the fact that the main attraction of quality management research has been dedicated to lossy media encoding; there is also a strong research investigating priority-based procedures in data routing and queuing. The current results propose many *absolute* quantitative quality evaluation methods, especially for images and video. These methods work fine for controlling the image encoding methods alone; for a generalised quality evaluation one should introduce a relative quality measure that involves all critical degradation factors in a system. In other words, a good service quality measure would refer to the best object presentation on a specific equipment rather than to a least mean square criterion based encoding with a degraded, sampled and quantised reference object.

3.1   System Descriptors

In order to provide a framework for a generalised type of service adaptivity, both mechanisms of adaptivity and standardised system descriptors are required. Descriptors of the system and information elements may be given any form of notation [40]. To allow interoperability among several systems, standardised notation is preffered; one of the dominant standards in this area is the descriptive language XML [40]. The structure of XML is dynamic, offering flexible notation with application-specific notation.

Metadata for service quality management

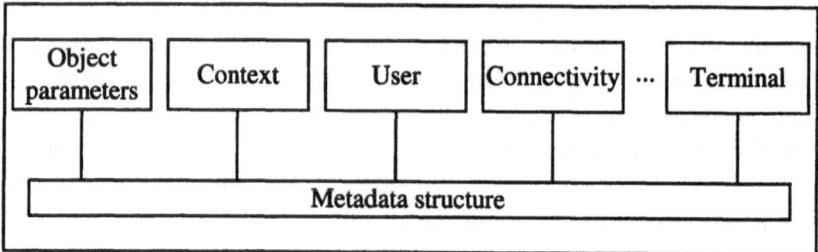

Figure 1.3.   Metadata structure.

This is why XML itself cannot assure application-specific data interchange at a system level. The dedicated structure should also be a subject to applicative standards, and there exist methods like DTD[11] or XML-Schema that make an appropriate framework for data structuring and validation. The data structure for our specific application should provide fundamental information about critical system components that either affect or prescribe the overall service quality. An analysis of system components has been made with some notation suggestions in XML.

3.1.1   Data context and origin.   The purpose of multimedia communications is not to achieve high scores in synthetical quality evaluation methods but to present selected materials in a most illustrative fashion. In this aspect, more data may not always mean better service. To achieve integrative quality measure, one must know that encoding methods and communication failures do not represent the only source of information loss. The mentioned effects should be measured relatively to unavoidable degradations such as sensor transfer functions, sampling and signal level quantisation.

The quality judgements also depend on contextual information such as regions of interest and current viewpoint. The type of information to be presented may also influence other requirements like minimum spatial resolution or observation window. For example, the text may not be readable unless a specified

---

[11]Document-Type-Definition.

resolution is reached; on the other hand, an observation window narrower than the column width may cause the same effect.

3.1.2   Media encoding.   There exist several compression techniques offering removal of redundant information, thus significantly reducing the number of bits used to encode the desired information. The encoding compression ratio may be severely improved at a certain level of quality degradation. The nature of human perception makes a foundation for standardised lossy compression schemes. However, the general assumptions of quality may differ with respect to a particular user within a particular type of imaging application.

A very popular set of compression systems is built upon block based linear transformations followed by coefficient quantisation that introduces irreversibility to the encoding process. Final data compression is achieved by means of the entropy-based encoding of the quantised coefficients [41, 42, 43]. Such an approach is a foundation for a range of industry-accepted standards like JPEG [44], MPEG1 [45], MPEG2 [46] and H.261. The operator of a linear transformation is there used to transform the signal to a space where more efficient compression can be achieved. Recently, subband-based methods such as Wavelet transform with improved performance at high compression ratios over block linear transform are evolving [47, 48]. In any case, it is not the decorrelation procedure but the coefficient quantisation that provides lossy data compression through manageable quality degradation.

3.1.3   Communication network.   To provide an optimal level of service, the application should adapt its performance to the available network resources. Of a particular interest is a class of applications that allow a certain level of user mobility. Wireless communications remain the area where specific care about the network bandwidth and responsiveness should be taken care of; the resources are more limited and the cost per bit transferred is higher. A specific user may be given access to a public network, local private network or even a wired connection (see Table 1.1), and the application itself should maintain the present resources with respect to personal user requirements in the best possible way in terms of cost and availability.

	Bandwidth (bps)	Round-trip latency (ms)
LAN	10-100 M	0.5-2
ISDN	$n \times 64K$	10-20
Analog modem	$14.4 - 33.6K$	350
Cellular	$n \times 9.6K$	100-500

Table 1.1.   Typical widely available network capacities.

3.1.4   Display equipment.   Technical limitations of a multimedia communication system are dominantly imposed by technical properties of a user terminal. Restrictions of the display size and resolution are accompanied by processing power and system support restrictions, that are especially critical in case of low-power mobile terminals. A list of typical display equipment properties is given in Table 1.2. These restrictions represent fixed constraints for the service optimisation procedure.

	Image size	Bits/pixel	Power consumption
Desktop PC	1600 × 1200	32, color	high
Portable PC	1024 × 768	24, color	moderate
Handheld computer	320 × 200	16, color	low
WAP/phone	96 × 65	2, gray	very low

Table 1.2.   Typical capabilities of an imaging terminal.

3.1.5   Personal preferences.   The final quantitative evaluation of the service quality can be assisted by introducing personal criteria. The user profile may include own parameters and even methods of quality evaluation. The user profile may therefore provide additional information about user requirements, such as transcoding methods that an individual prefers, regions of interest or criteria functions for quality evaluation. An illustration of user preference XML file is given in Figure 1.4.

```
<USER_PREFERENCE>
     <ENCODING_INFO>
          <REGION_TYPE Name="OTHER">
               <Impact> 0.1 </Impact>
          </REGION_TYPE>
          <REGION_TYPE Name="PERSON">
               <Impact> 1.0 </Impact>
          </REGION_TYPE>
          <RESIZE>
               FALSE
          </RESIZE>
          <CROP>
               TRUE
          </CROP>
     </ENCODING_INFO>
</USER_PREFERENCE>
```

Figure 1.4.   Example of a partial user profile in XML.

3.2 Quality Measures and System Approach

In order to manage system resources and data flow in a distributed communication service environment, a system is needed that will maintain communication connection and prepare multimedia data in a most suitable presentation form. The goals of such a system are clear; the application quality grade must fulfill at least minimum criteria, regardless of current conditions in the system.

The initial constraints are imposed by static system parameters. Due to restrictive operating environment, especially in case of low-bandwidth communication connections, active quality management may be required in order to fulfill minimum user expectations. The management procedure will mostly influence two major service quality parameters, namely

- degradation, imposed by lossy media encoding *and*

- service delay, proportional to the size of the media container.

3.2.1 Degradation imposed by lossy encoding.

The presentation quality measures arise from the experience in image encoding, however, if properly adapted, they may be further generalised in order to satisfy specific demands of lossy encoding of varied digital media.

There exist a number of degradation measures for media encoding quality evaluation. The majority of methods is general enough to be implemented on any type of signals. Classical error measurements are related to power or energy of the measured signal difference. The most known is the function of the *Mean Squared Error*[12] [43], $MSE = \frac{1}{n}\sum_{i=1}^{n}(X_i - \hat{X}_i)^2$, also referred to as l_2 norm. The alternative is to use the well known *Peak Signal-to-Noise Ratio*[13] [42], $PSNR = 10\log(\frac{M^2}{MSE})$. The expression is based on the same metrics. The result is scaled to the number of quantisation levels $M = 2^b$ and utilises the logarithmic property of human perception. In order to emphasize the perceptual characteristics, adjustments can be made using the *Weighted Mean-Square Error*[14] [43] approach. More sophisticated evaluation methods are dedicated to specific media encoding procedures [49, 50, 51, 52].

To allow real-time service adaptation, fast and efficient methods are required providing quantitative measures of service quality. The target encoding quality shall be combined with other quality criteria in order to achieve a balanced service operation.

3.2.2 Service latency as a measure of quality.

The responsiveness of an interactive communication service is also one of the critical quality ar-

---

[12]MSE
[13]PSNR
[14]WMSE

guments. It is most common to define the allowed service latency as a fixed threshold and to perform optimisation with constraints based on this value. Such an approach is good while latency thresholds are below the noticeable delay. From standards in the telecommunication industry, these values are typically below $T_{os} \leq 150ms$ [53, 54, 55].

In the area of interactive multimedia telecommunication services, these values are hard to be achieved, and the criterion of an *immediate response* (e.g. no delay perception) is often replaced by the value of a *tolerable delay*. Some authors [56, 57], suggest that a delay of $T_{os} \geq 10s$ leads to the loss of the user focus and seriously degrades the level of service. According to the same studies, a delay of $T_{os} = 1s$ ensures the feeling of interactivity and is generally acceptable.

To ensure an optimal service operation, latency thresholds should be replaced by a continuous cost function that is to be integrated in the quality optimisation procedure [58]. There exist some approaches based on economic models rather than on human perception. We set up a basic framework for a continuous latency evaluation, defining the impact function of responsiveness $F_{os} = f(t)$ as an increasing function of latency t. For the purpose of evaluation, we used a model based on the Steven's power law [59], which resulted in $F_{os} = k \cdot t^p$, $p \approx 3$ [58]. The suggested impact function of the system responsiveness was later used in an on-line service optimisation procedure.

3.2.3   Balance between latency and quality.

The use of lossy compression in multimedia communications is fundamental for service quality management. The relations that enable bitrate controlled encoding arise from the Shannon's work in the field of the information theory. The distortion caused by the reduced bitrate notation follows the boundaries given by the well known Rate-Distortion relation, $R(D)$.

The typical use of the above relations is known from many algorithms for optimal image encoding. There exist two well known optimisation strategies, namely to find the best possible encoding for:

- a predefined quality *or*

- a predefined bitrate.

In each case, the goal of the procedure is to select encoding parameters to be as close as possible to the theoretical boundaries of the $R(D)$ (or $D(R)$) function relation. The idea can be seen from Figure 1.5, following the curve $D(R)$.

The procedure suggests no answer to what exactly is the optimal bitrate or optimal presentation quality. In a resource-restricted environment, the conditions of perceptually lossless encoding or, on the other hand, virtually immediate data delivery are not likely to be met. The idea is to combine both the encoding quality $D(R)$ and service latency measures F_{os} into a common cost function.

Generally, the combination of the two can be calculated using the expression

$$P(R) = (c_1 D(R)^r + c_2 F_{os}(R)^r)^{\frac{1}{r}}, \qquad (1.6)$$

using predetermined constant factors c_1 and c_2 at a given metric determined by r. An optimal operating point is then defined as a set of operation parameters minimising the given cost function. The strategy and the optimal encoding point can be given from Figure 1.5, where crosses mark achievable operational points of the encoder. The cost function $P(R)$ is used to determine optimal encoder rate which can later be achieved using classical encoding algorithms.

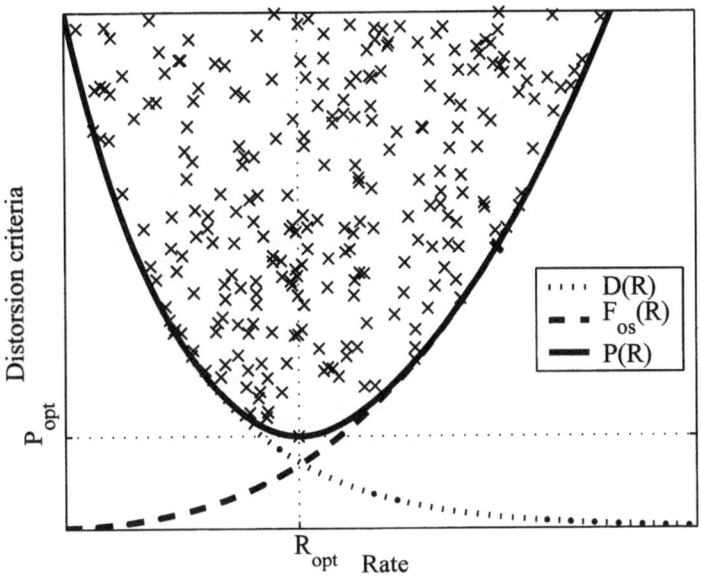

Figure 1.5. Optimal encoding.

4. System Architecture and Functionality

Two issues need to be taken into account when talking about the multimedia content usage: content selection and content presentation. Both are equally important and practically independent of each other. Figure 1.6 presents the metadata and object flow for mobile terminals. On the left-hand side of the figure is presented the content selection via the metadata store, while on the right-hand side is the content processing with regard to content presentation. Content selection is a subject to a number of issues and may therefore have different possible architectures. These issues, and issues regarding content presentation are discussed in the following sections.

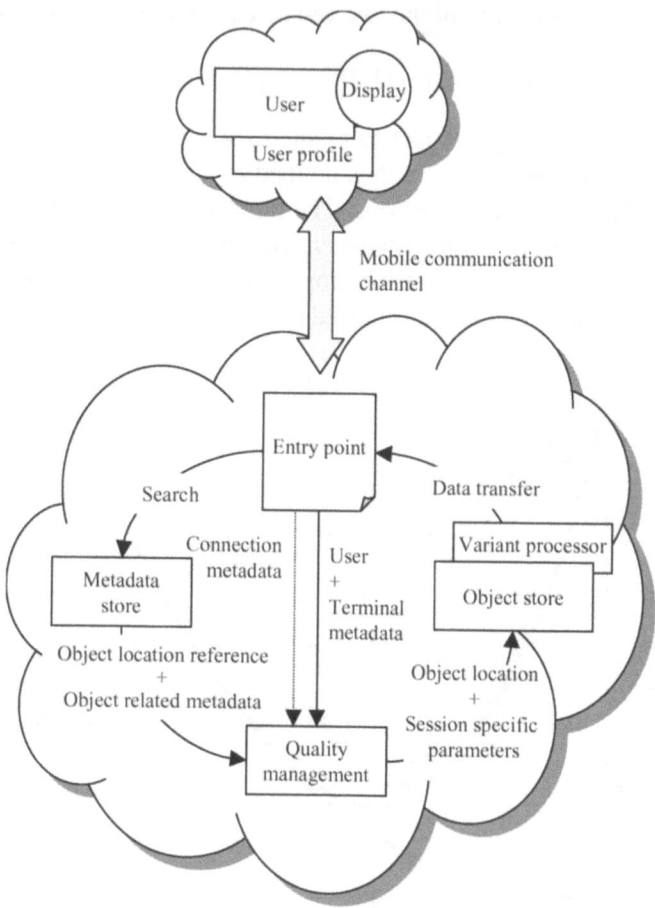

Figure 1.6.   Metadata and object flow for mobile terminals.

4.1   Personalised Information Retrieval

When designing the architecture of a personalised information system, one has more than one possibility. There are a number of criteria according to which one should choose this or some other solution. Among them are privacy (security) constraints, possibility of collaborative filtering, dispersion of the content (local database vs. the Web), etc.

As far as privacy is concerned, the best idea appears to be local storage and creation of profiles. First, in this way, the profile can be digitally encrypted with the user personal key, which prevents unauthorised use of the profile. Second, the user can determine what parts of the profile he or she wants to exchange with potential remote applications, if any at all. On the other hand, some systems do require a number of profiles stored in a central repository, for the purpose of collaborative filtering. In such cases, the privacy of profiles must be guaranteed by the service provider, which should not allow unauthorised access and use of the profile information, outside the scope of a signed agreement. Local availability of the content also influences the architecture. If the data is available from local databases, it is better to keep the profiles at the source of information, since the search process is performed at one location. When the information pool involves the Web, filtering of the content is usually performed on the client side or at least at a proxy server. In this case the profiles are stored on the client machine or at the proxy. Examples of the discussed architectures are presented in Figures 1.7, 1.8 and 1.9.

Examples of the client side user profiling are in [60] and [26]. In both cases the information is retrieved from databases, which contain a closed corpus of information. Collaborative filtering cannot be used, but this kind of architecture maintains the privacy of the user profile. The architecture is presented in Figure 1.7.

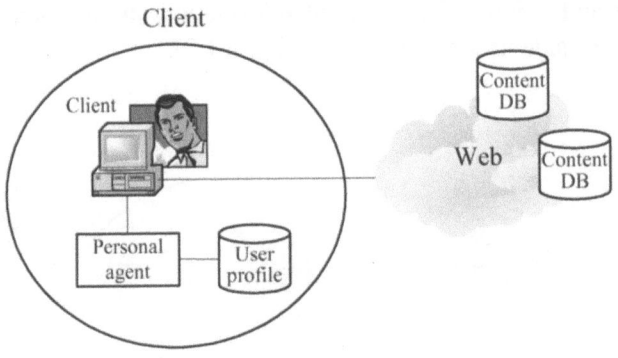

Figure 1.7. Client side user profiling.

The solution with a proxy server is also widely used [29, 32, 31, 34]. Users log-in the system through a browser and access the information selected by the personalisation module, running on the proxy. This architecture type enables remote access to the system from an arbitrary device with Internet access, since user profiles are stored at the proxy server. On the other hand, management and dissemination of the user information is not entirely in the hands of the user. Information items (news, Web documents) are gathered from an open corpus, that is, the entire Web [29, 32], or in the case of news retrieval, a number of news groups [31, 34]. This architecture is presented in Figure 1.8.

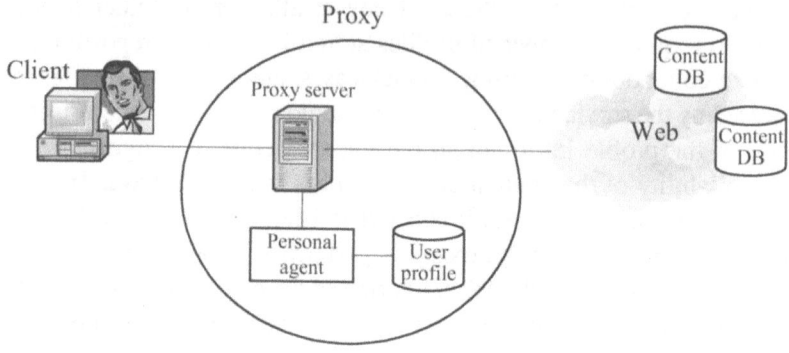

Figure 1.8.   User profiling on the proxy.

The third architecture also stores user profiles at a remote location in an information repository, together with information items or information metadata [23, 37]. Users similarly log in the system through a browser (proxy) and browse through a list of suggested query results. These systems provide both collaborative and individual filtering of information and as such use the full benefit of server-side user profiling (see Figure 1.9).

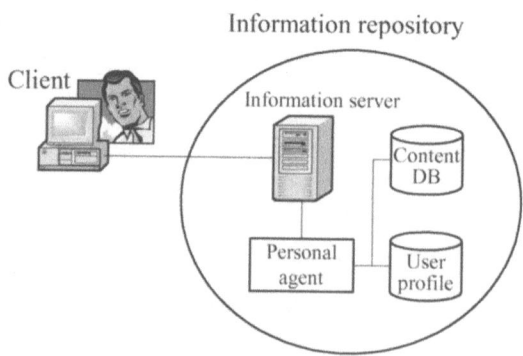

Figure 1.9.   Server side user profiling.

It is evident that the benefits of one approach are the drawbacks of another and that the design of the personalisation system relies heavily on the prioritised functionality. In [27] a user model server and protocols for communication and exchange of user profiles are proposed. The idea is to standardise storage, management and exchange of user profiles, which would allow interoperability among different systems involving personalisation and reuse of user profiles.

Similarly, [61] suggests the use of a protocol for expressing information about users and even evolves the idea of a user modelling language.

4.2   Personalised Content Presentation

In order to enable users having access to visual data information across various network-enabled devices, the applications should provide mechanisms that will adapt to the currently imposed constraints of the variable resources. The majority of the current services rely on the generic approach, where static applications are assisted by independent intelligent network resource management. We also believe that scalable proprietary applications may not represent the most appropriate solution to the problem. We recommend a modular concept, in which certain components of the entire system adapt themselves to restrictions imposed by the current status of the environment resources. Typical *environmental constraints* of an adaptive imaging application are first to be identified for a successful adaptive application implementation. These, for a typical imaging terminal, include network behavior, display capabilities and, for mobile devices, also power supply limitations.

The goals of the adaptive process may vary among different users and current types of service, and need to be observed within the context of a specific task. The description of goals, also referred to as *application semantics*, is more complex for generic, more generalised application frameworks. Maintenance of semantics will in any case cause the changes in the application behavior, and the choice of an equivalent mechanism that minimises degradation of service quality while fitting to a constrained environment is desired.

The choice of the most suitable image compression mechanism regarding current user requirements, object characteristics as well as available connection bandwidth is therefore required. Some concepts known from software engineering as *open implementation* [62] may be adopted as a technology providing a distributed modular application architecture. In such a model, the behavior of the system components should be described in a form of interface abstractions, while the internal implementation is not public. The component interfaces provide the mechanism with a choice of implementations that basically set up an orthogonal space. Each implementation is therefore represented as a point in optimisation space whose number of independent axes represents degrees of freedom of the application. The component behavior is therefore described as a trajectory in the named space.

The level of transparency at which the system operates should be assigned by the system designer and should be a compromise between automated operation and carefully designed light user interaction. The actual service grade estimates should basically refer at least to the image encoding quality, level of details, display resolution and service latencies. The system policies that control the

Figure 1.10.   High quality JPEG image: 98 KB.

application behavior can either be provided during the design phase of the image database, and dynamically during data access, based on the image context of use. To ensure maximum customer satisfaction, no system designer can provide the user with an ideal system behavior. Some user interaction that enables the user to control the adaptive behavior of the application has to be provided. Hence it is not expected that an average user is either willing or technically capable of controlling the application at the system level, the interaction should be performed at an abstract level based on the user experience.

Assuming that the server holds images of the best possible quality, it is the client hardware that limits color depth and image resolution. The terminal parameters as well as the current network connection setup may therefore be considered as the current session parameters. Additional optimisation constraints may be added to complement generic quality measures in order to adapt the system according to the current user, console and service profile.

The session management, following the above statements, demands a number of session parameters, related to terminal characteristics, user preferences, service demands and non-optimised image sources. Generic approaches to parameter gathering, which are required prior to start of service optimisation procedures, are possible. It has been shown that metadata-based systems may also be used for quality of service management [63]. We propose to use metadata to describe both system and user constraints and requirements. With a flexible metadata structure, optimisation policies themselves may also become part of the same data structure. Based on standardised metadata sets, a move towards standardised, cross-platform oriented mobile imaging services can be made. The entire process, including object and metadata flows, is shown in Figure 1.6.

It is up to the system designer to provide components either as elements placed on an image server or in a form of independent proxy-like services. The selected optimisation strategy on a real system requires moderate computational power and on a stationary server represents no significant issue in terms of speed, power and service latency [64]. Client terminals should rely on standardised image compression formats supported by the terminal hardware; formats that require intensive client-based computations are to be avoided. Because of communication and storage limitations of mobile terminals, lossy compression schemes that provide high compression ratios at an acceptable level of quality degradation are favorable. The use of DCT-based compression algorithms [41, 42, 43], where quantisation and entropy coding parameters are being modified to ensure optimal image compression, is still preferred as a result of to mobile hardware processing restrictions [65].

The results of the procedure will be shown on an example of image encoding to be displayed on a typical mobile terminal. In the initial preprocessing stage, the image cropping and rescalling will be made according to service and user preference settings. The results are shown in Figures 1.10, 1.11 and 1.12, respectively. This initial stage already reduces the quantity of data with no direct influence on the display quality on account of performance optimisation. In the given example, the data quantity was reduced from 96 KB dow to 10 KB, minimizing latency by a factor of 10.

The metadata on the available network services indicated the presence of a GSM data service. Since the network bandwidth in this case is critical, it was decided to reduce service latency by further encoding optimisation. For example, two typical bitrates were evaluated, the GSM data connection at $\bar{B}_1 = 9600bit/s$ and a more advanced HSCSD connection at the currently available $\bar{B}_2 = 28800bit/s$.

In an optimisation procedure, the minimum of the function $P(R)$ is to be found for given values of $\bar{B}_1$ and $\bar{B}_2$. The necessary functions can be obtained from service metadata profiles. For a classical GSM data interface, the optimal encoding is at $R = 3011B$, with delay time $T_z = 2,5s$. The resulting picture is shown in Figure 1.13.

Figure 1.11.   Adapted quality JPEG image: 20 KB.

In case of the HSCSD data connection, a new optimal point is to be determined, to give better picture quality at a reduced latency time. In this case, the optimal data rate equals to $R = 4989B$, with latency of $T_z = 1,35s$. The resulting picture is in Figure 1.14.

Let us remind that, with no transcoding strategies and by using GSM data connection, a delay of $T_z = R_{max}/\bar{B} = 81,92s$ would occur displaying the original image of $R = 96KB$. By adapting the image size and quality to the user terminal capabilities, the data size is reduced to $R = 10KB$, resulting in $T_z = R_{max}/\bar{B} = 8,16s$ latency. By using suggested methods, a true balanced operation is enabled, in terms of latency and presentation quality.

5.   Conclusions

The worldwide information availability is becoming a reality, especially with the development of wireless mobile data access. We are witnessing the growth of available information resources as well as the growth of wireless mobile devices. One of the characteristics of wide spreading mobile devices is the heterogeneity of their display properties and bandwidth connectivity, which

Figure 1.12. Adapted quality JPEG image, cropped to the user display size: 10 KB.

Figure 1.13. JPEG optimised encoding for $\bar{B} = 9600 \; bps$, $R = 3011B$, $T_z = 2, 5s$.

Figure 1.14.   JPEG optimised encoding for $\bar{B} = 28800\ bps$, $R = 4898B$, $T_z = 1,35s$.

requires certain adaptations in the communications infrastructure. This includes adaptation of transferred content in terms of image size and quality, based on display capabilities and user preferences regarding the price-performance ratio. The other aspect of accounting for user preferences is the selection of the appropriate content. It is becoming more and more difficult for users to find appropriate content, since the extent and number of available information sources is increasing with a tremendous rate. Automatic content search and analysis, based on identified user preferences is the step in the right direction.

The agent-based techniques already known from computer-assisted data mining and gathering can be used in order to maintain a networked user and application profile. Such profiles, combined with the current status of the available connection, processing and display resources, may be used to control the process of user specific information flow and encoding. The optimisation needs to be performed on-line and should adjust the quality of the related data. The main goal is to keep latencies below a certain threshold of acceptance, while providing best possible service quality. The computationally demanding optimisation and recompression tasks are to be provided either by data servers or in a form of public transcoding services.

Undoubtedly, the field of personalised approach to selection and presentation of information is prosperous and brings new challenges. Future systems may give us fully automated systems, adaptive interfaces with multiple sensors, gathering information not only about our explicit requests, but also about our current mood, fatigue, environment etc. Regardless of the fact how advanced the systems will be, we should take especially one aspect into consideration: The security and privacy of information about users. User profiles, containing the information about user preferences regarding the device capabilities and especially content preferences, are extremely delicate pieces of information. They contain a lot of confidential information, which should by no means be available to unauthorised entities, humans or machines. But, even though the risk of a "big brother" is evident, we believe that the advantages, brought by personalisation techniques, are having significant impact on the quality of the information retrieval. Therefore they should be used, with adequate precaution and security measures.

References

[1] S. Weibel, J. Kunze, C. Lagoze, and M. Wolf, "The Dublin Core initiative home page." http://www.ietf.org/rfc/rfc2413.txt, September 1998.

[2] G. N. Venkat, R. V. Vijay, G. I. William, and K. Rajesh, "Information retrieval on the World Wide Web," *IEEE Internet Computing*, vol. 1, no. 5, pp. 58–68, 1997.

[3] A. Pretchner, "Ontology based personalised search," Master's thesis, USA, State of Kansas, Lawrence, University of Kansas, Faculty of the graduate school, Departement of electrical engineering, 1998.

[4] P. Edwards, D. Bayer, C. L. Green, and T. R. Payne, "Experience with learning agents which manage Internet-based information," in *AAAI 1996 Stanford Spring Symposium on Machine Learning in Information Access, Stanford, California, USA* (M. A. Hearst and H. Hirsh, eds.), pp. 31–40, AAAI, 1996.

[5] G. Salton and M. J. McGill, *An introduction to modern information retrieval.* McGraw-Hill, 1983.

[6] K. Aas, "A survey on personalised information filtering systems in the world wide web," Tech. Rep. 922, Norwegian Computing Center, Oslo, Norway, December 1997.

[7] S. B. Palmer, "The semantic web: An introduction." http://infomesh.net/2001/swintro. Last accessed in Jan. 2003.

[8] A. Swartz, "The semantic web in breadth." http://logicerror.com/SemanticWeb-long. Last accessed in Jan 2003.

[9] M. Flickner *et al.*, "Query by image and video content: The QBIC system," *Computer Magazine IEEE*, vol. 28, pp. 23–32, September 1995.

[10] J. M. Martínez, "Overview of the MPEG-7 standard (version 6.0), ISO/IEC JTC1/SC29/WG11 - coding of moving pictures and audio." http://mpeg.telecomitalialab.com/standards/mpeg-7/mpeg-7.htm, December 2001.

[11] S. Chang, J. R. Smith, H. J. Meng, H. Wang, and D. Zhong, "Finding images/video in large archives," *D-Lib Magazine*, vol. 3, February 1997.

[12] N. Vasconcelos and A. Lippman, "Embeded mixture modeling for efficient probabilistic content-based indexing and retrieval," in *SPIE Multimedia Storage and Archiving Systems III*, vol. 3527, pp. 134–143, 1998.

[13] S. Belongie, C. Carson, H. Greenspan, and J. Malik, "Color- and texture-based image segmentation using em and its application to content-based image retrieval," in *Proceedings of the Sixth International Conference on Computer Vision*, (Bombay, India), pp. 675–682, 1998.

[14] E. J. Posnak, G. Lavender, and M. Harrick, "An adaptive framework for developing multimedia software components," *Communications of the ACM*, 1997.

[15] K. Shah and A. Sheth, "Infoharness: Managing distributed heterogeneous information," *IEEE Internet Computing*, vol. 3, pp. 18–28, November-December 1999.

[16] X. Zhu, H. Zhang, and L. Wu, "New query refinement and semantics integrated image retrieval system with semiautomatic annotation scheme," *Journal of electronic imaging*, vol. 10, pp. 850–860, October 2001.

[17] A. Benitez, J. Smith, and S. Chang, "Medianet: A multimedia information network for knowledge representation," in *Proceedings of SPIE* (J. Smith, C. Le, S. Panchanathan, and C. Kuo, eds.), vol. 4210, pp. 1–12, October 2000.

[18] G. A. Miller, "WordNet: A lexical database for English," *Proceedings of the ACM*, vol. 38, no. 11, pp. 39–41, 1995.

[19] U. Srinivasan and G. Riessen, "A video data model for content based search," in *8th International Workshop on Database and Expert Systems Applications (DEXA '97)*, (Toulouse, France), p. 178, September 1997.

[20] P. Faudemay and C. Seyrat, "Intelligent delivery of personalised video programmes from a video database," in *8th International workshop on database and expert systems applications (DEXA97)*, (Toulouse, France), p. 172, 1997.

[21] G. Amato, G. Mainetto, and P. Savino, "An approach to a content-based retrieval of multimedia data," *Multimedia Tools and Applications*, vol. 7, no. 1, pp. 9–36, 1998.

[22] G. Anhanger and D. C. Little, "Data semantics for improving performance of digital news video system," *IEEE Transactions on knowledge and data engineering*, vol. 13, pp. 352–360, May-June 2001.

[23] M. Pogačnik and J. F. Tasič, "Interactive and personalised television of the future," in *Proceedings of the First COST 276 workshop on Information and Knowledge Management for Integrated Media Communication (CD-ROM)*, (Florence, Italy), October 2001.

[24] http://www.tv-anytime.org.

[25] Sambits Consortium, "The usage of MPEG-7 metadata in a broadcast application," 2001.

[26] M. Turpeinen, "Agent mediated personalised multimedia services," Master's thesis, Helsinki University of Technology, Helsinki, Finland, 1995.

[27] P. Brusilovsky, *Adaptive hypermedia, User modeling and user adapted interaction*, pp. 87–110. Kluwer Academic publishers, 2001.

[28] M. Pazzani and D. Billsus, "Learning and revising user profiles: The identification of interesting web sites," *Machine Learning*, vol. 27, pp. 313–331, 1997.

[29] D. Mladenić, "Personal webwatcher: Implementation and design," Tech. Rep. IJS-DP-7472, Carnegie Mellon University, Pittsburgh, 1996.

[30] B. Crabtree and S. J. Soltysiak, "Identifying and tracking changing interests," *International journal of digital libraries*, vol. 2, no. 1, pp. 38–53, 1998.

[31] D. Billsus and M. J.Pazzani, "A hybrid user model for news story classification," in *Proceedings of the Seventh International Conference on User Modeling (UM99)*, (Banff, Canada), pp. 99–108, Springer-Verlag, 1999.

[32] A. Stefani and C. Strapparava, "Personalizing access to Web sites: The SiteIF project," in *Proceedings of the 2nd Workshop on Adaptive Hypertext and Hypermedia HYPERTEXT'98*, (Pittsburg, USA), pp. 69–74, June 1998.

[33] J. Mirković *et al.*, "Genetic algorithms for intelligent internet search." B.Sc. thesis, University of Belgrade, Belgrade, Yugoslavia, 1998.

[34] B. Sheth, "A learning approach to personalized information filtering," Master's thesis, Massachusetts Institute of Technology, 1994.

[35] J. Konstan, B. Miller, D. Maltz, J. Herlocker, L. Gordon, and J. Riedl, "Grouplens: Applying collaborative filtering to usenet news," *Communications of the ACM*, vol. 40, no. 3, pp. 77–87, 1997.

[36] T. Malone, K. Grant, and F. Turbak, "Intelligent information sharing systems," *Communications of the ACM*, vol. 30, pp. 390–402, 1987.

[37] P. Cotter and B.Smyth, "PTV: Intelligent personalised TV guides," in *Proceedings of the 12th Innovative Applications of Artificial Intelligence (IAAI-2000) Conference, Austin, Texas, USA*, pp. 957–964, AAAI Press/The MIT Press, 2000.

[38] "TiVo." http://www.tivo.com. Last accessed in Jan. 2003.

[39] "ReplayTV." http://www.replaytv.com. Last accessed in Jan. 2003.

[40] S. Spainhour and R. Eckstein, *Webmaster in a nutshell, Second edition.* 101 Morris Street, Sebastopol, CA, US: O'Reilly & Associates, Inc., 1999.

[41] R. J. Clarke, *Transform Coding of Images.* London: Academic Press, 1985.

[42] R. J. Clarke, *Digital Compression of Still Images and Video.* London: Academic Press, 1995.

[43] K. R. Rao and J. J. Hwang, *Techniques and Standards for Image Video and Audio Coding.* New Jersey: Prentice Hall, 1996.

[44] ISO/IEC, "TC1 10918-1 ITU-T REC. T.81 information technology: Digital compression and coding of continuous- time still images: requirements and guidelines," 1994.

[45] ISO/IEC, "11172-5 information technology: Coding of moving pictures and associated audio for digital storage media at up to about 1 Mbit/s: Part1: Systems; part 2: Video; part 3: Audio; part4: Conformance testing," 1993.

[46] ISO/IEC, "JTC1/SC29/WG11: preliminary working draft MPEG 92/086 document AVC-212," 1992.

[47] M. Vetterli and J. Kovačević, *Wavelets and Subband Coding.* Englewood Cliffs, New Jersey: Prentice Hall, 1995.

[48] N. S. Jayant and P. Noll, *Digital Coding of Waveforms.* Prentice Hall, 1984.

[49] A. A. Webster *et al.*, "An objective video quality assessment system based on human perception," in *Proceedings of SPIE, Human Vision, Visual Processing and Digital Display*, vol. 1913, (San Jose, California, USA), pp. 15–26, 1993.

[50] W. Xu and G. Hauske., "Picture quality evaluation based on error segmentation," in *Proceedings of SPIE, Visual Communications and Image Processing*, vol. 2308, (Chicago, IL, USA), pp. 1454–1465, 1994.

[51] M. Miyahara, K. Kotani, and V. R. Algazi, "Objective picture quality scale (PQS) for image coding," tech. rep., Technical Center for Image Processing and Integrated Computing, University of California, Davis, CA, USA, 1996.

[52] A. B. Watson, "DCT quantization matrices visually optimized for individual images," in *Proceedings of SPIE, Human Vision, Visual Processing, and Digital Display IV*, (San Jose, CA, USA), pp. 202–216, February 1993.

[53] A. Shah, D. Staddon, I. Rubin, and A. Ratkovic., "Multimedia over FDDI," in *Proceedings of 17th Conference on Local Computer Networks*, (Mineapolis, USA), pp. 110–124, 1992.

[54] C. Partridge, *Gigabit Networking*. Reading, MA: Adison Wesley, 1994.

[55] G. Karlsson, "Video over ATM networks." http://citeseer.nj.nec.com/15810.html, 1997.

[56] J. Nielson, *Usability Engineering*. Boston, US: AP Professional Press, 1994.

[57] N. T. Bhatti, A. Bouch, and A. Kuchinsky, "Integrating user-perceived quality into web server design," *WWW9 / Computer Networks*, vol. 33, no. 1-6, pp. 1–16, 2000.

[58] U. Burnik, *Optimalno kodiranje za prenos slik po ozkopasovnih prenosnih medijih (Optimal encoding for image transmission over narrowband communication media)*. PhD thesis, Univerza v Ljubljani, Fakulteta za elektrotehniko, 2002.

[59] D. Westen, *Psychology: mind, brain & culture*. US: John Wiley & Sons, 2nd ed., 1999.

[60] I. Akoulchina and J.-G. Ganascia, "SATELIT-Agent: An adaptive interface based on learning interface agents technology," in *User Modeling: Proceedings of the Sixth International Conference, UM97* (A. Jameson, C. Paris, and C. Tasso, eds.), (Vienna, New York), pp. 21–32, Springer Wien New York, 1997. Available from http://um.org.

[61] J. Orwant, "For want of a bit the user was lost: Cheap user modeling," *IBM Systems Journal*, vol. 35, no. 3, pp. 398–416, 1996.

[62] G. Kiczales and X. Parc, "Beyond the black box: Open implementation," *IEEE Software*, vol. 13, no. 1, pp. 8–11, 1996.

[63] B. Kerherve, A. Pons, G. V. Bochmann, and A. Hafid, "Metadata modeling for quality of service management in distributed multimedia systems," in *Proceedings of the First IEEE Metadata Conference*, (Silver Spring, Maryland, USA), 1996.

[64] U. Burnik and J. F. Tasič, "Network optimisation for remote multimedia imaging applications," in *Proceedings of COST #254 Workshop*, (Toulouse, France), pp. 147–151, 1997.

[65] U. Burnik and J. F. Tasič, "Efficient image-intensive communication system supporting personalised user requirements," in *Proceedings of X Eu-*

ropean Signal Processing Conference (EUSIPCO 2000) CD-ROM, vol. 1, (Tampere, Finland), September 2000.

Chapter 2

ANNOTATION, STORAGE, RETRIEVAL AND ANALYSIS OF DIGITAL VIDEO

Honglin Li, Yi Zhao, José-Luis Sancho-Gómez, and Stanley Ahalt

Department of Electrical Engineering, The Ohio State University,
2015 Neil Avenue, Columbus, Ohio 43210, United States
{lih,sca}@ee.eng.ohio-state.edu

Abstract   The volume of multimedia data that is generated everyday motivates a growing need for efficient and effective methods to index, organize, retrieve, and analyze video data. This chapter investigates techniques that are needed to access video data effectively. These techniques can be classified into three categories: frame-grouping, semantic association, and story/structure construction. To demonstrate our method of grouping we introduce a video query system based on similarity measures of low-level video features. We then demonstrate how predictive models can be built on the low-level video features in order to predict mid-level semantic characteristics associated with individual video units. Finally, by adopting syntactic pattern analysis, we show how high-level complex video patterns can be analyzed and syntactic models can be built to represent high-level video events that capture the structure of the video. We argue that all levels of video information –video meta-data– needs to be organized so that it is convenient for subsequent data storage, exchange, transmission, and analysis. As in the case with MPEG-7, we describe the use of an XML-based video meta-data for these purposes.

Keywords: Video indexing, video annotation, video retrieval, extensible markup language (XML), syntactic video analysis.

1.   Introduction

The past decade has witnessed significant advances in multimedia technology. Increasing computing power, broadband networks, new compression techniques, cheaper storage, and inexpensive and ubiquitous acquisition devices make multimedia data -image, video, audio, speech, and graphics- virtually omnipresent. Through Video CD (VCD) and Digital Versatile Disc (DVD)

J.F. Tasič et al. (eds.), Intelligent Integrated Media Communication Techniques, 43-87.

players, MPEG Audio Layer 3 (MP3) players, and the Internet, everyone is becoming a multimedia consumer. Moreover, because of the popularity of digital cameras and camcorders everyone is becoming a potential multimedia content producer, and is capable of publishing the multimedia data using the web. Add to this mixture emerging multimedia data sources such as digital libraries, distance learning, telemedicine, and video surveillance, and one can see that there is an explosion of multimedia data occurring - and which will continue to grow for the foreseeable future. It is a reality that the sheer volume of digital multimedia data that is being generated all over the world, everyday, threatens to overwhelm multimedia consumers. Indeed, it can be argued that - without efficient means to access this vast data resource - our global society may be overwhelmed and never make appropriate use of this unprecedented resource. Thus we argue that there is a growing need to investigate how we access multimedia data.

We currently have relatively mature methods to access traditional data. For example, for books and articles, we have library catalogs of which most are now electronic, various indices, abstracts (e.g. Chemical Abstracts), and Table Of Contents (TOCs). For textual information on the Internet, we have search engines like Google [1] and Yahoo [2]. For database information, we have the widely used Structured Query Language (SQL) to do various queries.

Unfortunately, in contrast, we have relatively poor tools to access multimedia data. For most multimedia we can FastForward (FF), Rewind (REW), or we can linearly browse through data streams in their entirety, which is a painful and frustrating experience. And the nature of the ever-growing corpus of multimedia data only makes this limited-mode access more difficult. While the pile of hay is getting bigger, finding the needle is only getting harder. Further, both consumers and producers are and will encounter the problem of how to identify and access multimedia content more efficiently.

A natural solution would be to manually analyze the multimedia data and annotate the data by text, which can then be accessed in traditional ways. For example, a text-based search engine could then be applied on the annotations. In fact, Google [1] provides image search based on key words. However, this approach has some inherent drawbacks. First of all, it is very expensive to manually analyze and annotate multimedia data. Considering the sheer volume of multimedia data, especially video data, it will continue to be impossible to manually annotate the data in all cases. Second, different people naturally have different interpretations for the same multimedia data. For example, Figure 2.1 shows one simple image. One interpretation would say the image is about a street, another would say it is about the buildings, and a third might say it is about two men.

There are no unique ways to annotate images using text, and annotating the image with all possible interpretations produces annotations too extensive to be

Figure 2.1. Figure shows one simple image. One interpretation would say the image is about a street, another would say it is about the buildings, and a third might say it is about two men.

manageable. Thus, although manually annotating multimedia data using text might have utility in some specialized domains, it is expected that automatic, or at least semiautomatic, non-text based audiovisual features will play important roles in accessing multimedia data. Currently, automatically generated features are limited to low-level features, such as color, motion, and shape. However, these non-text features are believed to provide content consumers richer and more powerful information for identifying, filtering, retrieving, and analyzing multimedia data. Augmented by various content-based features, text or non-text, multimedia data is expected to become much more easily accessed and analyzed.

In this chapter, we will focus on video data instead of general multimedia data. Video is becoming more and more popular and there is consequently a significant and growing need for efficient and flexible video access methodologies. Consumers would like to see video systems that provide means of easily browsing, retrieving, and analyzing video. To browse video, we might need TOCs, visual abstracts and efficient methods of playback. To retrieve video, we need to provide users with flexible approaches such as query by key words, query by attributes (color, luminance, motion, etc.), query by sketch, query by event, and query by example. Also, users might need to retrieve video based on different granularities such as at the image level, or at the shot level, or at the scene level. To analyze video, the video systems should provide users mechanisms to detect special patterns, anomalies, and events. In the light of the above discussion, a typical video system would take the form illustrated in Figure 2.2.

The Information Extraction block is responsible for extracting information from the raw video data. In this chapter we define two different data forms: raw video data (or reproduction data) and video meta-data (or descriptive data). Video meta-data, the data that describes the video content, is the information extracted from the raw video data. Meta-data can be text or non-text, low-level or high-level, automatically generated or manually annotated. Video access will, in most cases, be conducted based on video meta-data.

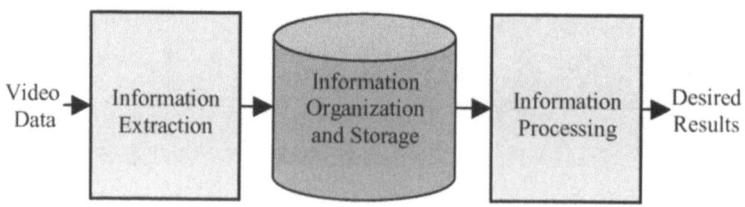

Figure 2.2.   A typical video system.

The Information Organization and Storage block deals with how to organize and store the video data, including both the raw data and meta-data. The video meta-data might be stored with the raw video data, or it could be stored separately from the raw video data. In any case, the organization of the data should be conducive to the subsequent video accessing.

The Information Processing block represents video access, retrieval, or analysis operations, each of which operates on the video meta-data. It is expected that future video content providers will provide the video meta-data along with the raw video data and at the same time there will be effective video analysis systems working on legacy video data.

There is considerable research underway on all of the above three blocks. Video segmentation, video indexing, video query, video storage and video analysis all fall into this model. The essential issues that researchers are trying to address are numerous. What kinds of video features are important? How can we extract various video features? How should the extracted meta-data or video features be organized? How can we store the meta-data? How can we efficiently retrieve video data, and how can we perform pattern analysis on video data to find specific images, shots, or scenes? Finally, what kind of potential applications are most critical and economically feasible? For an overview of the key ideas currently being studied there are numerous review articles [3],[4], [5],[6],[7], most of which focus on the big picture and the various techniques of video feature extraction.

For example, by associating similarity measures to the various video features, image/video retrieval or query systems have been developed. Indeed, there are examples of working systems such as QBIC [8], Virage [9], VideoQ [10], and Photo-book [11], for specific video domains. Other researchers [12] have demonstrated a video annotation engine in which the video meta-data is organized using SGML (Standard Generalized Markup Language). In September 2001, ISO released a new international standard, MPEG-7 [13], [14], formally called Multimedia Content Description Interface, which specifies some standard video descriptors -representations of video features- covering color, motion, shape and texture. The core of MPEG-7 is the use of XML (eXtensible Markup Language) as the language for the Description Definition Language (DDL) used to organize the meta-data. With respect to video analysis, there are also many attempts using various techniques including a proposed three-level video event detection methodology using neural networks [15]. Alternatively, [16] describes a probabilistic syntactic approach to the detection and recognition of temporally extended video activities. There is also a body of research employing string analysis working on the automatic labeling of video shots using time-constrained clustering to detect specific basic events [17].

We observe that although there is a lot of research working on various aspects of video indexing, annotation, storage, and analysis, there is no consistent framework. In this chapter, we propose a systematic approach that can structure the various components of video analysis, and we provide examples of the proposed components.

This chapter is organized as follows. Section 2 discusses the proposed system at a relatively detailed level. Sections 3, 4 and 5 each deals with a subcomponent of the system. In Section 3, a video query system applied on fixed-view motion imagery is described. An example of automatic video annotation is presented in Section 4. Section 5 presents our prototype work on syntactic video analysis. Finally, Section 6 summarizes this chapter.

2. Proposed System Architecture

The proposed system architecture is shown in Figure 2.3. The raw video data first undergoes a video segmentation process, which in this case is a shot detection operation that divides the video data into basic units -shots. Information extraction then operates on the shots in order to produce video meta-data that describes the video content. Next the video meta-data is organized using XML. Various video operations can subsequently be performed on the video meta-data. In this section, we will describe these steps at a relatively detailed level.

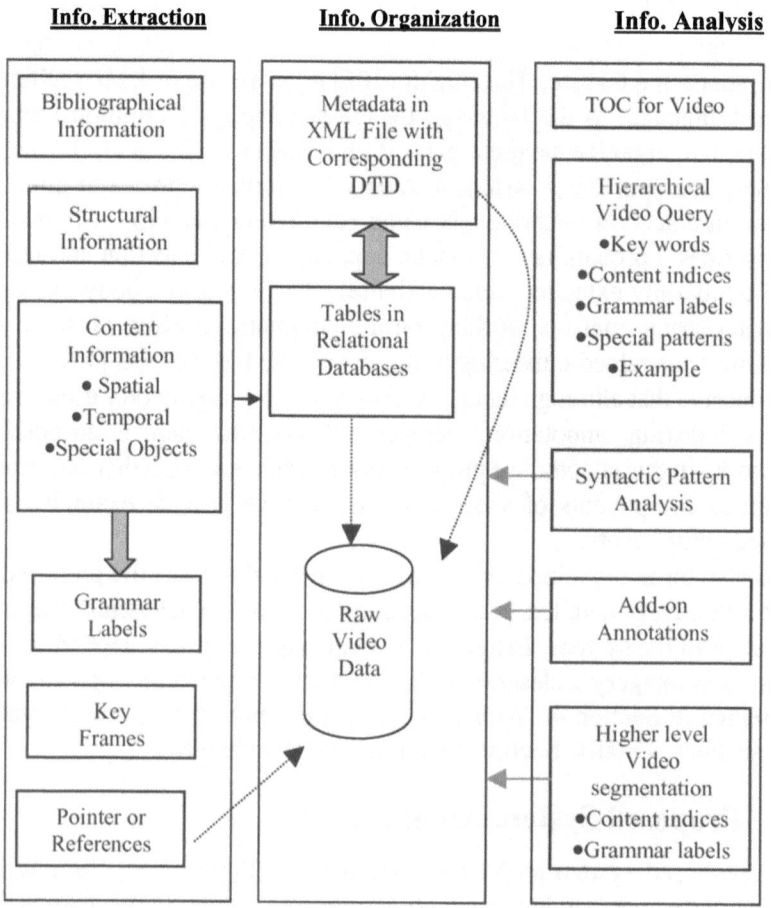

Figure 2.3.   Proposed system architecture.

2.1   Shot Detection

The structure within a video stems from the fact that video streams are formed by editing different video segments known as shots. A shot is a sequence of frames generated during a continuous camera operation and it represents continuous action in time and space. Shots can be joined together in either an abrupt transition mode, in which two shots are simply concatenated, or through gradual transitions, in which additional frames may be introduced using editing operations such as dissolve, fade-in, fade-out, and wipe. The purpose of shot detection is to detect the shot boundaries, either abrupt transition or gradual transition, and so segment the video stream into shots, which are the basic video units for subsequent processing.

There are a number of shot detection algorithms that fall into two categories: those that work in the uncompressed domain and those that operate in the compressed domain.

2.1.1   Shot detection in the uncompressed domain.

Pixel-wise comparison [18], [19].   Shot detection can be detected by emphasizing the spatial similarity between two frames. The simplest way to measure the spatial similarity is to compare every pixel in one frame with the corresponding pixel in the next frame. A shot boundary is declared if the sum of the pixel value difference exceeds a pre-specified threshold. This algorithm is sensitive to noise, object motion, and camera operations.

Histogram-based techniques [20],[21].   The intensity/color histogram of a gray/color image f is an N-dimensional vector $\{H(f,i); i = 1, 2, ..., N\}$ where N is the number of levels/colors and $H(f,i)$ is the number of pixels of level/color i in the image f. The histogram-based approach tries to compare the histograms of two consecutive frames and claims a shot boundary if the difference exceeds a certain threshold. This technique is less sensitive to camera operations and object motion compared to the pixel-wise comparison since histograms are invariant to image rotation and change slowly under the variations of viewing angle, scale, and occlusion.

Block-based techniques [18],[22],[23].   The previous two techniques work on global attributes that are not sufficient in some scenarios. For example, if a white object moves from one region to another region with a fixed black background, the color histogram cannot catch the difference. It is a natural extension to consider local attributes instead of global ones. In block-based techniques, each frame is partitioned into a set of blocks, each of which will be compared between two consecutive frames. The difference of two frames will be assessed on the differences of corresponding blocks.

Twin-comparison [24].   This algorithm has been proposed for the detection of gradual scene changes using a dual threshold. If the histogram difference between two consecutive frames exceeds a relatively high threshold, an abrupt shot boundary is declared. If the histogram difference exceeds the lower threshold we assume that a gradual shot transition is started and we begin to compute the accumulated difference of the following frames from the start of the assumed gradual transition and keep track of the difference between two consecutive frames. If the accumulated difference exceeds the high threshold, a gradual transition is declared. If the difference between two consecutive frames drops below the low threshold for more than two frames, we declare that there is no graduation transition.

Model-based segmentation [25],[26].   In model-based techniques, the problem of video segmentation is viewed as the process of locating the edit boundaries within the video sequence. Different edit types, such as cuts, translates, wipes, fades, rotations and dissolves are modeled by mathematical functions. The consecutive frames are checked against those mathematical functions to see whether there is a certain shot boundary between them.

2.1.2   Shot detection in the compressed domain.

Compression differences [27].   One simple technique [27] uses differences in the size of JPEG compressed files to detect shot boundaries.

DCT coefficients [28], [29], [30], [31].   The popular image/video compression standards, JPEG, MPEG and H.261, are based on Discrete Cosine Transform (DCT). The DCT coefficients in the frequency domain are related to the pixel values in the spatial domain. As a result, the DCT coefficients can be used for shot detection in the compressed domain. The fundamental principle is to investigate the difference between the DCT coefficients of two consecutive frames.

Motion vectors [29], [32].   For H.261 and MPEG compression standards, the motion information is encoded in the form of motion vectors, which can then be employed to detect shot boundaries. Recent work [29] has proposed a technique for shot detection using motion vectors in MPEG. This approach is based on the number of nonzero motion vectors.

Subband decomposition [33].   Due to the popularity of wavelet techniques, which has, for example, been incorporated into the new compression standard JPEG2000 [34], shot detection based on wavelet coefficients has drawn much attention. Lee and Dickinson [33] have presented a shot detection algorithm applied on the lowest subband coefficients. Again, the difference between the co-efficients of two consecutive frames is investigated to locate the shot boundaries. We note that the previously discussed pixel-wise comparison, histogram-based techniques, and block-based techniques are also well suited for this purpose.

2.2   Information Extraction

After the video is segmented into shots, the video content information can be extracted from shots in the form of various video features. By analyzing the shot level video features, one can group shots into higher level units –scenes. Additionally, one can go further to group scenes into story-units that compose the whole video. This hierarchical relation is illustrated in Figure 2.4.

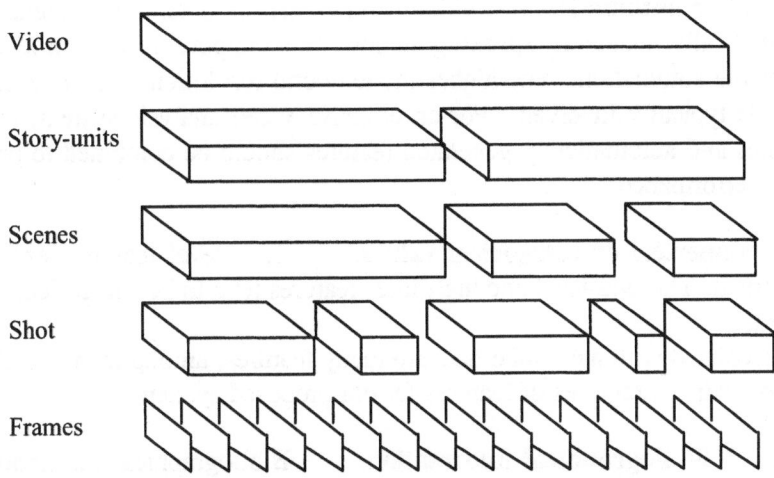

Figure 2.4.   Video structure.

Video features have the following properties:

Abstraction level.   Low-level features are typically quantitative values calculated directly from the pixel value of the image(s). For instance, the color histogram and motion vector are two low-level features. Mid-level features are those that describe the objects in the video stream and the relationships among the objects, such as the appearance and locations of persons, or the relative spatial relationships among players on a team. High-level features are generally considered to contain those features that describe the overall activity, or story, contained in the video. Generally speaking, the higher level features require more manual work. In most cases, automatic video indexing does not work well for high-level video features.

Temporal resolution level.   Video features can also have different levels of temporal resolution, e.g., at the frame level, the shot level, the scene level, the story-unit level, and the entire video level [35]. Generally, video data will be segmented -shot detection or/and scene detection- prior to extracting the features.

Aggregation level [35].   If a feature is calculated for each frame in the shot, such a feature description is called non-aggregated. If a feature is derived from a set/sequence of frames, such a feature description is called partially aggregated. If at the very most, only one feature is derived for the whole shot, it is called completely aggregated. The same concepts apply to the other different temporal resolution levels.

Manual or automatic.   Video features can be determined manually or automatically. Generally speaking the lower the temporal resolution level, the easier the automation. The higher the temporal resolution, the more manual work is typically involved. For an effective video analysis system, manual features and automatically generated features should be combined to provide good performance.

Text, numerical or categorical values.   Low-level features tend to be numerical. The mid-level and high-level features tend to be categorical values or text.

Associated with any video data are many features, among them we choose the following categories of features for our proposed system.

2.2.1   Bibliographical information.   Bibliographical information is textual indices that consist of knowledge of the production source, the abstract, the frame rate, the video resolution, the compression format, the category, the frame numbers and so forth. This information is mostly annotated manually and mainly for high temporal lever units.

2.2.2   Structural information.   Structural information encodes the hierarchical structure of the video, in which a whole video is divided into many story-units that in turn are divided into scenes, shots, and frames. This structural information easily yields a TOC of the video that expedites the video browsing process.

2.2.3   Low-level content-based information.   Low-level content-based information consists of various features that fall into two categories: spatial, and temporal.

Spatial features. A set of key frame(s) can be selected to represent each shot, and image indexing techniques are then applied to the key frame(s) to derive spatial features for the shot.

Color features. Color features are one of the most widely used visual features in image/video applications. In image retrieval, the color histogram is the most commonly used color feature. Besides the color histogram, several other color feature representations have been proposed, including color moments [36] and color sets. By taking into account the location of the colors, the spatial information can be incorporated into the color feature. For example, a local color histogram [37], which will be explained more in later sections, captures the locality of colors in an image.

Texture features. Texture refers to the visual patterns that have properties of homogeneity that do not result from the presence of a single color or intensity. It is an important feature of a visible surface where repetition or quasi-repetition of fundamental pattern occurs. Texture features such as contrast, uniformity, coarseness, roughness, regularity, frequency, density, and directionality provide significant information for scene interpretation and image classification [38].

Shape feature. The shape of an object refers to its profile and physical structure. Shape features are fundamental to systems such as medical image databases, where the color and texture of objects are similar. In general, shape features can be represented using traditional shape analysis methods such as invariant moments [39], Fourier descriptors [40], autoregressive models [41], and geometry attributes [42].

Special objects. By applying computer vision techniques, special objects such as human faces can be recognized and stored with their locations.

Temporal features. The apparent motions in video data can be attributed to object motions and camera motions, which can be captured as important video indices.

General motion. Motion information can be predicated using Block Matching Algorithms (BMAs) or Optical Flow Fields (OFFs) [43]. In fact, H.261 and MPEG compression standards incorporate block-based motion vectors into the compressed files and so it is possible to extract the motion vectors in the compressed domain. Of course, feature representations other than motion vectors can also be chosen to capture the motion information. One example is the difference of color histograms.

Object motion. If we can successfully segment and recognize special objects, every object can be associated with motion information. Object trajectories can then be derived, which are of great importance in video analysis.

Camera motion. Camera motion is another common source to producing motion. The seven basic camera operations are fixed, panning (horizontal rotation), tracking (horizontal transverse), tilting (vertical rotation), booming (vertical transverse), zooming (varying the focusing distance), and dolly (horizontal lateral movement). The category and intensity of camera motion can be used as motion indices. Akutsu et al. [44] have used motion vectors and their Hough transforms to identify the seven basic camera operations.

2.2.4   Grammar labels.

Video shots can be classified based on the low-level video features and one or more labels can then be assigned to each shot. In this chapter, we call these labels grammar labels since they can be used in syntactic video analysis. The grammar labels may have practical real-life meaning or may not, depending on the potential applications. In Section 4, a supervised automatic annotation model is discussed in which the grammar label has a specific meaning – the sky condition associated with the satellite cloud cover data. In Section 5, an unsupervised labeling scheme is presented in which the grammar labels do not have specific meaning, they are just symbols used to represent and differentiate different shot clusters. In [17], the video shots are labeled using time-constrained clustering and the labels are employed to detect dialogues, actions and story-units. Again, the labels do not have practical meanings.

Most of the time, in order to generate the grammar labels, a predictive model must be constructed based on the low-level video features. In Section 4 and Section 5, we will demonstrate both supervised model and unsupervised model.

2.2.5   Key frames extraction.

A set of key frames can be extracted from each shot in order to a) represent the shot and make the presentation of the shot easier and b) extract spatial features from the key frames instead of the whole shot. Different algorithms have been proposed for key frames extraction, based on visual differences or on the motion patterns.

Predefined positions. In [45] one key-frame is selected at predefined position for each shot. This is the simplest method, but not sufficient for many applications.

Local minima of motion. An algorithm using optical flow analysis [46] to measure the motion in a shot has been presented, and this algorithm selects key frames at the local minima of motion. First the optical flow field is computed and the sum of optical flow magnitudes for each frame is calculated as follows.

$$M(t) = \sum_{i,j} |O_x(i,j,t)| + |O_y(i,j,t)|, \qquad (2.1)$$

where $O_x(i,j,t)$ and $O_y(i,j,t)$ are the horizontal and vertical components of the optical flow at pixel (i,j). Then the key frames are selected at the local minima of $M(t)$.

Significant pauses. A proposed algorithm [47] to select key frames by detecting significant pauses in a video stream has been discussed in the literature.

Their approach is based on a χ^2-test between the intensity distribution of the frame representing the possible start of a pause and the subsequent frames. Good results have been obtained even in the presence of slow camera motion.

Dissimilarity measure based on normalized difference of luminance. The algorithm proposed in [48] uses the following dissimilarity measure based on the normalized difference of luminance projections,

$$
d_{l_p}(f_i, f_j) \quad = \quad \frac{1}{255 \cdot (J + K)} \tag{2.2}
$$

$$
\cdot \left(\frac{1}{J} \sum_{n=1}^{K} |l_n^r(f_i) - l_n^r(f_j)| + \frac{1}{K} \sum_{m=1}^{J} |l_m^c(f_i) - l_m^c(f_j)| \right),
$$

where $l_k^r(f_i), l_k^c(f_i)$ represent the luminance projection for the kth row and kth column respectively. Initially, the first frame is selected as one key frame. A new frame is selected as a key frame whenever its dissimilarity from the last selected key frame is greater than or equal to a given threshold ε. Since the dissimilarity used are normalized to the interval [0,1], when $\varepsilon=0$ all frames are selected as key frames, while when $\varepsilon=1$ only the first frame is selected as key frame.

Seek and spread strategy. Xiong et. al. [49] proposed a "seek and spread strategy" to locate the key frames within one shot, which consists of two stages. In the first stage, the first frame is compared with the following ones until a frame that is different enough is found or the end of the shot is reached. The frame before the found frame is selected as a key frame. In the second stage, the current key frame is compared with the following frames until finding a frame different enough or reaching the end of the shot. The newly found frame is the start frame for computing the next key frame.

2.2.6   Pointers and references.   The video meta-data has to be associated with the raw video data by pointers and references. By adding pointers, it is easy to locate the physical raw video data from the meta-data. By adding references, the relevant raw video data, such as similar shots, can also be located. In this way, the video meta-data can be stored separately from the raw video data.

2.3   Information Organization

After a number of video features are extracted, the problem becomes how to organize the video meta-data. A scheme is needed to describe and organize the video information so that it is easy for data storage, exchange, transmission, and operation. There are requirements for any description scheme. It should be hierarchical in order to represent the various temporal resolution levels of video such as video, story unit, scene, shot, and frame. The scheme also should be self-descriptive so that every application can understand what the meta-data means. Still, the scheme should be extensible since more information may be needed. Finally, the scheme should be able to easily inter-operate with other applications. Considering all the above factors, as is the case in MPEG-7 [14] we choose XML [50] to be the language used to describe video meta-data.

2.3.1   XML to organize video information.   XML stands for eXtensible Markup Language, which is a markup language much like HTML. However, there are some fundamental differences between XML and HTML. Here we list some key characteristics of XML to indicate why we choose XML as the language to organize video meta-data.

- XML was designed to describe data and to focus on what the data consists of. HTML was designed to display data and to focus on how data looks. HTML is focused on information display, and XML is about focused on information description. Thus XML is not a replacement for HTML, but a complement to HTML.

- XML tags are not predefined in XML. You can define your own tags in XML while HTML tags are predefined.

- XML focuses on data structure.

- XML is inherently hierarchical.

- With DTDs (Document Type Definitions), XML document is totally self-descriptive and can be validated against the specified DTD.

- XML is extensible. People are free to define their own tags.

- XML is neutral with regard to databases. More and more databases support XML.

- XML is increasingly popular, with more and more applications supporting XML.

XML documents use a self-describing and simple syntax. Figure 2.5 shows a toy XML example. The first line in the document - the XML declaration - defines the XML version of the document. In this case the document conforms to the 1.0 specification of XML. The next line describes the root element of the document (like it was saying: "this document is a note"). The next 4 lines describe 4 child elements of the root (to, from, heading, and body). XML elements can have attributes in the start tag, just like HTML. Attributes are used to provide additional information about elements.

```
<?xml version = "1.0"?>
<note>
<to>Tove</to>
<from>Jani</from>
<heading>Reminder</heading>
<body>Don't forget me this weekend</body>
</note>
```

Figure 2.5.   A toy XML example.

A DTD (Document Type Definition) defines the legal elements of an XML document. The purpose of a DTD is to define the legal building blocks of an XML document. It defines the document structure with a list of legal elements. A "Well Formed" XML document should conform to the rules of a DTD. By defining new elements in a DTD, one has more customized tags, which makes XML extensible. Of course, different applications need different sets of tags and consequently need different DTDs. For video data, different types of video may need different DTDs.

For our system, after the video information has been extracted, the resultant video meta-data is stored using XML, which is a convenient format for subsequent analysis and for data storage, exchange. Sometimes it is desired to map the XML file to a database schema so that the video meta-data can be stored in a database. Due to the strict syntax of XML, it is not difficult to parse an XML file and convert the XML content to a database schema. In fact, more and more databases have XML interfaces, which makes the data translation process between database and XML file even easier.

2.3.2   An example.   Figure 2.6 illustrates an example using XML to describe video meta-data. This example is a simple prototype used to demonstrate the basic capabilities of this format. In this example, the whole video is segmented into story-units, story-units are segmented into scenes, and scenes into shots, thanks to the hierarchical structure of XML. Some bibliography information data, mainly from manual annotation, are included in the higher temporal resolution level structure. On the other hand, the automatically generated low level spatial and temporal features are included in the lower temporal resolution level. Based on the low-level features, the grammar labels, part of the legal elements for this XML file, are assigned to the video shots. Furthermore, there are pointers associated with every temporal resolution level.

```xml
<?xml version="1.0"?> <Video ID = "001">
    <Title> NBA Final </Title>
    <Source> CNN </Source>
    <Resolution> CIF(352x288) </Resolution>
    <Format> MPEG-1 </Format>
    <FrameRate> 30 </FrameRate>
    <Abstract> The first game in NBA final stage between LAKE and 76ers
</Abstract>
    <Category> Sports </Category>
    <StartF> 1 </StartF>
    <EndF> 16,000 </EndF>
    <Story-unit ID = "001">
        <StartF> 1 </StartF>
        <EndF> 1012 </EndF>
        <Category> Actions </Category>
        <Scene ID = "001" >
            <StartF> 1 </StartF>
            <EndF> 300 </EndF>
            <Category> Lake Offense </Category>
            <Shot ID = "001">
                <StartF> 1 </StartF>
                <EndF> 98 </EndF>
                <R-Frame> 54 </R-Frame>
                <SpatialFeatures>
                    <ColorHistogram> 25,39,34,30 </ColorHistogram>
                    <EdgeOrientHist> 34,34, 5, 54 </EdgeOrientHist>
                </SpatialFeatures>
                <TemporalFeatures>
                    <MotionVectors> 34,45, 54 ,34 </MotionVectors>
                    <DiffColorHist> 3,4,5, 3 </DiffColorHist>
                </TemporalFeatures>
                <AudioFeatures>
                    <Pitch> 34 </Pitch>
                </AudioFeatures>
                <ClosedCaptions>
                    <HostA> Ivanson is flying! </HostA>
                    <HostB> I bet he will be the MVP this year!</HostB>
                </ClosedCaptions>
                <Label> B </Label>
            </Shot>
            <Shot ID = "003">
            </Shot>
            <Shot ID = "004">
            </Shot>
        </Scene>
        <Scene ID = "002">
        </Scene>
        <Scene ID = "003">
        </Scene>
    </Story-unit>
    <Story-unit ID = "002">
        <StartF> 1013 </StartF>
        <EndF> 2056 </EndF>
        <Category> Commercials </Category>
    </Story-unit>
    <Story-unit ID = "003">
    </Story-unit>
    <Story-unit ID = "004">
    </Story-unit>
</Video>
```

Figure 2.6. An example XML file organizing the video information.

2.4   Information Analysis

Once we have the available video information and the meta-data organized using XML, there are various video analysis operations that can be performed. Here, several important applications of the video meta-data are identified.

Table of contents (TOCs).   Like a TOC for books, a video TOC is an effective tool to assist people in accessing video information. Because of the hierarchical structure of the video meta-data organized using XML and the pointers associated with each temporal resolution level, a TOC is naturally and directly drawn from the meta-data. With the help of key frames, the video TOC can be made visually, which expedites the browsing process.

Hierarchical video query.   Due to the extensive information contained in the video meta-data, various video queries can be applied to the video data. For example, we can perform query by key words, query by low-level features, query by grammar labels, query by special patterns, and query by example. Nevertheless, the hierarchical structure of the video meta-data enables us to retrieve video according to different time scales, which can answer questions such as retrieving the shots that are similar to an example shot; retrieving the scenes that consist of shots that are similar to the example shot; or retrieving the story-units that consist of shots that are similar to the example shot.

Syntactic pattern analysis.   The grammar labels provide the possibility that syntactic pattern analysis can be performed on the video meta-data. Syntactic pattern analysis is an appropriate tool for analyzing structured pattern, which cannot be easily accomplished by traditional statistical pattern analysis methods. Many video patterns are structured, which makes syntactic pattern analysis a good choice for video pattern analysis. In our proposed system the grammar labels and the hierarchical video structure generate words, sentences, paragraphs and articles. Special video patterns can be identified using syntactic pattern analysis methods, and grammars can be deduced for various video patterns, which is of great help for further pattern analysis. Moreover, the results of syntactic pattern analysis can be augmented with the video meta-data and consequently added to the meta-data.

High level video segmentation.   To have a clear hierarchical video structure is one of our goals for video information extraction. However, in the Information Extraction block in Figure 2.3 we do not include significant higher-level video segmentation operations. Indeed, it is possible that we do not have a complete notion of the video structure from the information extraction block, even though some higher-level video segmentation operations could be incorporated. In other words, the video meta-data we have may not represent the complete video structure, and some parts of the structure may be missing. One possibility is to use the video meta-data and perform higher-level video segmentation, and then augment the meta-data with the results. For example, syntactic pattern analysis can be employed to group shots into scenes.

3.   Video Query Applied to Fixed-View Motion Imagery

Typical video indexing and annotation techniques have been applied to many applications such as news and interview programming [43], movie analysis [51], and sports video analysis [9], [43], [52]. However, there have been very few investigations of fixed-view motion imagery such as registered surveillance imagery, radar imagery, hyperspectral imagery, pollution monitoring, and satellite imagery. In this section, a novel scheme is advocated to apply video indexing on these types of data. By using satellite cloud cover data as a typical example, we show that content-based query can be implemented effectively on fixed-view motion imagery database. Considering the peculiarities of satellite cloud cover data, local color histograms and local motion vectors are chosen as the indexing features. Although the features are rather straightforward, the experimental results have shown that the system is quite effective. Readers can investigate this web-based video query system on satellite cloud cover data.[1]

The rest of this section is organized as follows. Subsection 3.1 presents an overview of the system we have prototyped. In Subsection 3.2, the various components of video indexing process are described in a relatively detailed level. Subsection 3.2 discusses the video clip matching problem. Our experimental results are presented in Subsection 3.4 and Subsection 3.5 identifies possible future research directions.

3.1   System Overview

In this subsection we provide an overview, illustrated in Figure 2.7, of the fixed-field system we have constructed. We emphasize that this is a prototype system. Again, a specific example -satellite cloud cover data— will be used to illustrate the key components of the system. However, please note that the same framework is applicable to many other forms of fixed-view motion imagery. We have chosen to use satellite cloud imagery because it is readily available and relatively easy to analyze.

Video query can be performed using different types of inputs, which result in different kinds of video query, such as query by key words [8], query by attributes [8], query be sketch [10], [53], and query by example [43]. As shown in Figure 2.7, we are interested in query by example due to the potential applications of such a system, for example coarse weather predication. The

---

[1]http://eepc269.eng.ohio-state.edu/matlab/[Jan. 2003]

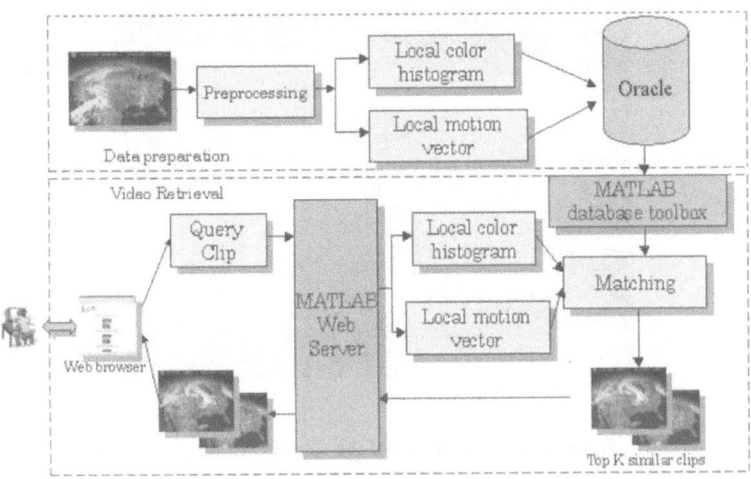

Figure 2.7. The system framework.

system divides naturally into two parts, one for data preparation (or database population), and the other for video retrieval. For data preparation the database video clip is gathered, and after some preprocessing (explained in Section 3), the features of the data – in this case, local color histograms and local motion vectors – are extracted in the video indexing process. The features are then stored in a database, in this case an Oracle database. During video retrieval a query clip is presented to the system, its features are also extracted, and then subsequently compared with the features of the stored clips in the database. The most similar clips are then output as the query result.

The user interface of this system is web-based. Due to the popularity of MATLAB in engineering and scientific community, the MATLAB Web Server was chosen for use in this system. Similarly, the Matlab Database Toolbox serves as the bridge between MATLAB and Oracle database.

3.2 Content-Based Video Indexing

Content-based video indexing is the heart of any video query system. Video indexing uses various extracted features drawn from a video clip and the features are used to measure the similarity between two video clips. In the following subsections, we provide a more detailed description of the video indexing process on fixed-view motion imagery, using satellite cloud cover data as a typical example.

Figure 2.8.   One frame of the satellite cloud picture.

3.2.1   Satellite cloud cover data.   The satellite cloud cover pictures
were collected by Yahoo Weather.[2] They are grouped into 6-picture sets. Each
is referred as a video clip in this paper, and covers three hours of observation.
Figure 2.8 shows one example frame of such a video clip.

Satellite cloud cover pictures have some special properties. First of all these
pictures are one kind of fixed-view motion imagery, which implies that there
is no camera motion. Hence, as for motion-based features, we only need to
consider the object motion, i.e. the cloud movement. Second, since the camera
does not move and the cloud is always moving, the concepts of shot, scene, and
story unit don't directly apply to satellite cloud pictures. Therefore, there is
no shot detection or any other kind of video segmentation. We simply use the
6-picture set as the unit of the video clips and do video query based on these
video clips.

3.2.2   Outline of video indexing.   Figure 2.9 illustrates the video index-
ing process, which serves as a key component in both the data preparation and
video retrieval. The input video clip first undergoes a series of preprocessing
actions, which includes decompression and background elimination (explained
later). Subsequently, the two important video features -local color histogram
and local motion vector- will be extracted using image and video processing
techniques. Furthermore, a key frame is extracted from the video clip and the
decimated version of the key frame is used as an icon to represent the video
clip. For this paper the method of extracting the key frame is trivial – as we
always choose the last frame as the key frame.

---

[2]http://weather.yahoo.com [Nov. 2001].   This   data   originated   from   Weathernews,   Inc.
(http://us.weathernews.com/)

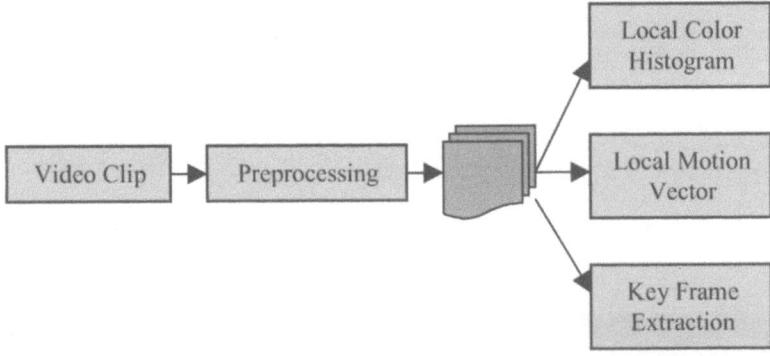

Figure 2.9. Video indexing process.

The feature vector, which consists of local color histograms and local motion vectors, is stored in the Oracle database.

3.2.3   Preprocessing.   As described in the previous section, the preprocessing step has two actions, decompression and background elimination. The video clips we have are in GIF89a format. Decompression is done because we extract the video features in uncompressed domain. We eliminate the fixed, mainly the land and the sea, as shown in Figure 2.10(b). By doing this, we will have only the clouds left in each frame.

There are two motivations for doing background elimination. One advantage is that it can significantly reduce the range of the pixel color value and thus simplify the subsequent processing. Indeed, only the gray pixel values instead of RGB color values will work well. As shown in Figure 2.10(c), the image after background elimination has primarily two color values, white or black. Another reason for doing background elimination is that the process improves the performance of subsequent steps. This occurs since the background does not move for fixed-view motion imagery, and if we keep the background in the images, the variations of different sub areas of the background may have some deleterious effect when we calculate the local motion vector.

Assume $O(i,j)$, $B(i,j)$, $N(i,j)$ are the pixel values for the pixel located at the ith row and j^{th} column in the original image, the background image and the image after background elimination, respectively. Background elimination results in

$$N(i,j) = \begin{cases} O(i,j), & \text{if } O(i,j) \neq B(i,j) \\ 0, & \text{if } O(i,j) = B(i,j) \end{cases} \qquad (2.3)$$

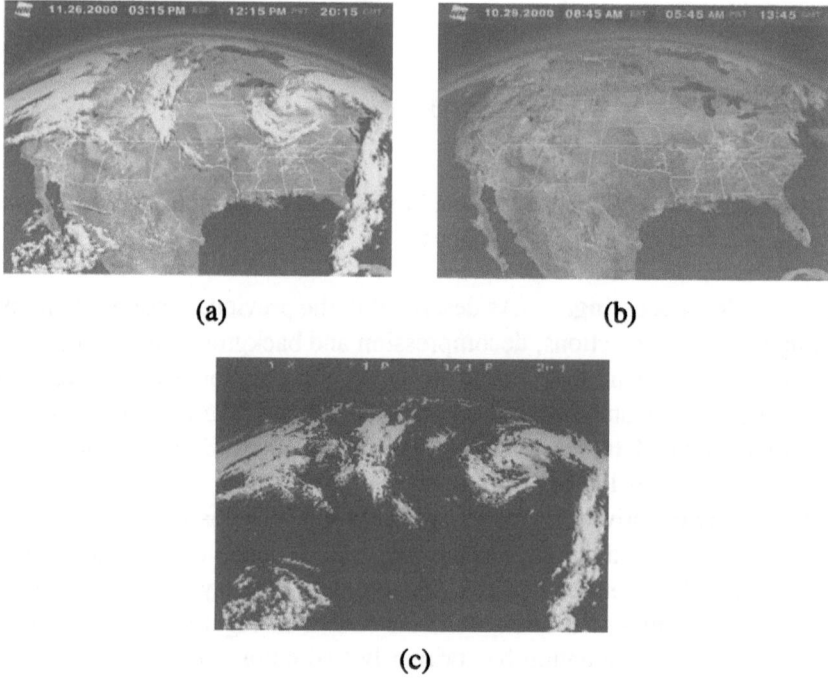

Figure 2.10.   Background elimination: (a) The original image, (b) The background image, (c) The image after background elimination.

3.2.4 Local features instead of global features.

Video features can be either global or local. Global features are calculated based on the whole image, while local features are calculated by dividing the whole image into sub areas, and then having feature values for each sub area. In this paper, we choose to use local features instead of global ones. The reason for this choice is that local features give better accuracy and thus better performance. By using local features, the spatial information is incorporated into the video features. Assume that we have two cloud images consisting of clouds with the same areas but different locations. Under this scenario, a global color histogram will give the exactly same results, while local color histogram can easily emphasize the difference and the same reasoning holds for local motion vectors.

Figure 2.11. Local features are based on the subblocks.

However, calculating local features involves more computation than calculating global features. Consequently, it is common to strike a compromise between computational complexity and spatial or temporal accuracy. In this chapter, we divide one video frame into 40 square blocks, each of which produces one color histogram and one motion vector. Figure 2.11 illustrates one frame, and how it is divided into sub areas.

3.2.5 Local color histogram and its distance measure.

In image and video databases, color is an important factor in classification, clustering, and retrieval. In order to characterize color features of individual image or video clips, different approaches have been investigated.

Some authors [54] have proposed the division of the whole color space into a certain number of bins, and then use the bin with the most pixels as the dominant color in a frame –which becomes a color feature. Other authors [10] have computed the mean color value for all pixels in a frame, usually referred as a color space center for the frame, and this mean is used as a color feature. Although the dominant color and the color space center are simple to compute and can result in good results in image and video retrieval, they are not appropriate for use with fixed view motion imagery because the image clips are typically very similar. Even if the second order information, i.e., the variance, is taken into account, the color feature description is still ambiguous for many types of fixed-view imagery.

In image indexing systems, global color/intensity histograms [4], [7] are often used because of their invariance to viewing angle, image rotation and translation, etc. However, for fixed view motion imagery, this virtue is also not desirable, due to the fact that the object location is typically critical, as discussed in Section 3.4. Consequently, in this paper, we propose the use of local color histograms as one color feature for fixed view motion imagery.

Local Color Histogram Computation.   Generally speaking, to compute local color histograms, each frame is first divided into B blocks; then the color histogram for each small block is computed using N bins, where B and N are empirical parameters. In this paper, $C_{V_m}(f_n, b_i)$ denotes the local histogram, a vector of length N, for block b_i in frame f_n of clip V_m.

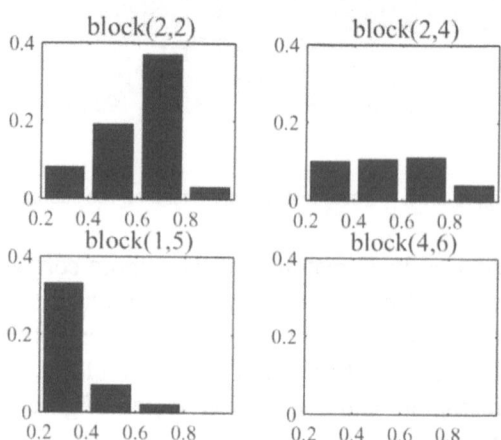

Figure 2.12.   Local histogram for four different blocks of the image shown in Figure 2.11.

After background elimination, the satellite cloud image does not have too many color variations so we can convert the original color images to grayscale images without losing accuracy. Furthermore, the intensity histogram can be used to simplify the problem. Since the grayscale values of most pixels after background elimination are zero, we additionally exclude the very low intensity pixels so that the local histograms only reflect cloud distribution information. For the two parameters B and N, we empirically choose $B = 40$ and $N = 4$.

Figure 2.12 shows the local histograms of four different blocks for the image frame shown in Figure 2.11.

Distance measure of local color histogram.   The local histogram $C_{V_m}(f_n, b_i)$ is also the approximate probability mass function (PMF) of color for block b_i in frame f_n of clip V_m, and one of the standard methods to measure the distance of two PMFs is the Kullback Leibler distance [55] (abbreviated as K-L distance below). Hence in this paper we use K-L distance to measure the similarity in the color feature of two cloud clips.

The K-L distance of two histograms H_1 and H_2 is

$$d_{KL}(H_1, H_2) = \sum_{j=1}^{N} H_1(j) \log_{10} \frac{H_1(j)}{H_2(j)} \tag{2.4}$$

where N is the number of bins in the histogram.

We define the corresponding blocks at the same location in two cloud clips to be a block pair. Thus the K-L distance of two cloud clips can be defined as the average over the K-L distances for all block pairs in the two clips:

$$d_C(C_{V_1}, C_{V_2}) = \frac{1}{6B} \sum_{n=1}^{6} \sum_{i=1}^{B} d_{KL}(C_{V_1}(f_n, b_i), C_{V_2}(f_n, b_i)) \tag{2.5}$$

3.2.6   Local motion vector and its distance measure.   As shown in Section 3.2.4, every block in the image produces one motion vector, which is also called the block-based motion vector. In other words, all pixels in a block are assumed to have the same motion. The motion estimate for the current block can therefore be obtained by searching the most similar block in the previous frame around the position of the current block. Figure 2.13 shows the motion vector for one block. The light gray block is the block in the current frame and the dark block is the most similar block in the previous frame. The arrow corresponds to the motion vector. Video compression standard H.261 [41] and MPEG [42] use this kind of block-based motion vector to perform motion compensation.

The biggest problem in calculating motion vectors –equivalent to finding the most similar block in the previous frame- is its formidable computational complexity. Instead of doing an exhaustive search, we use forward motion prediction with early stopping, which is explained in the following subsection.

Forward motion prediction with early stopping.   The algorithm of forward motion prediction with early stopping can be described in terms of the following steps.

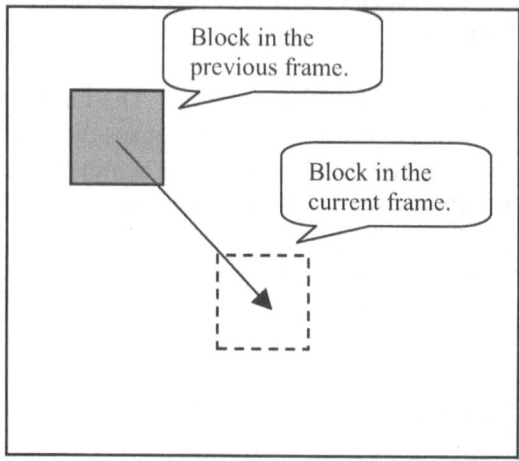

Figure 2.13. Block-based motion vector.

- The first frame in the video clip does not have motion vectors since it does not have a previous frame.

- For the second frame in the video clip, we use exhaustive search to find the motion vector for every block. Exhaustive search means that all the blocks that fall in the relatively large neighborhood of the current block are searched. Assume the motion vector is (d_x, d_y), the exhaustive search can be formulated as

$$(d_x, d_y) = \underset{\substack{-M_x \leq d_x \leq M_x \\ -M_y \leq d_y \leq M_y}}{\arg\min} \sum_{i,j} [I_1(i,j) - I_2(i - d_x, j - d_y)]^2$$

(2.6)

where $I_1(i,j)$ is the pixel value for the ith row and j^{th} column pixel –which falls in the current block- in the current frame and $I_2(i,j)$ is the pixel value for the ith row and j^{th} column pixel –which falls in the neighbor block– in the previous frame. M_x and M_y defines the size of the neighborhood. In this chapter, we use $M_x = 30$ and $M_y = 15$. M_y is less than M_x because the vertical motion of cloud is much less than the horizontal motion.

- For the subsequent frames we use the motion vectors of the previous frame as the motion estimation of the current frame. We then search a much smaller neighborhood than we would using an exhaustive search. Furthermore, the searching process stops when it finds a block that is similar enough to the current block, but not necessarily the most similar one. By doing this, we again are making compromise between computational complexity and accuracy. That is why this algorithm is called forward motion prediction with early stopping. Assume (d_x^n, d_y^n) is the motion vector for the current block and (d_x^{n-1}, d_y^{n-1}) is the motion vector for the corresponding block in the previous frame, the forward motion prediction can be formulated as

$$(d_x^n, d_y^n) = \mathop{\arg\min}_{\substack{d_x^n = d_x^{n-1} + \delta_x \\ d_y^n = d_y^{n-1} + \delta_y \\ -m_x \leq \delta_x \leq m_x \\ -m_y \leq \delta_y \leq m_y}} \sum_{i,j} [I_1(i,j) - I_2(i - d_x^n, j - d_y^n)]^2$$

(2.7)

and this minimization process stops if

$$[I_1(i,j) - I_2(i - d_x^n, j - d_y^n)]^2 \leq \delta \qquad (2.8)$$

where $m_x = 10$, $m_y = 5$, and δ is the threshold for early stopping.

Our experimental results have shown that the computational savings for forward motion prediction with early stopping can be up to 90% without significant loss of accuracy –for the applications we have investigated to date, which largely contain slow-moving objects.

Distance measure of local motion vector. By aggregating the motion vectors for each frame within the same video clip, a large motion-based feature vector M can be constructed. The problem then becomes how to measure the distance between two motion-based feature vectors M_1, M_2. Blindly using Euclidean distance is not satisfying since the horizontal motion and the vertical motion have different statistics. For example, the horizontal motion is generally larger than vertical motion, which is no surprise for satellite cloud cover picture. Also, the horizontal motion is most likely to be positive, which means the cloud generally moves from west to east. Furthermore, vertical motion is most likely to be negative, which means that the clouds generally move from south to north. The latter case is impacted by the angle of the satellite camera, which cannot be generalized to other different views of the satellite cloud cover picture. If we just measure the Euclidean distance between two vectors, M_1 and M_2, the horizontal motion and vertical motion will contribute to the distance by different amounts.

Considering the different statistics for horizontal motion and vertical motion, the horizontal and vertical parts of motion feature vector M are all normalized to $N(0, 1)$, which is a Gaussian distribution with zero mean and unit variance. Assume N_1, N_2 are the normalized version for M_1, M_2. The distance between M_1, M_2 is given by

$$d_M(M_1, M_2) = d_E(N_1, N_2) \qquad (2.9)$$

where d_E means Euclidean distance.

3.3   Video Clip Matching

Video clip matching is the process of locating the clips similar to the input clip in a database. Essentially, matching measures the distances between the input clip and the database clips, and then determines the clips with the smallest distances. Let V_1, V_2 denote two video clips ,C_1, C_2 be their color feature vectors M_1, M_2 be their motion feature vectors. The distance between V_1, V_2 is defined as

$$d(V_1, V_2) = w_C * d_C(C_1, C_2) + w_M * d_M(M_1, M_2) \qquad (2.10)$$

where d_C, d_M are defined in Section 3 and $w_C, w_M \in [0, 1]$ are the weights assigned to the color feature and motion feature.

Currently we use exhaustive search over all the clips in the database. We then retrieve the first k clips that are closest to the input clip.

3.4   Experimental Results

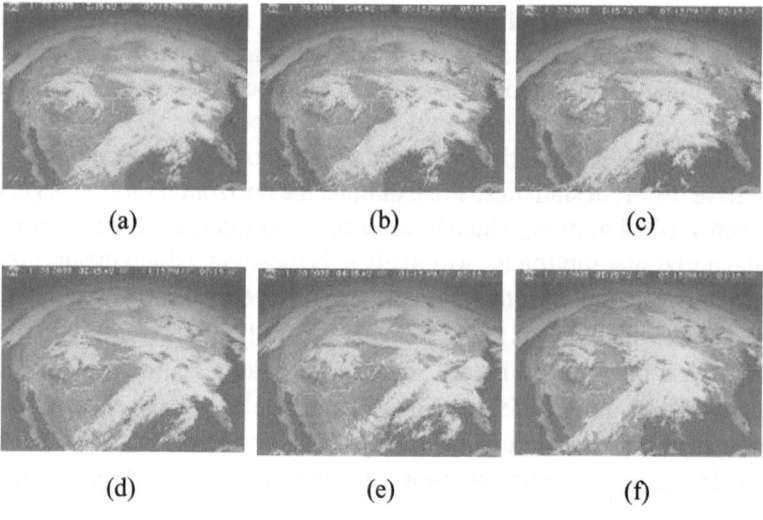

(a)   (b)   (c)

(d)   (e)   (f)

Figure 2.14.   A typical output of the system.

Currently, one thousand satellite cloud cover picture clips are archived in our database. Figure 2.14 shows a typical result of the system. Part (a) is the key frame of the input clip. Part (b), (c), (d), (e), (f) are the key frames of the first 5 most similar clips retrieved from the database. For this example, the input clip is drawn from the database. Consequently, part (b) is exactly the same clip as part (a). Absolute objective performance evaluation criteria are still not clear. However, subjectively, the system is relatively effective as we can see the retrieved clips are indeed very similar to the input clip.

3.5 Future Research Directions

The video query system we have currently implemented is far from perfect. We have identified the following potential research directions:

- Local motion vectors are not accurate enough. In particular, such vectors do not capture rotational motion very well. However, the rotational motion in cloud cover pictures is quite important since such patterns can imply potential severe weather patterns such as cyclones, tornadoes, etc. Therefore, one possible research direction is to develop better motion-based feature.

- The feature vectors calculated from the video clip are of very high dimensionality. In this simple example the dimensionality of the color feature is 960, while the dimensionality of the motion feature is 400. However, there exists strong correlation within the feature vectors. Hence, it appears to be a natural extension to perform dimensionality reduction or principal component analysis (PCA) on these feature vectors.

- With respect to matching algorithm, exhaustive search is the simplest but the most expensive method, and clearly a better search algorithm is needed. Accordingly, better data organization scheme is needed. One possible way is to perform video clustering on the database video clips. Then, the search algorithm can first locate the most similar cluster, and then search the most similar clips within that cluster instead of searching the whole database.

- A method for combining the different video features into a single distance measure is still a problem that needs to be solved.

- Objective performance evaluation criteria are needed in order to make better and quantitative evaluations.

4. Automatic Video Annotation – One Supervised Annotation Example

Based on the low-level spatial and temporal features, automatic video annotation can be performed for obtaining the grammar labels or for other purpose. In this section, a supervised automatic video annotation system is developed based on the video query system discussed in the previous section. For convenience, a specific example, again satellite cover data, will be used to illustrate the key components of the system. In this example, the sky condition is used as the spatio-temporal pattern we want to annotate.

4.1 Overview

Figure 2.15 illustrates the proposed framework, of which a preliminary version has been implemented [56] to demonstrate the viability of our approach to video annotation. The system divides naturally into two parts, one for training data preparation, and the other for automatic video annotation.

For training data preparation, a set of imagery data (video clips and images) is gathered. We refer to this original selected data set as the template imagery data. The features of this data are extracted in the video indexing process. Furthermore, a manual annotation process is carried out by domain experts, in which high level information needed by specific applications is added. From these two operations, video indexing and manual annotation, an annotated template data set is generated and used to train a model which will be the main tool of the automatic video annotation system. As further explained below, the automatic video annotation process, where a new video clip is annotated with the information provided by the template annotated data together with the training model process, is a critical task of the overall system.

The imagery database has two components, the template component and the general component. The template component data (raw data, features, and annotated information) are stored in the template component as an element of the database. When a new motion imagery clip is presented to the system, the system automatically annotates the data using the automatic annotation block. Then the imagery clip, its features, and the annotation information are stored in the general part of database.

4.2 Training Data Preparation

During the process of training data preparation, a collection of typical video clips (Template Imagery Data) is fed into the system. Features of these video clips are extracted and the annotation information is determined by domain experts, and then attached to the video clips. Subsequently the annotation information along with the indexing information and the raw video data (Template

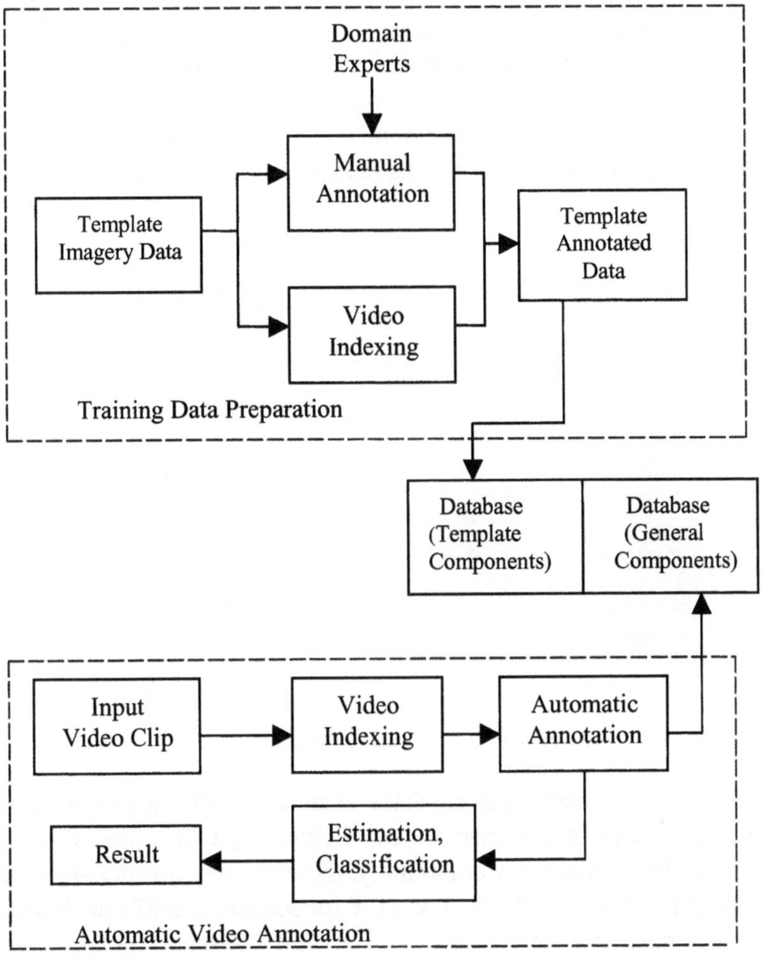

Domain
Experts
Manual
Annotation
Template
Imagery Data
Template
Annotated
Data
Video
Indexing
Training Data Preparation
Database
(Template
Components)
Database
(General
Components)
Input
Video Clip
Video
Indexing
Automatic
Annotation
Result
Estimation,
Classification
Automatic Video Annotation

Figure 2.15. The system architecture for automatic video annotation.

Manual Annotation
of Symbols
Template
Imagery Data
Database
Template Clips
with Annotation
Video Indexing
(Feature Extraction)

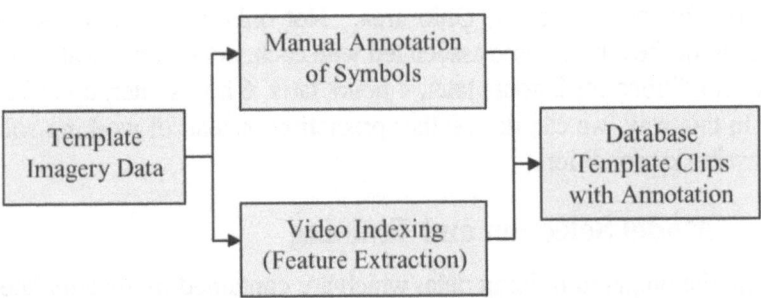

Figure 2.16. Training data preparation.

Annotated Data) are stored in the template components of the database. Figure 2.16 gives an overview of this process.

During the manual annotation process domain experts will annotate the template video clips according to their knowledge of common temporal events that are to be expected in the particular domain. In our example, the annotated information would be the sky conditions associated with the satellite cloud cover pictures.

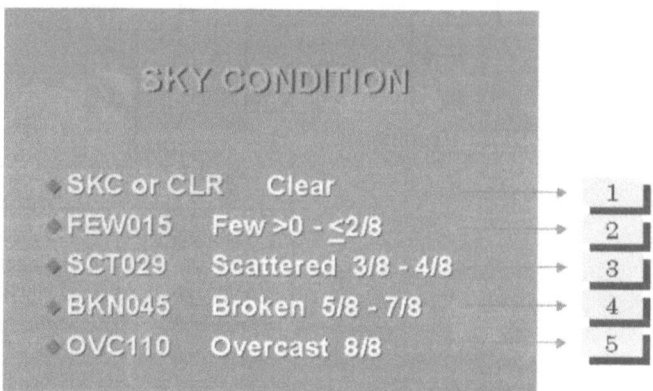

Figure 2.17.   The five sky conditions.

We choose to annotate each database video clip with the sky condition for a specific area, and in this case we use Columbus, Ohio. According to the standards of the meteorology community, there are five distinct sky condition codes: CLR for clear, FEW for few, SCT for scattered, BKN for broken, and OVC for overcast. For convenience, we use digit 1 to represent CLR, digit 2 to represent FEW and so forth, as shown in Figure 2.17.

By using the weather observation data from National Oceanic & Atmospheric Administration (NOAA), we annotate the video clips we collected with the sky condition for the Columbus, Ohio area. Not only do we annotate the sky condition for the current time associated with each video clip, we also annotate the sky conditions for 2 hours later, 4 hours later, 6 hours later, up to 72 hours later. In this way, we can do weather prediction instead of working solely on the current sky condition.

4.3   Model Selection and Training

Using the prepared training data, which are contained in the template component of the database, a model will be selected and trained in order to realize the automatic video annotation. This process is shown in Figure 2.18(a). The model will embody the underlying correlation between the video features and the annotation information. For example, the satellite cloud cover picture may have the correlation based on the described weather grammar as Figure 2.18(b) shows.

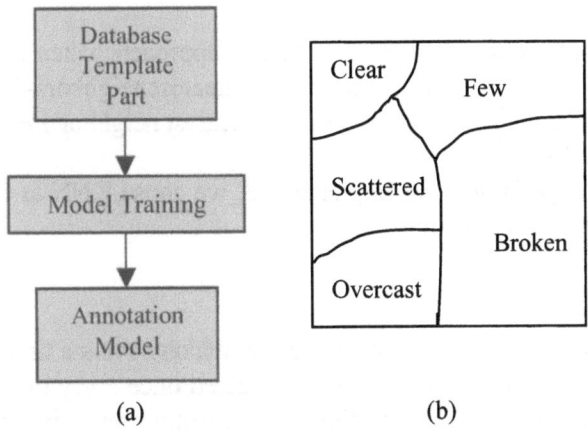

(a)   (b)

Figure 2.18.   (a) Model training process; (b) Underlying correlation.

Usually, the feature extraction method produces an overlapped data set. This means that, in general, a powerful model will be required in order to get appropriate generalization capabilities. In this project we propose to use two different models to do the automatic annotation. One is the parametric procedure embodied in Support Vector Machines (SVM); the other is the non-parametric method referred to as k-Nearest Neighbors (k-NN).

4.3.1   Support vector machines.   Support Vector Machines (SVMs) provide a new approach to the problem of machine learning. SVMs have clear connections to an underlying statistical learning theory [57]. Originally developed for pattern recognition [58], [59], SVMs represent the decision boundary in terms of a typically small subset of all training examples called the Support Vectors [60]. SVM can be considered as a universal learning procedure like Neural Networks (NN). However, SVM have some relevant advantages for this application. First, SVM training always finds a global minimum, and thus avoids becoming trapped in a local minimum. Second, unlike neural networks and other conventional statistical procedures, SVMs do not control model complexity by keeping the number of features small. Instead, with SVMs the model complexity is controlled independently of dimensionality. This fact allows SVMs to solve problems with a reasonably small number of patterns independent of their dimensionality.

There are other reasons why SVMs present a special interest for this work. SVMs are largely characterized by the choice of their kernel, and SVMs thus link the problems they are designed for with a large body of existing works on kernel based methods. Moreover, SVMs have a simple geometric interpretation that provides fertile ground for further investigation.

4.3.2   K nearest neighbors.   In this approach, when a query clip is presented, the most similar clips –the so called nearest neighbors- in the database are located. Then the pattern symbols of the nearest neighbor are the prediction for the query clip.

In the prototype system we implemented, we chose KNN as the algorithm to do the automatic annotation.

4.4   Rejection Window

The data we used has a total of 944 clips, which covers a time period from Nov. 2000 to Feb. 2001. The data was collected once every two hours. If we randomly divide the data into training set and testing set, it is very likely that the nearest neighbor of a certain clip could be the clip that is closest in time. If this happens, the prediction performance for 72 hours later will be almost the same as that for the current time because the two clips are close in time. In order to make the prediction more reasonable for the limited amount of data, we introduce a rejection window into our analysis system.

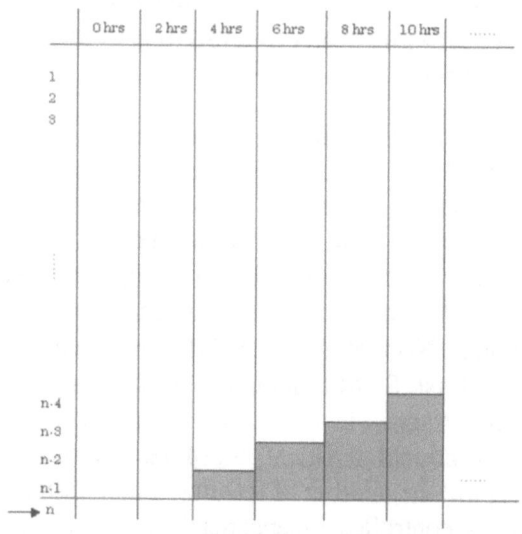

Figure 2.19.   Rejection window.

The problem is very specific: given the previous $(n-1)$ clips, how to make a prediction for the nth clip. As for the prediction of the current time, we can use all the previous $(n-1)$ clips because the current sky condition for all the previous $(n-1)$ clips are available. The same reasoning holds for the prediction of 2-hours-later sky condition. However, for the prediction of 4-hours-later sky condition, the $(n-1)^{th}$ clip is not eligible since the 4-hours-later sky condition for $(n-1)^{th}$ clip is still not available at the current time of nth clip. Likewise, for the prediction of 6-hours-later sky condition, the $(n-2)^{th}$ and $(n-1)^{th}$ clips are not eligible, and so forth. That is the reason we impose a rejection window, which is illustrated by the shadow areas in Figure 2.19.

4.5   Experiment Result

In the 944 clips, we choose the first 614 clips as a training set and the last 330 clips as a testing set. Figure 2.20 shows the result by using k-NN algorithm with rejection window. The solid line corresponds to the result for the exactly matched case. The dash dot line corresponds to the result for matching within 1 matching error, which means the predicted result is either exactly the correct answer or the next closest case. For example, "FEW" can be predicated as "CLR", "FEW" or "SCT", etc.

From Figure 2.20, we can see the result is good for prediction within ten hours. Indeed, the prediction result is not very bad for prediction made within 46 hours. However, our results deteriorate significantly for prediction of situations more than 46 hours later. This is reasonable since the cloud in the left side of the image usually takes two days to arrive at Ohio area. Better results are expected as we collect more data.

5.   Syntactic Video Analysis

Pattern analysis techniques may be grouped into two general categories: the statistical approach and the syntactic approach [61]. In the statistical approach, a set of features are extracted and the recognition of each pattern is then made by partitioning the feature space into some sup-spaces, each of which corresponds to one pattern. Many of the major developments in pattern recognition research fall into the statistical category. For example, k-nearest neighbor, neural networks, and support vector machines can be included in this category. However, in some domains the structural information of each pattern is more important and the goal is more likely to be recognizing structurally defined relationships of primitives. Purely statistical approaches are not sufficient since semantically equivalent patterns possess radically different statistical properties. The good news is syntactic approach can play an important role in recognizing the structural information. In syntactic approach, suitable and distinct grammars are defined to reflect the structure of each pattern class. The syntactic approach

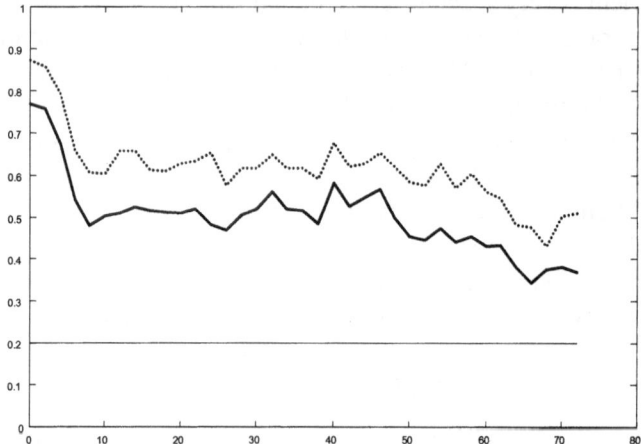

Figure 2.20. Pattern prediction result. The horizontal axis is the time delay in hours, and the vertical axis is the prediction rate. The solid line corresponds to the result for the exactly matched case. The dotted line corresponds to the result for matching within one matching error, which means the predicted result is either exactly the correct answer or the next closest one. The horizontal line in the figure is the performance for randomly guess.

provides a capability for describing a large set of complex patterns by using small sets of simple pattern primitives and grammatical rules.

In the video analysis domain some semantically similar events may have significantly different statistical properties and it is even possible that they cover very different time intervals. However, they may share very similar structural information. In such kind of cases, syntactic pattern analysis enters the picture with effective performance. In this section, we will use an example to show the basic idea of syntactic video analysis. Basketball video is used and our goal is try to identify and locate the free-throw events.

5.1   Overview

Figure 2.21 illustrates the architecture of the syntactic video analysis system we are prototyping. The basketball video undertakes shot detection, followed by video feature extraction. The shots then are clustered and assigned with grammar labels. The video features and grammar labels together are organized using XML and the resultant XML file is fed into a syntactic pattern analysis subsystem, which will be explained in Subsections 5.4 and 5.5. The locations of the free-throws in the basketball video will be the output.

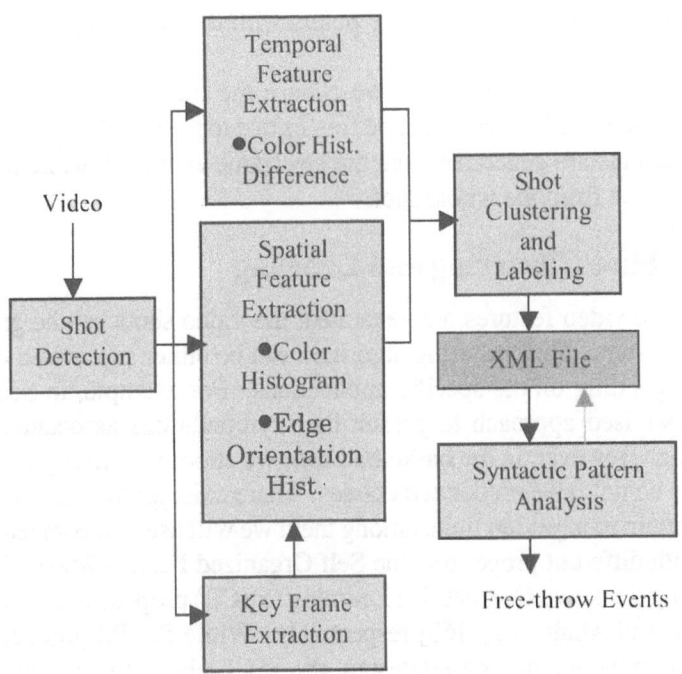

Figure 2.21.   The big picture of syntactic video analysis on basketball video.

5.2   Video Indexing

Shot detection segments the basketball video into elementary units, each of them being then processed by the information extraction block. Apart from key frame extraction, some typical spatial and temporal features will be extracted. In this case, the spatial features consist of local color histogram and edge orientation histogram [43]. Local color histogram has already been addressed in Section 3, while edge orientation histogram is defined as

$$H = \{H(i), 1 \leq i \leq n\}. \tag{2.11}$$

where $H(i)$ is the number of edge points with the edge angle falling into the ith bin.

As for the temporal features, we choose the difference between the local color histograms of two consecutive frames due to the simplicity. Of course, the spatial features are generated from the key frame extracted, while the temporal features come from the whole shot.

5.3   Shot Clustering and Labeling

After the video features are generated, the video shots can be grouped into a set of clusters. The clustering algorithm can be either supervised or unsupervised, depending on the specific applications. For example, in Section 4, we used supervised approach to predict the sky conditions associated with each video clip. However, in the basketball case we choose unsupervised clustering since we do not have predefined classes. There exist many clustering methods in the pattern recognition field, among them we will use two competitive procedures with different properties: the Self Organized Feature Maps (SOFM) and Frequency Sensitive Competitive Learning (FSCL) proposed and developed by Kohonen and Ahalt, [62], [63] respectively. While SOFM produce topologic maps due to the vicinity-based design, the FSCL places the prototypes in such a way that the entropy is maximized, i.e. every prototype is roughly with equal probability.

Labels, one or more for each cluster, are assigned to each video shots and consequently a string will be generated for the whole video. Syntactic pattern analysis will be performed on this string.

5.4   Grammatical Inference

Grammars need to be constructed to represent the pattern classes. This problem of learning grammar(s) from a training set of sample patterns is called grammatical inference, which corresponds to the training phase in statistical pattern analysis. Formally speaking the problem of grammatical inference is concerned mainly with the procedures that can be used to infer the syntactic

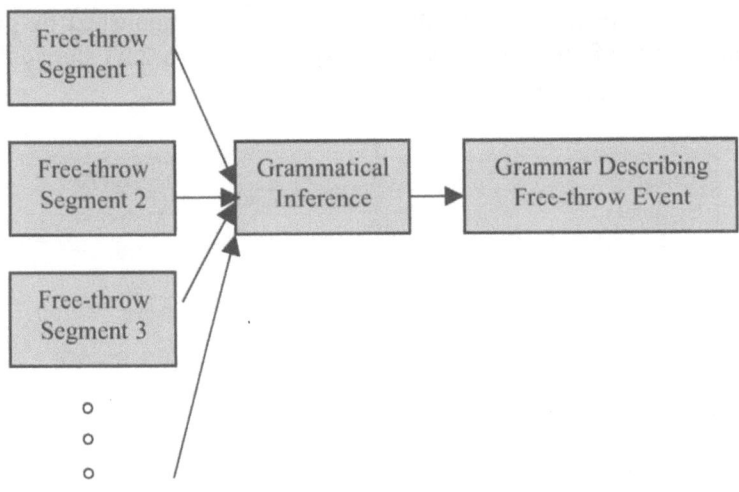

Figure 2.22.   Grammatical interface.

rules of an unknown grammar G based on a finite set of sentences or strings S_t from $L(G)$, the language generated by G, and possibly also on a finite set of strings from the complement of $L(G)$. The inferred grammar is a set of rules for describing the given finite set of strings from $L(G)$. A basic block diagram of a grammatical inference machine is shown in Figure 2.22. For more information, the readers are advised to check out [61].

5.5   Syntactic Pattern Recognition

The problem of syntactic pattern recognition is to determine which grammar one certain string belongs to, given several pattern grammars. There are two fundamental approaches.

String matching.   For some relatively simple patterns, one pattern class may correspond to just a few strings. In this case, the syntactic pattern recognition can be reduced to just string matching, which compares the input string with the pattern strings to see whether the input belongs to the pattern class or not. Of course, a distance can be calculated based on certain string distance measure. If the distances of the input string with every pattern class are computed, we even can use the nearest neighbor to classify the given input.

String parsing.   For relatively complex problems, the input string has to be checked against every grammar for each pattern class. Parsing is a fundamental concept related to the syntactic or structural approach, whose objective is to determine if the input pattern (string) is syntactically well formed in the context of one or more pre-specified grammars. Paring is accomplished by parsers, which are often referred to as syntax analyzers. Figure 2.23 shows the big picture of a parsing process.

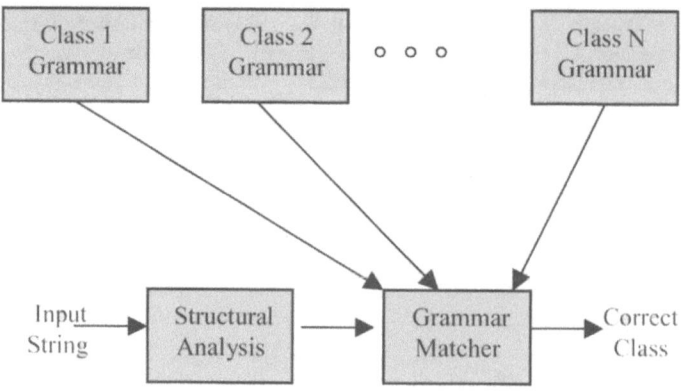

Figure 2.23.   Syntactic pattern recognition.

5.6   Potential Applications of Syntactic Video Analysis

There are many potential applications of syntactic video analysis, including:

- Video query based on grammar. Syntactic video analysis may help to answer questions like: retrieve the video segments whose structures are close to one certain given grammar, which corresponds to one certain spatio-temporal event.

- Detection and identification of events of interest. For example, the free-throws in basketball video.

- Prediction and detection of large-scale events.

- Prototypical sequence generation that can be used for meta-analysis and video indexing.

- Performing high level video segmentation based on grammar symbols.

6.   Conclusions

There is a significant and growing need for efficient indexing, storing, re-trieving, and analyzing video. In this chapter, we propose a systematic approach to video analysis, which incorporates various components together. In the pro-posed system, XML is chosen to be the language to describe the video meta-data.

Examples are given to further demonstrate some of the important aspects of the whole system. A video query system on fixed-view motion imagery is used to illustrate the process of video retrieval. For this simple problem, local color histogram and local motion vector are effective for video query on satellite cloud cover data. However, special techniques must be employed to deal with special problems. For fixed- view imagery, background elimination is performed to improve the performance. Based on the video query system we implemented, automatic video annotation can be readily achieved using predictive model. We have shown an example of doing sky condition prediction on satellite cloud cover data. Syntactic pattern analysis is a promising tool for video analysis and a prototype work of syntactic video analysis is shown. By applying syntactic pattern analysis, large-scale and more complex events can be analyzed.

References

[1] http://www.google.com/, Jan. 2003.

[2] http://www.yahoo.com/, Jan. 2003.

[3] Y. Rui, T. S. Huang, and S. Mehrotra, "Constructing table-of-content for videos," *Multimedia Systems*, vol. 7, no. 5, pp. 359–368, 1999.

[4] F. Idris and S. Panchanathan, "Review of image and video indexing techniques," *Journal of Visual Communication and Image Representation*, vol. 8, pp. 146–166, June 1997.

[5] G. Ahanger and T. D. C. Little, "A survey of technologies for parsing and indexing digital video," *Journal of Visual Communication and Image Representation*, vol. 7, pp. 28–43, March 1996.

[6] A. D. Bimbo, "Image and video database: visual browsing, querying and retrieval," *Journal of Visual Language and computing*, vol. 7, pp. 353–359, December 1996.

[7] R. Brunelli, O. Mich, and C. M. Modena, "A survey on the automatic indexing of video data," *Journal of Visual Communication and Image Representation*, vol. 10, pp. 78–112, June 1999.

[8] M. Flickner *et al.*, "Query by image and video content: the QBIC system," *IEEE Computer Magazine*, vol. 28, pp. 23–32, September 1995.

[9] A. Hampapur and R. Jain, "Virage video engine," in *Proceedings of the SPIE 1998 Conference on Storage and Retrieval for Image and Video Databases V* (R. C. J. Ishwar K. Sethi, ed.), vol. 3022, (San Jose, CA), pp. 188–197, February 1997.

[10] S. F. Chang *et al.*, "A fully automated content-based video search engine supporting spatiotemporal queries," *IEEE Transactions on Circuits and System for Video Technology*, vol. 8, pp. 602–615, September 1998.

[11] B. Gunsel, A. M. Tekalp, and P. J. L. van Beek, "Content-based access to video objects: temporal segmentation, visual summarization, and feature extraction," *Signal Processing*, vol. 66, pp. 261–280, April 1998.

[12] M. Carrer, L. Ligresti, G. Ahanger, and T. D. C. Little, "An annotation engine for supporting video database population," *Multimedia Tools and Techniques*, vol. 5, pp. 233–258, November 1997.

[13] F. Pereira and R. H. Koenen, *Multimedia Systems, Standards, and Networks*, ch. MPEG-7: status and directions. New York: Marcel Dekker, Inc., March 2000.

[14] J. M. Martinez, "Overview of the MPEG-7 standard.," Tech. Rep. N4509, ISO/IEC JTC1/SC29/WG11, December 2001.

[15] N. Haering, R. J. Qian, and M. I. Sezan, "A semantic event-detection approach and its application to detecting hunts in wildlife video," *IEEE Transactions on Circuits and System for Video Technology*, vol. 10, pp. 857–868, September 2000.

[16] Y. A. Ivanov and A. F. Bobick, "Recognition of visual activities and interactions by stochastic parsing," *IEEE Transactions on Pattern Analysis and Machine Intelligence*, vol. 22, pp. 852–872, August 2000.

[17] M. M. Yeung and B. L. Yeo, "Video content characterization and compaction for digital library applications," in *Proceedings of the SPIE 1998 Conference on Storage and Retrieval for Images and Video Databases V* (R. C. J. Ishwar K. Sethi, ed.), vol. 3022, (San Jose, CA), pp. 45–58, February 1997.

[18] R. Kasturi and R. Jain, *Computer Vision: Principles*, ch. Dynamic vision, pp. 469–480. Los Alamitos, CA: IEEE Computer Society Press, 1990.

[19] A. Nagasaka and Y. Tanaka, "Automatic video indexing and full video search for object appearance," in *Proc. of the IFIP: Visual Database Systems II*, pp. 113–127, 1992.

[20] U. Gargi *et al.*, "Evaluation of video sequence indexing and hierarchical video indexing," in *Proceedings of SPIE 1995 Conference on Storage and Retrieval for Images and Video Databases III* (W. Niblack and R. C. Jain, eds.), vol. 2420, pp. 144–151, 1995.

[21] Y. Tonomura, "Video handling based on structured information for hypermedia systems," in *Proc. ACM Int. Conf. Multimedia Information Systems*, (Singapore), pp. 333–344, 1991.

[22] S. Shahraray, "Scene change detection and content-based sampling of video sequences," in *Digital Video Compression: Algorithms Tech. 2419*, pp. 2–13, 1995.

[23] D. Swanberg, C. F. Shu, and R. Jain, "Knowledge guided parsing in video databases," in *Proceedings of SPIE 1993 Conference Storage and Re-*

trieval for Images and Video Databases (W. Niblack, ed.), vol. 1908, pp. 13–24, 1993.

[24] H. J.Zhang et al., "Automatic partitioning of full motion video," *ACM Multimedia Systems*, vol. 1, no. 1, pp. 10–28, 1993.

[25] A. Hampapur, R. Jain, and T. Weymouth, "Production model based digital video segmentation," *Multimedia Tools and Applications*, vol. 1, pp. 9–46, Mar. 1995.

[26] P. Aigrain and P. Joly, "The automatic real-time analysis of film editing and transition effects and its applications," *Comput. Graphics*, vol. 18, no. 1, pp. 93–103, 1994.

[27] T. D. C. Little et al., "A digital on-demand video service supporting content-based queries," in *First ACM International Conference on Multimedia*, (Anaheim, CA, USA), pp. 427–436, 1993.

[28] H. J. Zhang et al., "Video parsing and browsing using compressed data," *Multimedia and Tools Applications*, vol. 1, pp. 89–111, 1995.

[29] H. J. Zhang et al., "Video parsing compressed data," in *SPIE: Image Video Processing II*, (San Jose, CA, USA), pp. 142–149, 1994.

[30] F. Arman, A. Hsu, and M. Y. Chiu, "Image processing on compressed data for large video databases," in *First ACM International Conference on Multimedia*, (Anaheim, CA, USA), pp. 267–272, 1993.

[31] F. Arman, A. Hsu, and M. Y. Chiu, "Feature management for large video databases," in *Storage and Retrieval for Images and Video Databases* (W. Niblack, ed.), vol. 1908, pp. 2–12, 1993.

[32] H. C. H. Liuand and G. L. Zick, "Scene decomposition of MPEG compressed video," in *Digital Video Compression: Algorithms Tech*, (San Jose, CA, USA), pp. 26–37, 1995.

[33] J. Lee and B. W. Dickinson, "Multiresolution video indexing for subband coded video databases," in *Image Video Processing II*, (San Jose, CA, USA), pp. 321–330, 1994.

[34] D. Santa-Cruz and T. Ebrahimi, "An analytical study of JPEG 2000 functionalities," in *Proc. of the International Conference on Image Processing*, vol. 2, (Vancouver, Canada), pp. 49–52, 2000.

[35] R. Lienhart, W. Effelsberg, and R. Jain, "VisualGREP: A systematic method to compare and retrieve video sequences," *Multimedia Tools and Applications*, vol. 10, no. 1, pp. 47–72, 2000.

[36] M. Stricker and M. Orengo, "Similarity of color images," in *Storage and Retrieval for Images and Video Databases III* (W. Remesh and C. Jain, eds.), vol. 2420, pp. 381–393, 1995.

[37] Y. Gong *et al.*, "An image database system with content capturing and fast image indexing abilities," in *Proceedings of the International Conference on Multimedia Computing and Systems*, (Boston, MA, USA), pp. 121–130, 1994.

[38] M. Tuceryan and A. K. Jain, *Handbook of Pattern Recognition and Computer Vision*, ch. Texture analysis, pp. 235–276. World Scientific Publishing, 1993.

[39] G. Taubin and D. B. Coper, "Recognition and positioning of rigid objects using algebraic moment invariants," in *Geometric Methods Computer Vision* (B. C. Vemuri, ed.), vol. 1570, pp. 175–186, 1991.

[40] E. Persoon and K. S. Fu, "Shape discrimination using fourier descriptors," *IEEE Trans. Systems Man Cybernetics*, vol. 8, pp. 170–179, 1977.

[41] S. R. Rubois and F. H. Glanz, "An autoregressive model approach to dimensional shape classification," *IEEE Transactions on Pattern Analysis and Machine Intelligence*, vol. 8, pp. 55–66, 1986.

[42] A. K. Jain, *Fundamentals of Digital Image Processing*. Englewood Cliffs, NJ: Prentice Hall, 1989.

[43] A. K. Jain, A. Vailaya, and X. Wei, "Query by video clip," *Multimedia Systems*, vol. 7, pp. 369–384, 1999.

[44] A. Akutsu *et al.*, "Video indexing using motion vectors," in *Visual Communications and Image Processing'92* (P. Maragos, ed.), vol. 1818, pp. 1522–1530, 1992.

[45] H. Ueda, T. Miyataka, and S. Yoshizawa, "IMPACT: An interactive natural-motion-picture dedicated multimedia authoring system," in *Proc. Human Factors in Computing Systems CHI'91*, (New Orleans, Louisiana, USA), pp. 343–350, 1991.

[46] W. Wolf, "Key frame selection by motion analysis," in *Proceeding IEEE Int. Conf. Acoust., Speech and Signal Proc.*, (Atlanta, GA, USA), pp. 1228–1231, 1996.

[47] X. Liu, C. B. Owen, and F. Makedon, "Automatic video pause detection filter," Tech. Rep. PCS-TR97-307, Dartmouth College, Computer Science, Hanover, NH, USA, February 1997.

[48] M. M. Yeung and B. Liu, "Efficient matching and clustering of video shots," in *Proceedings of the IEEE Intl. Conference on Image Processing*, vol. 1, (Washington, D.C.), pp. 338–341, Oct. 1995.

[49] W. Xiong, J. C. M. Lee, and R. H. Ma, "Automatic video data structuring through shot partitioning and key-frame computing," *Machine Vision and Applications*, vol. 10, pp. 51–65, 1997.

[50] http://www.w3.org/XML, Jan. 2003.

[51] H. Jiang and A. K. Elmagarmid, "WVTDB - a semantic content-based video database system on the world wide web," *IEEE Transactions on Knowledge and Data Engineering*, vol. 10, no. 6, pp. 947–966, 1998.

[52] T. Kawashima *et al.*, "Indexing of baseball telecast for content-based video retrieval," in *Proceedings of International Conference on Image Processing*, (Chicago, IL, USA), pp. 871–874, 1998.

[53] T. Kato *et al.*, "A sketch retrieval method for full color image database query by visual example," in *Proceedings of 11th IAPR International Conference on Pattern Recognition*, (The Hague, The Netherlands), pp. 530–533, 1992.

[54] E. Ardizzone *et al.*, "Motion and color based video indexing and retrieval," in *Proc. Int. Conf. on Pattern Recognition (ICPR-96)*, vol. III, (Vienna, Austria), pp. 135–139, 1996.

[55] T. M. Cover and J. Thomas, *Elements of information theory*. John Wiley & Sons, Inc., 1991.

[56] H. Li and Y. Zhao. http://eepc269.eng.ohio-state.edu/matlab/, Jan. 2003.

[57] V. Vapnik, *The Nature of Statistical Learning Theory*. New York: Springer Verlag, 1995.

[58] B. E. Boser, I. M. Guyon, and V. N. Vapnik, "A training algorithm for optimal margin classifiers," in *Proceedings of the 5th Annual ACM Workshop on Computational Learning Theory*, (Pittsbugh, PA, USA), pp. 144–152, ACM Press, 1992.

[59] V. Vapnik and A. Chervonenkis, *Theory of Pattern Recognition [in Russian]*. Nauka, Moscow, 1974.

[60] B. Scholkopf, C. Burges, and V. Vapnik, "Extracting support data for a given task," in *Proceedings 1st International Conference on Knowledge Discovery & Data Mining*, (Menlo Park, CA), pp. 252–257, 1995.

[61] K. S. Fu, *Syntactic Pattern Recognition and Applications*. Englewood Cliffs: Prentice-Hall, 1982.

[62] T. Kohonen, *Self-Organizing Maps*. Spinger-Verlag, 1995.

[63] S. C. Ahalt *et al.*, "Competitive learning algorithms for vector quantization," *Neural Networks*, vol. 3, pp. 277–290, May 1990.

Chapter 3

SEGMENTATION TECHNIQUES FOR VIDEO SEQUENCES IN THE DOMAIN OF MPEG-COMPRESSED DATA

Irena Koprinska

School of Information Technologies, University of Sidney
Sydney, 2006 NSW, Australia
irena@it.usyd.edu.au

Sergio Carrato

Dept. of Electrical Engineering and Computer Science, University of Trieste
v. Valerio, 10, 34100 Trieste, Italy
carrato@univ.trieste.it

Abstract   Video segmentation into shots is the first step in content-based analysis of digital video. This chapter provides a comprehensive taxonomy and critical survey of the existing techniques for video segmentation operating on MPEG video stream. Their performance, relative merits and limitations are discussed and contrasted. The gradual development of the techniques and their similarities with the video segmentation methods operating on uncompressed video are also considered.

Keywords: Temporal video segmentation, shot change detection, cut detection, digital video archiving, MPEG compression.

1.   Introduction

Recent developments in computing performance, multimedia compression and communication technologies have made possible the creation of digital video archives. Applications such as digital libraries, video-on-demand, digital video broadcast, distance learning generate and use large collections of video data. It is also expected that the storage of digital video at home will soon overtake the current analog video systems [1, 2]. However, unlike the docu-

J.F. Tasič et al. (eds.), Intelligent Integrated Media Communication Techniques, 89-114.
© 2003 *Kluwer Academic Publishers.*

ment databases that use keywords to quickly access large quantities of data, video databases still lack techniques for efficient organization, searching and retrieval. Text-based video organization based on manual annotation is highly inefficient, tedious and time consuming. Hence, automatically organizing video according to its content, in a way that allows meaningful and rapid indexing, querying, browsing and retrieval of objects, has become increasingly important in recent years. Several content-based prototype systems [3, 4, 5, 6, 7] have been developed.

Temporal video segmentation is generally accepted as the first step in content-based analysis of video sequences. It breaks up the video stream into a set of meaningful and manageable segments called shots that are used as basic units for indexing and annotation. Each shot is represented by one or more key frames. The content of the shot is indexed by spatial features extracted from the key frame(s) (e.g. color, texture, shape, text). In addition, temporal features from the shot can be used (e.g. motion, camera operations, interframe relations, audio). Some other approaches further cluster the shots based on the similarity between their key frames [8, 9] and provide the user with a hierarchical representation of the video data [10]. Shots may be also concatenated into story units [11, 12]. These shot-based representations can be used as a video summary or to browse the content of the video sequence, to find quickly the subsequences of interest and view them.

A shot is defined as one or more frames generated and recorded contiguously, and representing a continuous action in time and space [13]. Video editing produces two general types of shot transitions: abrupt and gradual. Abrupt transitions (cuts) are the most common and they occur over a single frame by splicing two distinct scenes successively. Gradual transitions occur over multiple frames and are result of effects such as fade outs, fade ins and dissolves. A fade out is a slow decrease in brightness resulting in a black frame; a fade in is a gradual increase in intensity starting from a black image. Dissolves show one image superimposed on the other as the frames of the first shot get dimmer and these of the second one get brighter.

As the characteristics of the frames before and after an abrupt transition usually differ significantly, abrupt transitions are much easier to detect than gradual transitions. The recognition of the gradual transitions is further complicated by the presence of camera operations (e.g. zooms, pans, tilts, tracks, booms) and object movement as they exhibit temporal variances of the same order and cause false positives. It is particularly challenging to detect dissolves between sequences involving intensive motion [14, 15, 16].

Since MPEG was established as an international standard for compression of digital video that is optimised for efficient storage and retrieval, video is increasingly stored and moved in compressed format. Hence, it is highly desirable to develop methods that can operate directly on the encoded stream.

Working in the compressed domain offers the following advantages. Firstly, by not having to perform decoding/re-encoding, computational complexity is reduced and savings on decompression time and decompression storage are obtained. Secondly, operations are faster due to the lower data rate of compressed video. Last but not least, the encoded video stream already contains a rich set of pre-computed features, such as motion vectors and block averages, that are suitable for temporal video segmentation.

The goal of this chapter is to provide a comprehensive taxonomy and critical survey of the existing approaches for temporal video segmentation of compressed video. As most of these approaches bear a close resemblance to the approaches operating on uncompressed video, we start with a brief overview of temporal video segmentation in uncompressed domain. Then in Section 3 we first review the relevant parts of the MPEG compression standard and then survey segmentation approaches operating on compressed video. The gradual development of the techniques and how the uncompressed domain methods were tailored and imported into the compressed domain are also considered. Finally, Section 4 concludes this chapter.

2.   Temporal Video Segmentation in the Uncompressed Domain

A great variety of techniques for temporal video segmentation were reported. The majority of algorithms process uncompressed video. Usually, a similarity measure between successive frames is defined. When the two frames are sufficiently dissimilar, there may be a cut. Gradual transitions are found by using cumulative difference measures and more sophisticated thresholding schemes. Based on the metrics used to detect the difference between successive frames, the algorithms can be divided broadly into three categories: pixel, block-based and histogram comparisons. Other categories include clustering based, feature based and model driven temporal video segmentation. For more information see [17, 18, 19, 20].

2.1   Pixel Comparison

Pair-wise pixel comparison evaluates the differences in intensity or color values of corresponding pixels in two successive frames. The simplest way is to calculate the absolute sum of pixel differences and compare it against a threshold [21]. This method, however, is not able to distinguish between a large change in a small area and a small change in a large area. For example, cuts are misdetected when a small part of the frame undergoes a large, rapid change. A possible improvement is to count the number of pixels that change in value more than some threshold and to compare the total against a second threshold [22, 23]. Although some irrelevant frame differences are filtered out, these approaches are still sensitive to object and camera movements, illumination changes and noise. For example, if camera pans, a large number of pixels may be judged as changed, even though there is actually a shift of a few pixels.

2.2   Block-Based Comparison

Block-based approaches use local characteristic to increase the robustness to camera and object movement. Each frame i is divided into b blocks that are compared with their corresponding blocks in frame $i + 1$. In [24] the blocks are compared using a likelihood ratio based on the mean intensity values and variances. If the likelihood ratio is greater than a threshold, the block is considered as changed and a cut is declared when the number of changed blocks is large enough. Another block-based technique is proposed by Shahraray [25]. The frame is divided into 12 non-overlapping blocks. For each of them the best match in terms of intensity is found in the respective neighbourhoods in the previous image. A non-linear order statistics filter is used to combine the match values where the weight of a match value depends on its order in the match value list. Other block-based approaches are discussed in [26, 27]. Compared to template matching, block-based comparison is more tolerant to slow and small object motion from frame to frame.

2.3   Histogram Comparison

A step further towards reducing sensitivity to camera and object movements can be done by comparing the histograms of successive images. The idea behind histogram-based approaches is that two frames with unchanging background and unchanging (although moving) objects will have little difference in their histograms. In addition, histograms are invariant to image rotation and change slowly under the variations of viewing angle and scale [28]. Depending on whether the histogram comparison is conducted over the entire frame or blocks of the frame, the approaches can be classified into global or local histogram comparison.

2.3.1   Global histogram comparison.   In [22, 29, 23] the absolute sum of grey level histogram differences between two successive frames is compared. Another simple and very effective approach is to compare color histograms [23]. To reduce the bin number (3 colours x 8 bits create histograms with 2^{24} bins), only the upper two bits of each color intensity were used. The resulting 64 bins have been shown to give sufficient accuracy. To enhance the difference between two frames across a cut, several authors [22] propose the use of the χ^2 test to compare the color histograms of two successive frames. However, the results reported in [23] show that the χ^2 test not only enhances the difference between two frames across a cut but also increases the difference due to camera and object movements. Gargi et al. [30] evaluate the performance of three histogram based methods using six different color coordinate systems: RGB, HSV, YIQ, L*a*b*, L*u*v* and Munsell. The twin-comparison method [23] is a histogram-based technique for both abrupt and gradual transition recognition. In the first pass a high threshold T_h is used to detect cuts. In the second pass a lower threshold T_l is employed to detect the potential starting frame F_s of a gradual transition. F_s is than compared to subsequent frames. This is called an accumulated comparison as during a gradual transition this difference value increases. The end frame F_e of the transition is detected when the difference between consecutive frames decreases to less than T_l, while the accumulated comparison has increased to a value higher than T_h. If the consecutive difference falls below T_l before the accumulated difference exceeds T_h, then the potential start frame F_s is dropped and the search continues for other gradual transitions. In [31] several temporal video segmentation techniques were compared and was found that twin-comparison is a simple algorithm that works very well.

2.3.2 Local histogram comparison.

Global histogram based approaches are simple and more robust to object and camera movements but ignore the spatial information and, therefore, fail when two different images have similar histograms. On the other hand, block based comparison methods make use of spatial information. They typically perform better than pair-wise pixel comparison but are still sensitive to camera and object motion and are also computationally expensive. By integrating the two paradigms, false alarms due to camera and object movement can be reduced while enough spatial information is retained to produce more accurate results. For example, Nagasaka and Tanaka [22] compare several statistics based on gray-level and color pixel differences and histogram comparisons. The best results are obtained by breaking the image into 16 equal-sized regions, using χ^2 test on color histograms for these regions and discarding the largest differences to reduce the effects of noise, object and camera movements. Other local histogram based approaches are presented in [32, 33].

2.4 Clustering-Based Segmentation

The approaches discussed so far rely on suitable thresholding of similarities between successive frames. However, the thresholds are typically highly sensitive to the type of input video. This drawback is overcome in [34] by the application of unsupervised clustering algorithm. More specifically, the temporal video segmentation is viewed as a 2-class clustering problem ("scene change" and "no scene change") and the well-known K-means algorithm is used to cluster frame dissimilarities. Then the frames from the cluster "scene change" which are temporary adjacent are labelled as belonging to a gradual transition and the other frames from this cluster are considered as cuts. Two similarity measures based on color histograms are used. The main advantage of the clustering-based segmentation is that it is a generic techniques that not only eliminates the need for threshold setting but also allows multiple features to be used simultaneously to improve the performance. For example, in [35] both histogram difference and pair-wise pixel comparison are incorporated in the clustering method.

2.5   Feature-Based Segmentation

An interesting approach for temporal video segmentation based on features is described in [15]. It involves analysis of the intensity edges between consecutive frames. During a cut or a dissolve, new intensity edges appear far from the locations of the old edges. Similarly, old edges disappear far from the location of new edges. Thus, by counting the entering and exiting edge pixels, cuts, fades and dissolves are detected and classified.

2.6   Model Driven Segmentation

The video segmentation techniques presented so far are sometimes referred to as data driven, bottom-up approaches [14]. They address the problem from data analysis point of view. It is also possible to apply top-down algorithms that are based on mathematical models of video data. Such approaches allow a systematic analysis of the problem and the use of several domain-specific constraints that might improve the efficiency. In [14] a shot boundaries identification approach based on the mathematical model of the video production process is presented. Another model-based technique, called differential model of motion picture, is proposed in [36]. Gradual transitions detection based on a model of intensity changes during fade out, fade in and dissolve is discussed in [37]. Boreczky and Wilcox [38] use hidden Markov models (HMM) for temporal video segmentation.

3.   Temporal Video Segmentation in the Compressed Domain

Several algorithms for temporal video segmentation in the compressed domain have been reported. According to the type of information used (see Tab. 3.1), they can be divided into six groups - segmentation based on: 1) Discrete Cosine Transform (DCT) coefficients, 2) DC terms, 3) DC terms, macroblock (MB) coding mode and motion vectors (MVs), 4) DCT coefficients, MB coding mode and MVs, 5) MB coding mode and MVs, 6) MB coding mode and bitrate information. Before reviewing each of them, we present a brief description of the fundamentals of the MPEG compression standard.

Table 3.1. Six groups of approaches for temporal video segmentation in compressed domain based on the information used.

Information used	Group					
	1	2	3	4	5	6
DCT coefficients	★			★		
DC terms		★	★			
MB coding mode			★	★	★	★
MVs			★	★	★	
Bit rate						★

3.1 MPEG Stream

The Moving Picture Expert Group (MPEG) standard is the most widely accepted international standard for digital video compression. It uses two basic techniques: MB-based motion compensation to reduce temporal redundancy and transform domain block-based compression to capture spatial redundancy. An MPEG stream consists of three types of pictures - intra coded (I frames), predicted (P frames) and bi-directional (B frames). These pictures are combined in a repetitive pattern called group of picture (GOP). A GOP starts with an I frame. The I and P frames are referred to as anchor frames. B frames appear between each pair of consecutive anchor frames. Fig. 3.1 shows a typical GOP and the predictive relationships between the different types of frames.

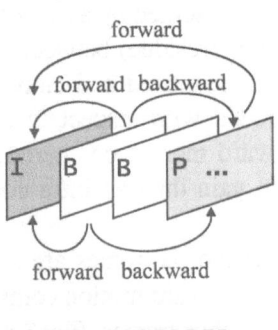

Figure 3.1. Typical GOP and predictive relationships between I, P and B pictures.

Each video frame is divided into a sequence of nonoverlapping MBs, typically 16×16 pixel. Each MB can be either intra coded or inter coded (i.e. coded with motion compensation). I frames are completely intra coded: every 8×8 pixel block in the MB is transformed to the frequency domain using the

Discrete Cosine Transform (DCT). The first DCT coefficient is called DC term and is 8 times the average intensity of the respective block, see Fig. 3.2. The 64 DCT coefficients are then (lossy) quantised and (lossless) entropy encoded using Run Length Encoding (RLE) and Huffman entropy coding. As I frames are coded without reference to any other video frames, they can be decoded independently and hence provide random access points into the compressed video.

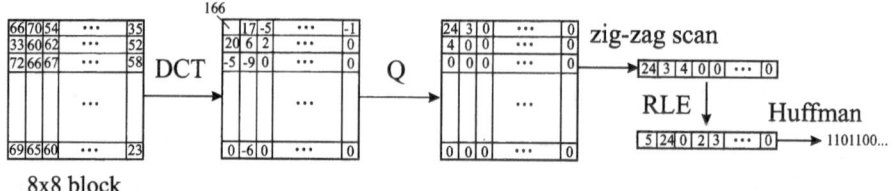

8x8 block

Figure 3.2. Intra coding.

P frames are predictively coded with reference to the nearest previous anchor frame (i.e. the previous I or P frame). For each MB in a P frame, the encoder searches the anchor frame and finds the best matching block in terms of intensity, see Fig. 3.3. The MB is then represented by a MV (which points to the position of the match) and the difference (residue) between the MB and its match. The residue is then DCT encoded, quantised and entropy coded while the MV is differentially and entropy coded with respect to its neighbouring MV. This is called encoding with forward motion compensation. An inter coded MB provides higher compression gain than an intra coded as the residue can be coded with fewer bits.

To achieve further compression, B frames are bi-directionally predictively encoded with forward and/or backward motion compensation with reference to the nearest past and/or future anchor frames, Fig. 3.4. As B frames are not used as reference for coding other frames, they can accommodate more distortion, and thus, provide higher compression gain compared to I and P frames. During the encoding process a test is made on each MB of P and B frame to see if it is more expensive to use motion compensation or intra coding. The latter occurs when the current frame does not have much in common with the anchor frame(s). As a result each MB of a P frame could be coded either intra or forward while for each MB of a B frame there are four possibilities: intra, forward, backward or interpolated. For more information about MPEG see [39].

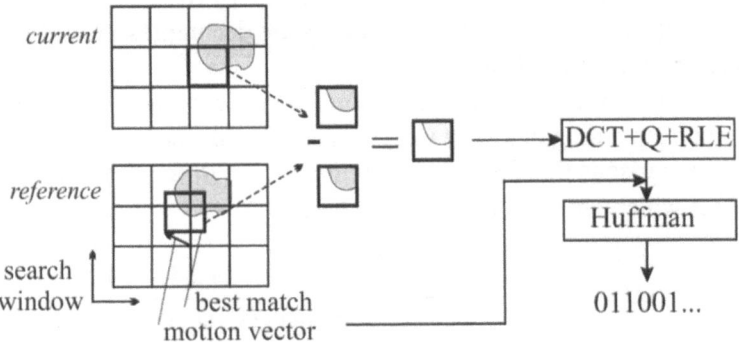

Figure 3.3.   Forward prediction for P frames.

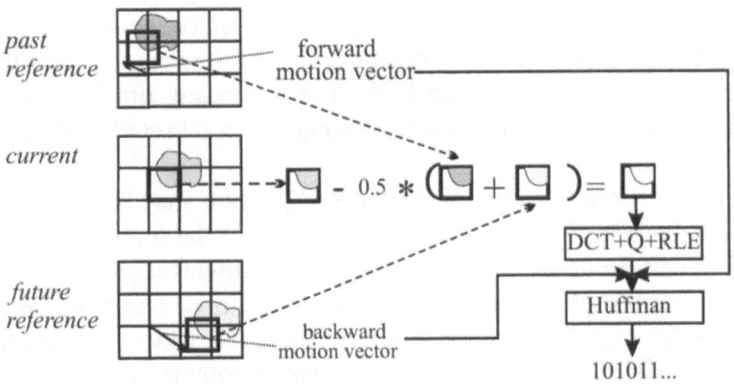

Figure 3.4.   Interpolated prediction for B frames.

3.2   Using DCT Coefficients

The pioneering work on video parsing directly in compressed domain is conducted by Arman, Hsu and Chiu [40] who proposed a technique for cut detection based on the DCT coefficients of I frames. For each frame a subset of the DCT coefficients of a subset of the blocks is selected in order to construct a vector $V_i = \{c_1, c_2 \ldots\}$. V_i represents the frame i from the video sequence in the DCT space. The normalised inner product is then used to find the difference between frames i and $i + \phi$:

$$D(i, i + \phi) = \frac{V_i V_{i+\phi}}{|V_i||V_{i+\phi}|} \tag{3.1}$$

A cut is detected if $1 - |D(i, i + \phi)| > T_1$, where T_1 is a threshold. In order to reduce false positives due to camera and object motion, video cuts are examined more closely using a second threshold T_2 $(0 < T_1 < T_2 < 1)$. If $T_1 < 1 - |D(i, i + \phi)| < T_2$, the two frames are decompressed and examined by comparing their color histograms.

Zhang et al. [16] apply a pair-wise comparison technique to the DCT coefficients of corresponding blocks of video frames. The difference metric is similar to pixel comparisons [22, 23]. More specifically, the difference of block l from two frames which are ϕ frames apart is given by:

$$DP(i, i + \phi, l) = \frac{1}{64} \sum_{k=1}^{64} \frac{|c_{l,k}(i) - c_{l,k}(i + \phi)|}{\max[c_{l,k}(i), c_{l,k}(i + \phi)]} \qquad (3.2)$$

where $c_{l,k}(i)$ is the k-th DCT coefficient of block l in the frame i, $k = 1 \ldots 64$, and l depends on the size of the frame. If the difference exceeds a given threshold T_1, the block l is considered to be changed. If the number of changed blocks is larger than another threshold T_2, a transition between the two frames is declared. The pair-wise comparison requires far less computation than the difference metric used by Arman [40]. The processing time can be further reduced by applying Arman's method of using only a subset of coefficients and blocks.

It should be noted that both of the above algorithms may be applied only to I frames of the MPEG compressed video, as they are the frames fully encoded with DCT coefficients. As a result, the processing time is significantly reduced but the temporal resolution is low. In addition, due to the loss of the resolution between the I frames, false positives are introduced and, hence, the classification accuracy decreases. Also, neither of the two algorithms can handle gradual transitions or false positives introduced by camera operations and object motion.

3.3   Using DC Terms

For temporal video segmentation in MPEG compressed domain the most natural solution is to use the DC terms as they are directly related to the pixel domain, possibly reconstructing them for P and B frames, when only DC terms of the residual errors are available. Then, analogously to the uncompressed domain methods, the changes between successive frames are evaluated by difference metrics and the decision is taken by complex thresholding.

For example, Yeo and Liu [41] propose a method where so called DC-images are created and compared. DC-images are spatially reduced versions of the original images: the (i, j) pixel of the DC-image is the average value of the (i, j) block of the image (Fig. 3.5).

As each DC term is a scaled version of the block's average value, DC-images can be constructed from DC terms. The DC terms of I frames are directly available in the MPEG stream while those of B and P frames are estimated using the MVs and DCT coefficients of previous I frames. It should be noted

Figure 3.5.   A full image (352 × 288 pixels) and its dc image (44 × 36 pixels).

that the reconstruction techniques are computationally very expensive—in order to compute the DC term of a reference frame (DC_{ref}) for each block, eight 8×8 matrix multiplications and 4 matrix summations are required. Then, the pixel differences of DC-images are compared and a sliding window is used to set the thresholds because the shot transition is a local activity.

In order to find a suitable similarity measure, the authors compare metrics based on pixel differences and color histograms. They confirm that when full images are compared, the first group of metrics is more sensitive to camera and object movements but computationally less expensive than the second one. However, when DC-images are compared, pixel differences based metrics give satisfactory results as DC-images are already smoothed versions of the corresponding full images.

Hence, similarly to the pixel comparison approaches, abrupt transitions are detected using a difference measure based on the sum of absolute pixel differences of two consecutive frames (DC-images in this case):

$$D(l, l+1) = \sum_{i,j} |P_l(i,j) - P_{l+1}(i,j)|, \qquad (3.3)$$

where l and $l+1$ are two consecutive DC-images and $P_l(i,j)$ is the intensity value of the pixel in l-th DC-image at the coordinates (i,j).

In contrast to the previous methods for cut detection that apply global thresholds on the difference metrics, Yeo and Liu propose to use local thresholds, as scene changes are local activities in the temporal domain. In this way false positives due to significant camera and object motions are reduced. More specifically, a sliding window is used to examine m successive frame differences. A cut between frames l and $l+1$ is declared if the following two conditions are satisfied:

- $D(l, l+1)$ is maximum within a symmetric sliding window of size $2m-1$

- $D(l, l+1)$ is n times the second largest value in the window

The second condition guards against false positives due to fast panning or zooming and camera flashes that typically manifest themselves as sequences of large differences or two consecutive peaks, respectively. The size of the sliding window m is set to be smaller than the minimum duration between two transitions, while the values of n typically range from 2 to 3.

Gradual transitions are detected by comparing each frame with the following k-th frame where k is larger than the number of frames in the gradual transition. A gradual transition g_n in the form of linear transition from c_1 to c_2 in the time interval (α_1, α_2) is modelled as

$$
g_n = \begin{cases}
c_1 & n < \alpha_1 \\
\dfrac{c_2 - c_1}{\alpha_2 - \alpha_1}(n - \alpha_2) + c_2 & \alpha_1 \le n < \alpha_2 \\
c_2 & n \ge \alpha_2
\end{cases}
\tag{3.4}
$$

Then if $k > \alpha_2 - \alpha_1$, the difference between frames l and $l+k$ from the transition g_n will be

$$
D_{g_n}(l, l+k) = \begin{cases}
0 & n < \alpha_1 - k \\
\dfrac{|c_2 - c_1|}{|\alpha_2 - \alpha_1|}[n - (\alpha_1 - k)] & \alpha_1 - k \le n < \alpha_2 - k \\
|c_2 - c_1| & \alpha_2 - k \le n < \alpha_1 \\
-\dfrac{|c_2 - c_1|}{|\alpha_2 - \alpha_1|}(n - \alpha_2) & \alpha_1 \le n < \alpha_2 \\
0 & n \ge \alpha_2
\end{cases}
\tag{3.5}
$$

As $D_{g_n}(l, l+k)$ corresponds to a symmetric plateau with sloping sides (see Fig. 3.6), the goal of the gradual transition detection algorithm is to identify such plateau patterns. The algorithm of Yeo and Liu needs 11 parameters to be specified.

In [42] shots are detected by color histogram comparison of DC term images of consecutive frames. Such images are formed by the DC terms of the DCT co-efficients for a frame. DC terms of I pictures are taken directly from the MPEG stream, while those for P and B frames are reconstructed by the following fast algorithm. First, the DC term of the reference image (DC_{ref}) is approximated using the weighted average of the DC terms of the blocks pointed by the MVs, Fig. 3.7:

$$
DC_{ref} = \frac{1}{64} \sum_{\alpha \in E} N_\alpha DC_\alpha
\tag{3.6}
$$

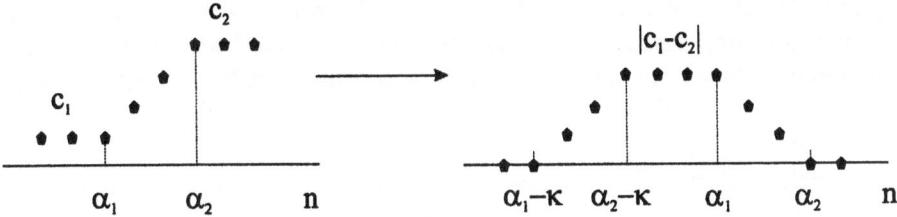

Figure 3.6. g_n and $D_{g_n}(l, l+k)$ in the dissolve detection algorithm of Yeo and Liu.

where DC_α is the DC term of block α, E is the collection of all blocks that are overlapped by the reference block, and N_α is the number of pixels in block α that is overlapped by the reference block.

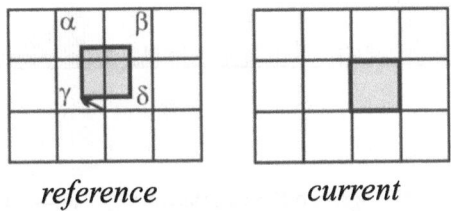

reference   current

Figure 3.7. DC term estimation in the method of Shen and Delp.

Then, the approximated DC terms of the predicted pictures are added to the encoded DC terms of the difference images in order to form the DC terms of P and B pictures:

$$DC = DC_{diff} + DC_{ref} \tag{3.7}$$

(only forward or only backward prediction) or

$$DC = DC_{diff} + (DC_{ref1} + DC_{ref2})/2 \tag{3.8}$$

(interpolated prediction).

In this way the computations are reduced to at most 4 scalar multiplications and 3 scalar summations for each block to determine DC_{ref}.

The histogram difference diagram is generated by calculating the absolute sum of histogram differences (see Section 2.3.1) between DC terms of successive images. As it can be seen from Fig. 3.8, a break is represented by a single sharp pulse and a dissolve entails a number of consecutive medium-heighted pulses. Cuts are detected using a static threshold. For the recognition of gradual transitions, the histogram difference of the current frame is compared with the average of the histogram differences of the previous frames within a window. If this difference is n times larger than the average value, a possible start of a gradual transition is marked. The same value of n is used as a soft threshold for the following frames. End of the transition is declared when the histogram difference is lower than the threshold. Since during a gradual transition not all of the histogram differences may be higher than the soft threshold, similarly to the twin comparison, several frames are allowed to have lower difference as long as the majority of the frames in the transition have higher magnitude than the soft threshold.

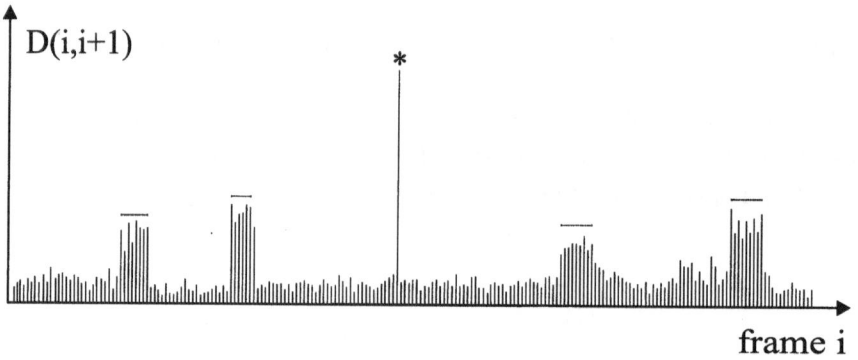

Figure 3.8.   Histogram difference diagram (*: cut;—: dissolve).

As only the DC terms are used, the computation of the histograms is 64 times faster than that using the original pixel values. The approach is not able to distinguish rapid object movement from gradual transition. As a partial solution, a median filter (of size 3) is applied to smooth the histogram differences when detecting gradual transitions. There are 7 parameters that need to be specified.

An interesting extension of the previous approach is proposed by Taskiran and Delp [43]. After the DC term image sequence and the luminance histogram for each image are obtained, a two dimensional feature vector is extracted from each pair of images. The first component is the dissimilarity measure based on the histogram intersection of the consecutive DC term images:

$$x_{1i} = 1 - \text{intersection}(H_i, H_{i+1}) \tag{3.9}$$

$$= \frac{\sum_{j=1}^{n} \min(H_i(j), H_{i+1}(j))}{\sum_{j=1}^{n} H_{i+1}(j)} \tag{3.10}$$

where $H_i(j)$ is the luminance histogram value for the bin j in frame i and n is the number of bins used. Note that the definition of the histogram intersection is slightly different from that used in [30].

The second feature is the absolute value of the difference of standard deviations for the luminance component of the DC term images, i.e. $x_{2i} = |\sigma_i - \sigma_{i+1}|$. The so called *generalised sequence trace* d for a video stream composed of n frames is defined as $d_i = ||x_i - x_{i+1}||, i = 1 \ldots n$.

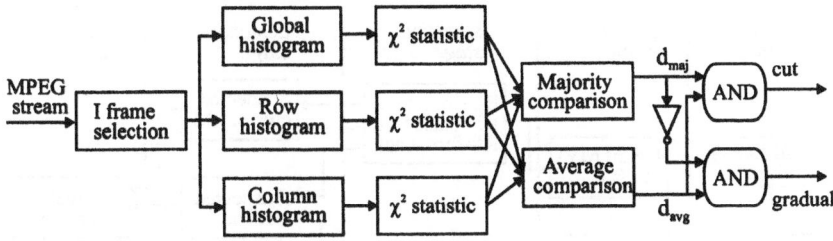

Figure 3.9.   Video shot detection scheme of Patel and Sethi.

These features are chosen not only because they are easy to extract. Combining histogram-based and pixel-based parameters makes sense as they complement some of their disadvantages. As it was discussed already, pixel-based techniques give false alarms in case of camera and object movements. On the other hand, histogram-based techniques are less sensitive to these effects but may miss shot transition if the luminance distribution of the frames do not change significantly. It is shown that there are different types of peaks in the generalised trace plot: wide, narrow and middle corresponding to a fade out followed by a fade in, cuts, and dissolves, respectively. Then, in contrast to the other approaches that apply global or local thresholds to detect the shot boundaries, Taskiran and Delp pose the problem as a one dimensional edge detection and apply a method based on mathematical morphology.

Patel and Sethi [44, 45] use only the DC components of I frames. In [45] they compute the intensity histogram for the DC term images and compare them using three different statistics: Yakimovski likelihood ratio, χ^2 test and Kolmogorov-Smirnov statistics. The experiments show that χ^2 test gives satisfactory results and outperforms the other techniques. In their consequent paper [44], Patel and Sethi compare local and global histograms of consecutive DC term images using χ^2 test, Fig. 3.9.

The local row and column histograms X_i and Y_j are defined as follows:

$$X_i = \frac{1}{M} \sum_{j=1}^{M} b_{0,0}(i,j), \quad Y_j = \frac{1}{N} \sum_{i=1}^{N} b_{0,0}(i,j), \qquad (3.11)$$

where $b_{0,0}(i,j)$ is the DC term of block (i,j), $i = 1, \ldots, N, j = 1, \ldots M$. The outputs of the χ^2 test are combined using majority and average comparison in order to detect abrupt and gradual transitions. As only I frames are used, the DC recovering is eliminated. However, the temporal resolution is low as in a typical GOP every 12th frame is an I frame and, hence, the exact shot boundaries cannot be labelled.

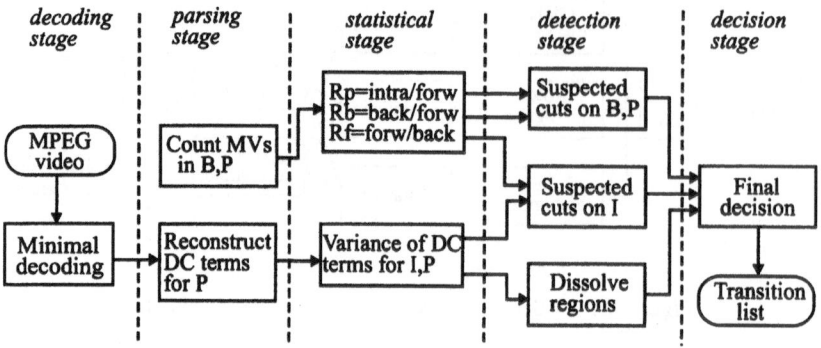

Figure 3.10.   Shot detection algorithm of Meng, Juan and Chang.

3.4   Using DC Terms and MB Coding Mode

Meng, Juan and Chang [46] propose an algorithm based on the DC terms and the type of MB coding, Fig. 3.10. DC components only for P frames are reconstructed. Gradual transitions are detected by calculating the variance of the DC term sequence for I and P frames and looking for parabolic shapes in this curve. This is based on the fact that if gradual transitions are linear mixture of two video sequences f_1 and f_2 with intensity variances σ_1 and σ_2, respectively, and are characterised by $f(t) = f_1(t)[1 - \alpha(t)] + f_2(t)\alpha(t)$ where $\alpha(t)$ is a linear parameter, then the shape of the variance curve is parabolic, $\sigma^2(t) = (\sigma_1^2 + \sigma_2^2)\alpha(t) - 2\sigma_1^2\alpha(t) + \sigma_1^2$. Cuts are detected by the computation of the following three ratios:

$$R_p = \frac{intra}{forw}, \quad R_b = \frac{back}{forw}, \quad R_f = \frac{forw}{back},$$

where $intra$, $forw$, and $back$ are the number of MBs in the current frame that are intra, forward and backward coded, respectively.

If there is a cut on a P frame, the encoder can not use many MBs from the previous anchor frame for motion compensation and as a result many MBs will be coded intra. Hence, a suspected cut on P frame is declared if R_p peaks. On the other hand, if there is a cut on a B frame, the encoding will be mainly backward. Therefore, a suspected cut on B frame is declared if there is a peak in R_b. An I frame is a suspected cut frame if two conditions are satisfied:

- there is a peak in $\Delta\sigma^2$ for this frame

- the B frames before I have peaks in R_f.

The first condition is based on the observation that the intensity variance of the frames during a shot is stable, while the second condition prevents false positives due to motion.

This technique is relatively simple, requires minimum encoding and produces good accuracy. The total number of parameters needed to implement this algorithm is 7.

3.5   Using DCT Coefficients, MB Coding Mode and MVs

A very interesting two-pass approach is taken by Zhang, Low and Smoliar [16]. They first locate the regions of potential transitions, camera operations and object motion, applying the pair-wise DCT coefficients comparison of I frames (Eq. 3.2) as in their previous approach (see Sec. 3.2). The goal of the second pass is to refine and confirm the break points detected by the first pass. By checking the number of MVs for the selected areas, the exact cut locations are detected. If M denotes the number of MVs in P frames and the smaller of the numbers of forward and backward nonzero MVs in B frames, then $M < T$ (where T is a threshold close to zero) is an effective indicator of a cut before or after the B and P frame.

Gradual transitions are found by an adaptation of the twin comparison algorithm utilising the DCT differences of I frames. By MV analysis, though using thresholds, false positives due to pan and zoom are detected and discriminated from gradual transitions.

Thus, the algorithm only uses information directly available in the MPEG stream. It offers high processing speed due to the multipass strategy, good accuracy and also detects false positives due to pan and zoom. However, the metric for cut detection yields false positives in the case of static frames. Also, the problem of how to distinguish object movements from gradual transitions is not addressed.

3.6   Using MB Coding Mode and MVs

In [47] cuts, fades and dissolves are detected only using MVs from P and B frames and information about MB coding mode. The system follows a two-pass scheme and has a hybrid rule-based/neural structure. During the rough scan, peaks in the number of intra coded MBs in P frames are detected. They can be sharp (Fig. 3.11) or gradual with specific shape (Fig. 3.12) and are good indicators of abrupt and gradual transitions, respectively.

The solution is then refined by a precise scan over the frames of the respective neighbourhoods. The simpler boundaries (cuts and black fade edges) are recognized by the rule-based module, while the decisions for the complex ones (dissolves and non-black fade edges) are taken by the neural part. The precise scan also reveals cuts that remain hidden for the rough scan, e.g. B_{24}, I_{49}, B_{71}

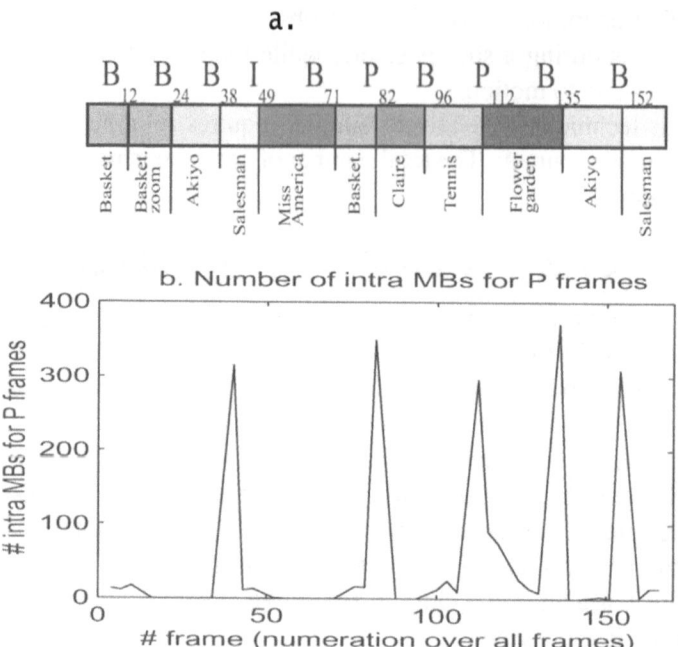

Figure 3.11.   Cuts: (a) Video structure, (b) Number of intra-coded MBs for P frames.

and B_{96} in Figure 3.11. The rules for the exact cut location are based on the number of backward and forward MBs while those for the fades black edges detection use the number of interpolated and backward coded MBs. There is only one threshold in the rules that is easy to set and not sensitive to the type of video.

The neural network module learns from pre-classified examples in the form of MV patterns corresponding to the following 6 classes: stationary, pan, zoom, object motion, tracking and dissolve. It is used to distinguish dissolves from object and camera movements, find the exact location of the complex boundaries of the gradual transition and further divide shots into sub-shots. The well-known supervised neural algorithm Learning Vector Quantisation (LVQ) [48] is used. Given a set of pre-classified feature vectors (training examples), LVQ creates a few prototypes for each class, adjusts their positions by learning and then classifies the unseen examples by means of the nearest-neighbour principle.

The MVs from both P and B frames are used to generate a 22-dimensional feature vector for each frame. The first component is calculated using the number of zero MVs in forward, backward and interpolated areas. Then, the forward MV pattern is sub-divided in 7 vertical strips for which the following 3 parameters are computed: the average of the MV direction, the standard

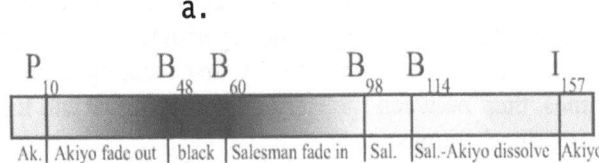

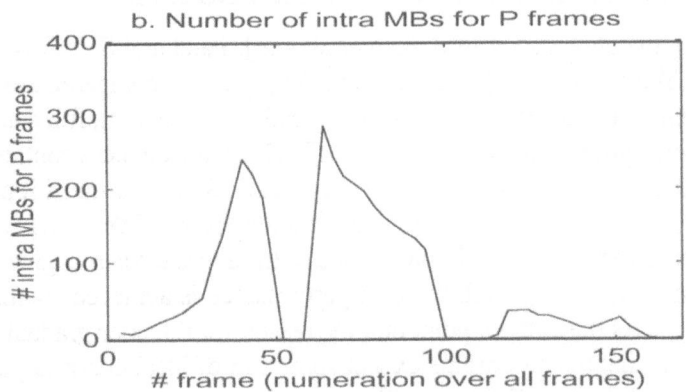

Figure 3.12.   Fade out, fade in, dissolve: (a) Video structure, (b) Number of intra-coded MBs for P frames.

deviation of the MV direction and the average of MV modulus. A technique that deals with the discontinuity of angles at $0/360^o$ is proposed for the calculation of the MV direction. To build the LVQ classifier, the MV patterns of 1200 P and B frames (200 for each class) were visually examined and manually labelled.

The approach is simple, fast, robust to camera operations and very accurate when detecting the exact locations of cuts, fades and simple dissolves. However, sometimes dissolves between busy sequences are recognized as object movement or their boundaries are not exactly determined.

3.7   Using MB Coding Mode and Bitrate Information

Although limited only to cut detection, a simple and effective approach is proposed in [49]. It only uses the bitrate information at MB level and the number of various motion predicted MBs. A large change in bitrate between two consecutive I or P frames indicates a cut between them. Similarly to [46], the number of backward predicted MBs is used for detecting cuts on B frames. Here, the ratio is calculated as

$$R_b = back/mc$$

where *back* and *mc* are the number of backward and all motion compensated MBs in a B frame, respectively. The algorithm is able to locate the exact cut locations. It operates hierarchically by first locating a suspected cut between two I frames, then between the P frames of the GOP and finally (if necessary) by checking the B frames.

3.8   Comparison of Algorithms for Temporal Video Segmentation in Compressed Domain

In [50] the approaches of Arman et al. [40], Patel and Sethi [44], Meng et al. [46], Yeo and Liu [41] and Shen and Delp [42] are compared along several parameters: classification performance (recall and precision), full data use, ease of implementation, source effects. Ten MPEG video sequences containing more than 30 000 frames connected with 172 cuts and 38 gradual transitions are used as an evaluation database. It is found that the algorithm of Yeo and Liu performs best when detecting cuts. Although none of the approaches recognizes gradual transitions particularly well, the best performance is achieved by the last two algorithms. As the authors point out, the reason for the poor gradual transition detection is that the algorithms expect some sort of ideal curve (a plateau or a parabola) but the actual frame differences are noisy and either do not follow this ideal pattern or do not do this smoothly for the entire transition. Another interesting conclusion is that not processing of all frame types (e.g., like in the first two methods) does decrease performance significantly. It was also found that the different measures used on the three frame types have different performance, e.g. in the approach of Meng et al. the P frame differences are more reliable than the B ones. The algorithm of Yeo and Liu is found to be easiest for implementation as it specifies the parameter values and even some performance analysis is already carried out by the authors. The dependence of the two best performing algorithms on bitrate variations is investigated and it is shown that they are robust to bitrate changes except at very low rates. Finally, the dependence of the algorithm of Yeo and Liu on two different software encoder implementations is studied and significant performance differences are reported.

4.   Conclusions

Temporal video segmentation is the first step towards automatic organization of digital video for browsing and retrieval. It is an active area of research gaining attention from several research communities including image processing, computer vision, pattern recognition and artificial intelligence.

In this chapter we reviewed and classified several existing approaches for temporal video segmentation in MPEG compressed domain, discussing their relative advantages and disadvantages. While early work focus on cut detection,

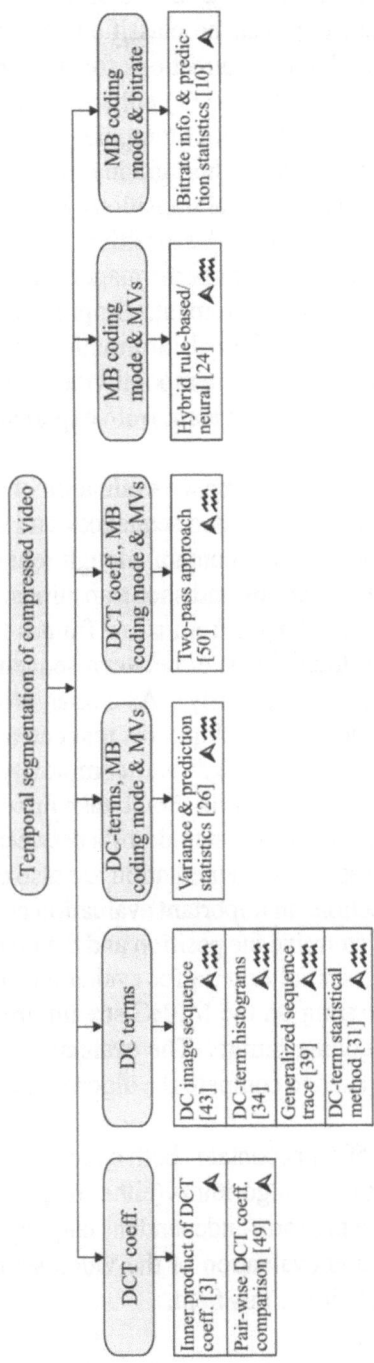

Figure 3.13.   Taxonomy of techniques for temporal video segmentation that process compressed video (▲ detect cut, ≋ detect gradual transitions).

more recent techniques deal with gradual transition detection. Based on the type of information used, they can be classified into six groups, Fig. 3.13. Their limitations, that highlight the directions for further development, can be summarised as follows. Most of them (1) either require the reconstruction of DC terms of P or P&B frames, or sacrifice temporal resolution and classification accuracy, (2) process unrealistically short gradual transitions and are unable to recognize the different types of gradual transitions, (3) involve many adjustable thresholds, and (4) do not handle false positives due to camera operations. Moreover, all of them have rather poor performances while trying to distinguish gradual transitions from object movement. Improvements may be possibly obtained thanks to the use of additional information (e.g., audio features and text captions), integration of different temporal video segmentation techniques and development of methods that can learn from experience how to adjust their parameters.

Consistent comparison and performance evaluation of the various techniques requires availability of benchmark video sequences and unified evaluation criteria. Benchmark sequences should contain enough representative data for the possible types of camera operations and shot transitions. They should include gradual transitions varying in length from a few frames to hundreds of frames and also complex gradual transition (i.e. between sequences involving motion or other effects occurring simultaneously). As human ground truthing is a subjective process there is a need to develop tools that can automatically generate ground truthed video of arbitrary length and complexity by splicing together video sequences. The evaluation criteria should take into consideration the type of application that may require different trade-offs between recall and precision. Threshold selection and ease of implementation are also critical issues. In case of gradual transition detection, an important evaluation criterion is the ability of the algorithm to exactly determine the position and the type (dissolve, fade, etc.) of the transition. Other essential criteria for evaluation of the shot boundaries detection algorithms operating on the MPEG stream are the sensitivity to the encoder used and the encoding bitrate. The evaluation process needs to state explicitly the parameters used to compare the algorithm output with the ground truth.

Building a repository [50] that contains both benchmark sequences and Web-based executable versions of the algorithms (either by providing a Web interface or implementing them in a platform-independent language such as Java) would allow better comparison and evaluation of the video segmentation algorithms and boost the research activity in this field.

References

[1] H. Okamoto, M. Nakamura, Y. Hatanaka, and S. Yamazaki, "A consumer digital VCR for advanced television," *IEEE Transactions on Consumer Electronics*, vol. 39, pp. 199–204, 1993.

[2] "SMASH project." http://www.extra.research.philips.com/euprojects/smash/, Feb. 2003.

[3] J. Y. Chen, C. Taskiran, E. Delp, and C. A. Bouman, "ViBE: A new paradigm for video database browsing and search," in *IEEE Workshop on Content-Based Access of Image and Video Libraries*, (Santa Barbara, USA), pp. 96–100, 1998.

[4] H. J. Zhang, J. Wu, D. Zhong, and S. Smoliar, "An integrated system for content-based video retrieval and browsing," *Pattern Recognition*, vol. 30, no. 4, pp. 643–658, 1997.

[5] S. F. Chang, W. Chen, H. J. Meng, H. Sundaram, and D. Zhong, "VideoQ: An automated content based video search system using visual cues," in *ACM Multimedia Conf.*, (Seattle, USA), pp. 313–324, 1997.

[6] W. Niblack, X. Zhu, J. L. Hafner, T. Breuer, D. B. Ponceleon, D. Petkovic, M. D. Flickner, E. Upfal, S. I. Nin, S. Sull, B. E. Dom, B. L. Yeo, S. Srinivasan, D. Zivkovic, and M. Penner, "Updates to the QBIC system," in *IS&T/SPIE Conf. Storage and Retrieval for Image and Video Databases VI*, vol. 3312, pp. 150–161, 1997.

[7] M. Smith and T. Kanade, "Video skimming and characterization through the combination of image and language understanding," in *Proc. of the 1998 IEEE International Workshop on Content-Based Access of Image and Video Databases (ICCV'98)*, (Bombay, India), pp. 61–70, 1998.

[8] D. Zhong, H. Zhang, and S. F. Chang, "Clustering methods for video browsing and annotation," in *IS&T/SPIE Storage and Retrieval for Still Image and Video databases IV*, vol. 2670, pp. 239–246, 1996.

[9] M. Yeung and B. L. Yeo, "Time-constrained clustering for segmentation of video into story units," in *Proceedings of the 13th International Conference on Pattern Recognition*, vol. 3, (Los Alamitos, USA), pp. 375–380, IEEE Comput. Soc. Press, 1996.

[10] M. Yeung and B.-L. Yeo, "Video visualization for compact presentation and fast browsing of pictorial content," *IEEE Transactions on Circuits and Systems for Video Technology*, vol. 7, no. 5, pp. 771–785, 1997.

[11] A. Hanjalic and R. L. Lagendijk, "Automated high-level movie segmentation for advanced video-retrieval systems," *IEEE Transactions on Circuits and Systems for Video Technology*, vol. 9, no. 4, pp. 580–588, 1999.

[12] Q. Huang, Z. Liu, and A. Rosenberg, "Automated semantic structure reconstruction and representation generation for broadcast news," in

IS&T/SPIE Conference on Storage and Retrieval for Image and Video databases VII, vol. 3656, pp. 50–62, 1999.

[13] G. Davenport, T. A. Smith, and N. Pincever, "Cinematic primitives for multimedia," *IEEE Transactions on Computer Graphics Applications*, vol. 11, no. 4, pp. 67–74, 1991.

[14] A. Hampapur, R. Jain, and T. E. Weymouth, "Production model based digital video segmentation," *Multimedia Tools and Applications*, vol. 1, no. 1, pp. 9–46, 1995.

[15] R. Zabih, J. Miler, and K. Mai, "A feature-based algorithm for detecting and classifying production effects," *Multimedia Systems*, vol. 7, no. 2, pp. 119–128, 1999.

[16] H. J. Zhang, C. Y. Low, and S. W. Smoliar, "Video parsing and browsing using compressed data," *Multimedia Tools and Applications*, vol. 1, pp. 89–111, 1995.

[17] G. Ahanger and T. D. C. Little, "A survey of technologies for parsing and indexing digital video," *Journal of Visual Communication and Image Representation*, vol. 7, no. 1, pp. 28–43, 1996.

[18] F. Idris and S. Panchanathan, "Review of image and video indexing techniques," *Journal of Visual Communication and Image Representation*, vol. 8, no. 2, pp. 146–166, 1997.

[19] R. M. Ford, C. Robson, D. Temple, and M. Gerlach, "Metrics for shot boundary detection in digital video sequences," *Multimedia Systems*, vol. 8, pp. 37–46, 2000.

[20] A. Dailianas, R. B. Allen, and P. England, "Comparisons of automatic video segmentation algorithms," in *Integration Issues in Large Commercial Media Delivery Systems*, no. 2615, pp. 2–16, 1995.

[21] T. Kikukawa and Ś. Kawafuchi, "Development of an automatic summary editing system for the audio-visual resources," *Transactions on Electronics and Information*, pp. 204–212, 1992.

[22] A. Nagasaka and Y. Tanaka, *Visual Database Systems II*, ch. Automatic video indexing and full-video search for object appearances, pp. 113–127. Elsevier, 1995.

[23] H. J. Zhang, A. Kankanhalli, and S. W. Smoliar, "Automatic partitioning of full-motion video," *Multimedia Systems*, vol. 1, no. 1, pp. 10–28, 1993.

[24] R. Kasturi and R. Jain, *Computer Vision: Principles*, ch. Dynamic vision, pp. 469–480. Washington DC, USA: IEEE Computer Society Press, 1991.

[25] B. Shahraray, "Scene change detection and content-based sampling of video sequences," in *Digital Video Compression: Algorithms and Technologies* (R. J. Safranek and A. A. Rodriquez, eds.), vol. 2419, pp. 2–13, Feb. 1995.

[26] W. Xiong, J. C. M. Lee, and M. C. Ip, "Net comparison: a fast and effective method for classifying image sequences," in *Storage and Retrieval for Image and Video Databases III*, vol. 2420, (San Jose, USA), pp. 318–328, 1995.

[27] W. Xiong and J. C. M. Lee, "Efficient scene change detection and camera motion annotation for video classification," *Computer Vision and Image Understanding*, vol. 71, no. 2, pp. 166–181, 1998.

[28] M. J. Swain, "Interactive indexing into image databases," in *SPIE Conf. Storage and Retrieval in Image and Video Databases*, pp. 173–187, 1993.

[29] Y. Tonomura, "Video handling based on structured information for hypermedia systems," in *ACM Int. Conf. on Multimedia Information Systems*, (Singapore), pp. 333–344, 1991.

[30] U. Gargi, S. Oswald, D. Kosiba, S. Devadiga, and R. Kasturi, "Evaluation of video sequence indexing and hierarchical video indexing," in *SPIE Conf. Storage and Retrieval in Image and Video Databases*, pp. 1522–1530, 1995.

[31] J. S. Boreczky and L. A. Rowe, "Comparison of video shot boundary detection techniques," in *IS&T/SPIE Intern. Symposium Electronic Imaging: Storage and Retrieval for Image and Video Databases*, (San Jose, USA), pp. 170–179, 1996.

[32] D. Swanberg, C. F. Shu, and R. Jain, "Knowledge guided parsing in video databases," in *Proc. of the Int. Conf. on Storage and Retrieval for Image and Video Databases*, (San Jose, USA), pp. 13–24, 1993.

[33] C. M. Lee and D. M. C. Ip, "A robust approach for camera break detection in color video sequences," in *IAPR Workshop Machine Vision Appl.*, (Kawasaki, Japan), pp. 502–505, 1994.

[34] B. Gunsel, A. M. Ferman, and A. M. Tekalp, "Temporal video segmentation using unsupervised clustering and semantic object tracking," *Journal of Electronic Imaging*, vol. 7, no. 3, pp. 592–604, 1998.

[35] A. M. Ferman and A. M. Tekalp, "Efficient filtering and clustering for temporal video segmentation and visual summarization," *Journal of Visual Communication and Image Representation*, vol. 9, no. 4, pp. 336–351, 1998.

[36] P. Aigrain and P. Joly, "The automatic real-time analysis of film editing and transition effects and its applications," *Computers and Graphics*, vol. 18, no. 1, pp. 93–103, 1994.

[37] H. Yu, G. Bozdagi, and S. Harrington, "Feature-based hierarchical video segmentation," in *Int. Conf. on Image Processing (ICIP97)*, (Santa Barbara, USA), pp. 498–501, 1997.

[38] J. S. Boreczky and L. D. Wilcox, "A hidden Markov model framework for video segmentation using audio and image features," in *Int. Conf. Acoustics, Speech, and Signal Proc. 6*, (Seattle, USA), pp. 3741–3744, 1998.

[39] ISO/IEC, "13818 Draft Int. Standard: Generic Coding of Moving Pictures and Associated Audio, Part 2: video."

[40] F. Arman, A. Hsu, and M.-Y. Chiu, "Image processing on compressed data for large video databases," in *First ACM Intern. Conference on Multimedia*, pp. 267–272, 1993.

[41] B. Yeo and B. Liu, "Rapid scene analysis on compressed video," *IEEE Transactions on Circuits & Systems for Video Technology*, vol. 5, no. 6, pp. 533–544, 1995.

[42] K. Shen and E. Delp, "A fast algorithm for video parsing using MPEG compressed sequences," in *Intern. Conf. Image Processing (ICIP'96)*, (Lausanne, Switzerland), pp. 69–72, 1996.

[43] C. Taskiran and E. Delp, "Video scene change detection using the generalized sequence trace," in *IEEE Int. Conf. Acoustics, Speech & Signal Processing*, (Seattle, USA), pp. 2961–2964, 1998.

[44] N. V. Patel and I. K. Sethi, "Video shot detection and characterization for video databases," *Pattern Recognition*, vol. 30, pp. 583–592, 1997.

[45] I. K. Sethi and N. V. Patel, "A statistical approach to scene change detection," in *IS&T/SPIE Conf. Storage and Retrieval for Image and Video Databases III*, vol. 2420, (San Jose, CA, USA), pp. 2–11, 1995.

[46] J. Meng, Y. Juan, and S. F. Chang, "Scene change detection in a MPEG compressed video sequence," in *IS&T/SPIE Int. Symp. Electronic Imaging*, vol. 2417, (San Jose, CA, USA), pp. 14–25, 1995.

[47] I. Koprinska and S. Carrato, "Detecting and classifying video shot boundaries in MPEG compressed sequences," in *IX Eur. Sig. Proc. Conf. (EUSIPCO)*, (Rhodes, Greece), pp. 1729–1732, 1998.

[48] T. Kohonen, "The self-organizing map," *Proc. of the IEEE*, vol. 78, no. 9, pp. 1464–1480, 1990.

[49] J. Feng, K. T. Lo, and H. Mehrpour, "Scene change detection algorithm for MPEG video sequence," in *Int. Conf. Image Processing (ICIP96)*, (Lausanne, Switzerland), pp. 821–824, 1996.

[50] U. Gargi, R. Kasturi, and S. Antani, "Performance characterization and comparison of video indexing algorithms," in *Conf. Computer Vision and Pattern Recognition (CVPR)*, (Santa Barbara CA, USA), pp. 559–565, 1998.

II

ROBUST VIDEO DATA PROTECTION AND WA-TERMARKING

Chapter 4

DIGITAL WATERMARKING FOR THE COPYRIGHT PROTECTION OF COMPRESSED VIDEO

Dimitrios Simitopoulos, Sotirios A. Tsaftaris, Nikolaos V. Boulgouris, Georgios A. Triantafyllidis, and Michael G. Strintzis

Information Processing Laboratory

Department of Electrical and Computer Engineering

Aristotle University of Thessaloniki

Thessaloniki 540 06, Greece

and

Informatics and Telematics Institute

1st Km Thermi-Panorama Road

57001 (PO Box 361)

Thermi-Thessaloniki, Greece

{dsim,stsaft,nblg,gatrian,strintzi}@iti.gr

Abstract   In this chapter, a new technique for the watermarking of MPEG-1 and MPEG-2 compressed video streams is proposed. The watermarking scheme operates directly in the domain of MPEG-1 system streams and MPEG-2 program streams (multiplexed streams). Perceptual models are used during the embedding process in order to preserve video quality. The watermark is embedded in the compressed domain and is detected without the use of the original video sequence. Experimental evaluation demonstrates that the proposed scheme is able to withstand a variety of attacks. The resulting watermarking system is very fast and reliable, and is suitable for the copyright protection of video content.

Keywords: Video watermarking, MPEG, perceptual analysis, data hiding, error correcting codes.

1.   Introduction

In parallel with the development and the introduction of Digital Versatile Disc (DVD) as the ultimate medium for the digital storage and distribution

J.F. Tasič et al. (eds.), Intelligent Integrated Media Communication Techniques, 117-151.

of audiovisual content, the MPEG-2 standard [1, 2] was established as the coding scheme for such content. These developments made the large-scale distribution and replication of multimedia very easy but at the same time also to a large extent uncontrollable. In order to protect multimedia content from unauthorized trading, digital watermarking techniques have been introduced [3]. Such techniques embed the digital signature of the copyright holder in image [4, 5, 6], audio [7] or video [8, 9, 10] signals. In the case of unauthorized copying of multimedia data, the copyright holder is able to prove the ownership of the data.

Especially for images and video, numerous watermarking techniques have been proposed that perform the watermark embedding and detection processes in either the spatial [11, 12], Fourier [13, 14], Discrete Cosine Transform (DCT) [4, 15] or wavelet [16, 17] domain. However, very few deal with the very important issue of compressed domain watermarking for video [18, 12, 19, 20].

The authors of [12] proposed a technique that decompresses the MPEG stream, watermarks the resulting DCT coefficients and re-encodes them into a new compressed bitstream. Chung et al. [19] applied a DCT domain embedding technique that also incorporates a block classification algorithm in order to select the coefficients to be watermarked. In [18] a faster approach was proposed, that embeds the watermark in the domain of quantized DCT coefficients but uses no perceptual models in order to increase the robustness of the watermark.

The important practical problem of watermarking of MPEG-1/2 multiplexed streams has not been addressed, up to date, in the literature. Multiplexed streams contain at least two elementary streams, an audio and a video elementary stream. Thus, it is necessary to develop a watermarking scheme that operates with multiplexed streams as its input.

In this chapter, a novel compressed domain watermarking scheme is presented which is suitable for MPEG-1/2 multiplexed streams. Embedding and detection are performed without fully de-multiplexing the video stream. During the embedding process, the data that are going to be watermarked are extracted from the stream, watermarked and placed back into the stream. This approach leads to a fast implementation which is necessary for real-time applications.

The watermark is embedded in the intraframes (I-frames) of the video sequence. In each I-frame, only the quantized AC coefficients of each discrete cosine transformed block of the luminance component are watermarked. In order to reach a satisfactory compromise between robustness and imperceptibility of the embedded watermark, a quantized DCT domain method for the application of perceptual analysis is introduced for the selection of the coefficients to be watermarked and of the watermark strength. The watermark detection strategy followed, however, operates in the DCT domain rather than the quantized domain. The resulting novel hybrid scheme is shown to withstand transcoding, as well as cropping and filtering.

The chapter is organized as follows. In Section 2, the requirements of a video watermarking system are analyzed. Section 3 describes the processing in the compressed stream. The proposed embedding scheme is presented in Section 4. Section 5 describes the detection process and in Section 6 two detector implementations for detection in the DCT domain are presented. Section 7 presents a data hiding scheme based on the proposed watermark embedding and detection scheme. Simulation results are discussed in Section 8 and finally, in Section 9, conclusions are drawn.

2.   Video Watermarking System Requirements

In all watermarking systems the watermark is required to be imperceptible and robust against attacks such as compression, cropping, filtering [9, 6, 21] and geometric transformations [13, 14]. Apart form the above, compressed video watermarking systems have the following additional requirements:

- **Fast embedding/detection.** A video watermarking system must be very fast due to the fact that a large amount of data has to be processed. Watermark embedding and detection procedures should be efficiently designed in order to offer fast processing times using a software implementation.

- **Blind detection.** The system should not use the original video for the detection of the watermark. This is necessary not only because of the important concerns raised in [21] about using the original data in the detection process (e.g. an attacker can create a fake original by subtracting his own watermark from a watermarked video frame so that the watermarked video frame appears to have the attacker's watermark), but also because it is often impractical to keep all original sequences in addition to the watermarked ones.

- **Preserve file size.** The size of the MPEG file should not be altered significantly. The watermark embedding procedure should take into account the fact that the total MPEG file size should not be significantly increased, because an MPEG file may have been created, so as to conform to specific bandwidth or storage constraints. This may be accomplished by watermarking only the DCT coefficients whose corresponding Variable Length Code (VLC) words after watermarking will have less than or equal length to the length of the original VLC words, as in [12, 19, 22, 18].

- **Avoid/Compensate drift error.** Due to the nature of the interframe coding applied by MPEG, alterations of the coded data in one frame may propagate in time and cause alterations to the subsequent decoded frames. Therefore, special care should be taken during the watermark embedding in order to avoid visible degradation in subsequent frames. Drift error of this nature was encountered in [12], where the watermark was embedded in all video frames (intra- and interframes) in the compressed domain; the authors of [12] proposed the addition of a drift compensation signal to compensate for watermark signals propagated from previous frames. Generally, either the watermarking method should be designed in a way such that drift error is imperceptible, or the drift error should be compensated, at the expense of additional computational complexity.

In the ensuing sections, an MPEG-1/2 watermarking system is described which meets the above requirements.

3.   Preprocessing of MPEG-1/2 Multiplexed Streams

For several reasons, it is often preferable to watermark video in the compressed rather than the spatial domain. In fact, it is very often impractical (due to high storage capacity requirements) or indeed entirely not feasible to decompress and then recompress the entire video data. In addition, a watermarking system that performs decoding and re-encoding of an MPEG stream would also significantly increase the required processing time, perhaps even prohibiting it from being used in real time applications. For these reasons, the video watermark embedding and detection methods presented in this chapter are carried out entirely in the compressed domain.

MPEG-2 program streams and MPEG-1 system streams are multiplexed streams that contain at least two elementary streams i.e. an audio and a video elementary stream. A fast and efficient video watermarking system should be able to cope with multiplexed streams. Specifically, it should be able to discern one elementary stream from the other, watermark the video stream (possibly also the audio stream), and then multiplex again the two elementary streams in one stream. An obvious approach to MPEG watermarking would be to use the following procedure. The original stream is de-multiplexed to its constituent elementary streams (video and audio). Subsequently the video elementary stream is processed in order to embed the watermark. Finally, the resulting watermarked video elementary stream and the audio elementary stream are again multiplexed to produce the final MPEG stream. The above process however is extremely costly computationally and very slow in implementation.

In order to keep complexity low, a technique was developed that does not fully de-multiplex the stream before the watermark embedding, but instead deals with the multiplexed stream itself. Specifically, first the video elementary stream packets are detected in the multiplexed stream. For the video packets that contain I-frame data, the encoded video data are extracted from the video packets and variable length decoding is performed in order to obtain the quantized DCT coefficients. The headers of these packets are left intact. This procedure is schematically described in Fig. 4.1. The quantized DCT coefficients are first

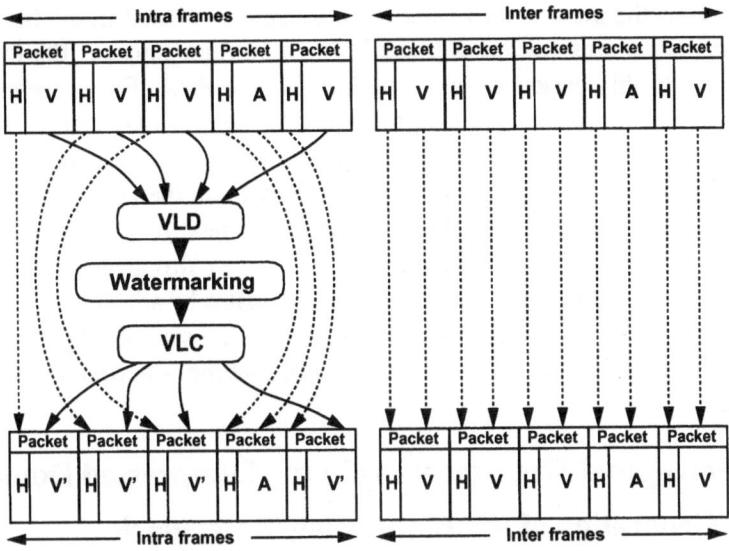

Figure 4.1. Operations performed on an MPEG multiplexed stream (V: encoded video data, A: encoded audio data, H: elementary stream packet header, Packet: elementary stream packet, V': watermarked encoded video data, VLC: Variable length coding, VLD: Variable length decoding).

watermarked. Then the watermarked coefficients are variable length coded. The video encoded data are partitioned so that they fit into video packets that use their original headers. Audio packets and packets containing interframe data are not altered. Basically, the stream structure remains unaffected and only the video packets that contain coded I-frame data are altered.

In order to perform the above described process on a multiplexed MPEG stream, a watermarking system must deal with two significant problems. The first involves the packet structure of MPEG, which permits the coexistence of intraframe and interframe data in a single packet. This means that it is possible that a packet contains data from a previous interframe along with data that correspond to a following I-frame. Also, the last data elements of an I-frame may coexist in the same packet with the beginning of interframe data.

Another important issue the implementation of our system had to confront, was the existence of video start codes that are split (cut) during the packetization process. While MPEG multiplexing is performed, the data bytes of such "cut start codes" are split into two segments, the one at the end of a packet and the other at the end of the next packet, making their detection (and consequently I-frame data tracing and substitution) more difficult.

The system proposed in the present chapter deals with both mentioned obstacles. Specifically, using information extracted during the decoding process, the position of the *picture start code* of an I-frame is located in the bitstream and the corresponding packet is traced. The algorithm replaces the original I-frame data in this packet by watermarked I-frame data. For all remaining video packets that contain only I-frame data (whose size may change as will be explained later) apart from the last one which requires special treatment, simple substitution occurs between the old original and the new watermarked data. In order to compete with the occurrence of a next *picture start code* in the last packet that contains I-frame data, the latter are dropped and only data from the next frame are kept. The procedure terminates by adjusting the packet size to its new length. As a result of this approach, the size of all packets that contained I-frame data is adaptively adjusted so that the watermarked I-frame data fit in.

4.   Perceptual Watermarking in the Compressed Domain

4.1   Generation of the Embedding Watermark

The values of the embedding watermark sequence W are either -1 or 1. This sequence is produced from an integer random number generator by setting the watermark coefficient to 1 when the generator outputs a positive number and by setting the watermark coefficient to -1 when the generator output is negative. The result is a zero mean, unit variance process. The random number generator is seeded with the result of a hash function. The MD5 algorithm [23] is used in order to produce a 128 bit integer seed from a meaningful message (owner ID). The watermark generation procedure is depicted in Fig. 4.2. As explained in [21], the watermark is generated so that even if an attacker finds a watermark sequence that leads to a high correlator output, he or she still cannot find a meaningful owner ID that would produce the watermark sequence through this procedure and therefore cannot claim to be the owner of the image. This is ensured by the use of the hashing function included in the watermark generation. The hashing procedure is a one way function that produces a unique seed for the random number generator when an owner ID is supplied. For example, *owner ID 1* produces *seed 1*. If another owner ID is supplied (*owner ID 2*), a different seed (*seed 2*) is produced. There is no way to produce *seed 1* using *owner ID 2*.

4.2   Perceptual Embedding

The proposed watermark embedding scheme (see Fig. 4.3) modifies only the quantized AC coefficients of a luminance block $X_{\kappa,\lambda}(m,n)$ (where κ is the index of the current macroblock, λ is the index of the block within the current macroblock and m, n are indices indicating the position of the current coefficient in an 8×8 DCT block) and leaves the chrominance block unal-

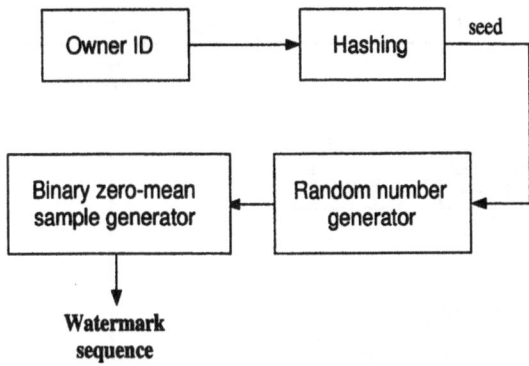

Figure 4.2. Watermark generation.

tered. In order to make the watermark as imperceptible as possible, a novel method is employed, combining perceptual analysis [9, 24] and block classification techniques [25, 19]. These are applied in the DCT domain in order to select which coefficients are best for watermarking. For each selected coefficient in the DCT domain, the product of the embedding watermark coefficient $W_{\kappa,\lambda}(m,n)$ with the corresponding parameters that result from the perceptual analysis (embedding mask $M_{\kappa,\lambda}(m,n)$) and block classification (classification mask $C_{\kappa,\lambda}(m,n)$) is added to the corresponding quantized coefficient, resulting in the quantized watermarked coefficient $X'_{Q\kappa,\lambda}(m,n)$

$$X'_{Q\kappa,\lambda}(m,n) = X_{Q\kappa,\lambda}(m,n) + C_{\kappa,\lambda}(m,n)M_{Q\kappa,\lambda}(m,n)W_{\kappa,\lambda}(m,n) \quad (4.1)$$

Initially, each discrete cosine transformed (DCT) luminance block is classified with respect to its energy distribution to one of the five possible classes: *low activity, diagonal edge, horizontal edge, vertical edge* and *textured block*. The calculation of energy distribution and the subsequent block classification are performed as in [19]. This procedure returns the class of the examined block. The binary mask values $C_{\kappa,\lambda}(m,n)$ corresponding to each class indicate the best coefficients to be altered (using an additive watermark whose strength is estimated by the perceptual analysis described in the sequel) without reducing the visual quality:

$$C_{\kappa,\lambda}(m,n) = \begin{cases} 0 & \text{no alteration} \\ 1 & \text{alteration} \end{cases}$$

where $0 \leq \kappa \leq$ total picture macroblocks, $0 \leq \lambda \leq 3$ and $m,n \in [0,7]$. For all block classes other than the *low activity* class, the binary mask C is one of those depicted in Fig. 4.4. In the case of *low activity* blocks the binary mask contains "ones" for all AC coefficients.

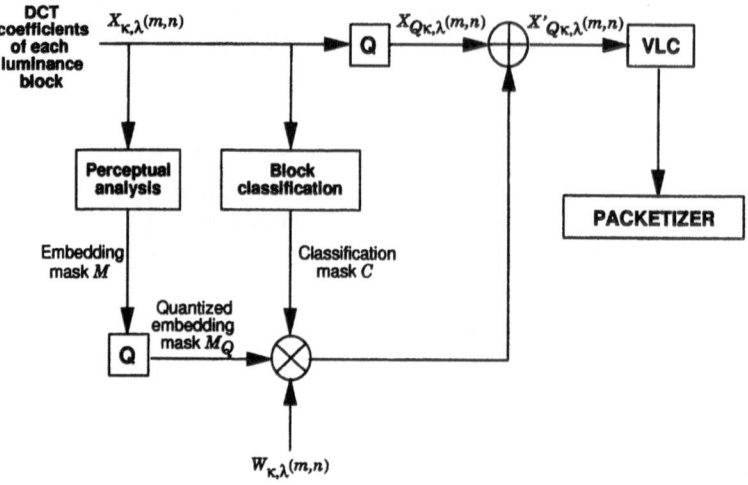

Figure 4.3.   Watermark embedding scheme.

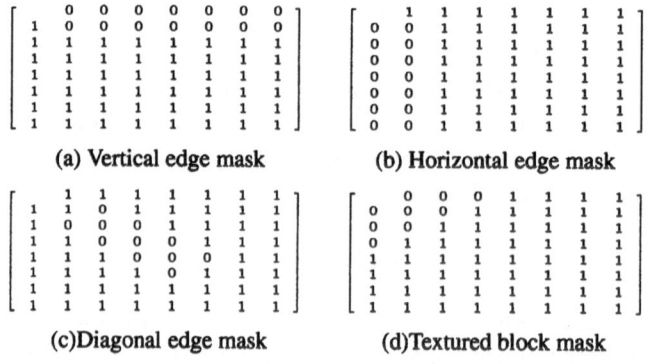

(a) Vertical edge mask   (b) Horizontal edge mask

(c)Diagonal edge mask   (d)Textured block mask

Figure 4.4.   Block classification masks.

The perceptual model that is used is a novel adaptation of the perceptual model proposed by Watson [24]. Specifically, a measure $T''_{\kappa,\lambda}(m, n)$ is introduced which determines the maximum Just Noticeable Difference (JND) for each DCT coefficient of a block and then this model is adapted in order to be applicable to the domain of quantized DCT coefficients.

Let w_x and w_y be the horizontal and vertical width of a pixel in 1/16 *pixels/degree* of visual angle for 48.7 cm viewing distance. Then the horizontal and vertical spatial frequencies are $f_{m,0} = \frac{m}{2Mw_x}$, $f_{0,n} = \frac{n}{2Mw_y}$, where m,n are the line and column indices respectively and $M \times M$ are the DCT block dimensions. The *frequency sensitivity* of the Human Visual System (HVS),

$T(m, n)$, is given by [24]:

$$\log T(m, n) = \log \left(\frac{T \min(f_{m,0}^2 + f_{0,n}^2)^2}{(f_{m,0}^2 + f_{0,n}^2)^2 - 4(1 - r)f_{m,0}^2 f_{0,n}^2} \right)$$

$$+ K \left(\log \sqrt{f_{m,0}^2 + f_{0,n}^2} - \log f_{\min} \right)^2$$

where $K = 1.728$, $f_{\min} = 3.68$ *cycles/degree*, $T_{\min} = 1.1548$. Then

$$T'_{\kappa,\lambda}(m, n) = T(m, n) \left(\frac{X_{\kappa,\lambda}(0,0)}{\overline{X(0,0)}} \right)^{a_T}$$

where $a_T = 0.749$, $T'_{\kappa,\lambda}(m, n)$ represents the *luminance sensitivity* of the HVS, $X_{\kappa,\lambda}(0,0)$ denotes the DC coefficient of each block and $\overline{X(0,0)}$ is the DC coefficient corresponding to the mean luminance of an 8×8 block. For the case of 8-bit color depth, $\overline{X(0,0)} = 1024$. For each 8×8 discrete cosine transformed block the values of $T'_{\kappa,\lambda}(m, n)$ compose a different 8×8 matrix. These values are compared with the absolute value of each DCT coefficient $|X_{\kappa,\lambda}(m, n)|$. In this manner, they are used as thresholds in order to determine which coefficients to watermark. Furthermore, for each 8×8 DCT block the measure $T''_{\kappa,\lambda}(m, n)$, which represents the *contrast masking* property of the HVS, is calculated as follows:

$$T''_{\kappa,\lambda}(m, n) = \max \left(T'_{\kappa,\lambda}(m, n), |X_{\kappa,\lambda}(m, n)|^w \, T'_{\kappa,\lambda}(m, n)^{(1-w)} \right)$$

where $X_{\kappa,\lambda}(m, n)$ denotes the DCT coefficient that belongs on the m row and n column of an 8×8 block and w is an exponent that lies between 0 and 1 usually chosen equal to 0.7 following [24]. The values of $T''_{\kappa,\lambda}(m, n)$ determine the embedding strength of the watermark. Specifically, a mask matrix $M_{\kappa,\lambda}(m, n)$ for each block in a macroblock is calculated. This mask contains the values of $T''_{\kappa,\lambda}(m, n)$ for the coefficients that exceed the $T'_{\kappa,\lambda}(m, n)$ thresholds and zeroes for the remaining coefficients i.e.

$$M_{\kappa,\lambda}(m, n) = \begin{cases} T''_{\kappa,\lambda}(m, n), & \text{if } |X_{\kappa,\lambda}(m, n)| > T'_{\kappa,\lambda}(m, n) \\ 0, & \text{otherwise} \end{cases}$$

In order to achieve efficient perceptual embedding in the quantized domain, the *quantized* values $M_{Q\kappa,\lambda}(m, n)$ of the perceptual mask values $M_{\kappa,\lambda}(m, n)$ are used as the embedding strength of the watermark as explained in Subsection 4.3.

4.3   Quantized Domain Embedding

Following the perceptual analysis described above, two options exist for the watermark embedding process. The watermark will be added either to the DCT coefficients, or to the quantized DCT coefficients. We shall first examine in detail the case where the watermark coefficient $W_{\kappa,\lambda}(m,n)$ is embedded in the DCT coefficient $X_{\kappa,\lambda}(m,n)$ before quantization is applied; then the watermark embedding equation is given by:

$$X'_{\kappa,\lambda}(m,n) = X_{\kappa,\lambda}(m,n) + C_{\kappa,\lambda}(m,n)M_{\kappa,\lambda}(m,n)W_{\kappa,\lambda}(m,n) \quad (4.2)$$

where $X'_{\kappa,\lambda}(m,n)$ denotes the corresponding watermarked DCT coefficient. Then, the MPEG coding algorithm quantizes the watermarked DCT coefficients using the *quant* function

$$quant\,[x(m,n)] = round\left(\frac{8x(m,n)}{q_s Q\,(m,n)}\right) \quad (4.3)$$

where $Q\,(m,n)$ denotes the (m,n) element of the quantization matrix used by MPEG [1] and q_s is the quantizer scale parameter (ranging from 1 to 31) that is selected by the MPEG encoding algorithm during the rate-control process in order to achieve a specific target bitrate for the entire video sequence. The *round* function performs rounding to the closest integer. The quantizer scale q_s has the same value for all discrete cosine transformed coefficients of an 8×8 block.

Similarly, the inverse mapping from quantized coefficients to DCT values is given by

$$quant^{-1}\,[x(m,n)] = \left\lfloor \frac{x(m,n)q_s Q\,(m,n)}{8} \right\rfloor \quad (4.4)$$

where the $\lfloor \cdot \rfloor$ operator denotes downward truncation.

The *quant* and *quant*$^{-1}$ functions in the above equations correspond to MPEG-1 quantization. MPEG-2 quantization is performed in the same way but the value of 8 is changed to 16 in both equations (4.3) and (4.4). Obviously, all analysis in the ensuing sections with obvious modifications applies to both MPEG-1 and MPEG-2.

Thus, using (4.2) and (4.3) the quantized watermarked coefficients are given by:

$$quant\,[X'_{\kappa,\lambda}(m,n)] = round\left(\frac{8X'_{\kappa,\lambda}(m,n)}{q_s Q\,(m,n)}\right) =$$

$$= round\left(\frac{8X_{\kappa,\lambda}(m,n)}{q_s Q\,(m,n)} + \frac{8C_{\kappa,\lambda}(m,n)M_{\kappa,\lambda}(m,n)W_{\kappa,\lambda}(m,n)}{q_s Q\,(m,n)}\right) \quad (4.5)$$

The decimal part of both fractions of equation (4.5) is uniformly distributed in [0,1). In case

$$\left| \frac{8C_{\kappa,\lambda}(m,n)M_{\kappa,\lambda}(m,n)W_{\kappa,\lambda}(m,n)}{q_s Q(m,n)} \right| < 1 \qquad (4.6)$$

there is a 50% probability that the second term of (4.5), which contains the watermark, vanishes altogether and the right-hand part of (4.5) simply yields

$$round \left(\frac{8X_{\kappa,\lambda}(m,n)}{q_s Q(m,n)} + \frac{8C_{\kappa,\lambda}(m,n)M_{\kappa,\lambda}(m,n)W_{\kappa,\lambda}(m,n)}{q_s Q(m,n)} \right) =$$

$$= round \left(\frac{8X_{\kappa,\lambda}(m,n)}{q_s Q(m,n)} \right)$$

which is identical to the quantized value as if no watermark had been embedded. Therefore, it is clear that if (4.6) is valid the embedded watermark may be entirely eliminated by the quantization process. More generally, it is clear that the damage to the watermark signal may be very severe, and that potentially, the watermark detection process may become unreliable.

Thus, in order to avoid reduced detection performance due to MPEG quantization, the second option will henceforth be employed where the watermark is embedded in the quantized DCT coefficients. Since the MPEG coding algorithm performs no other lossy operation after quantization (see Fig. 4.5), any information embedded as in Fig. 4.5 does not run the risk of being eliminated by the subsequent processing. Thus, the watermark exists intact in the quantized coefficients when the detection process is carried out and the quantized DCT coefficients $X_{Q\kappa,\lambda}(m,n)$ are watermarked in the following way (see Fig. 4.3):

$$X'_{Q\kappa,\lambda}(m,n) = X_{Q\kappa,\lambda}(m,n) + C_{\kappa,\lambda}(m,n)M_{Q\kappa,\lambda}(m,n)W_{\kappa,\lambda}(m,n) \quad (4.7)$$

where $M_{Q\kappa,\lambda}(m,n)$ is calculated by

$$M_{Q\kappa,\lambda}(m,n) = quant\,[M_{\kappa,\lambda}(m,n)] = round \left(\frac{8M_{\kappa,\lambda}(m,n)}{q_s Q(m,n)} \right) \qquad (4.8)$$

In addition, whenever the value of $M_{\kappa,\lambda}(m,n)$ is non-zero but the value of $M_{Q\kappa,\lambda}(m,n)$ becomes equal to zero due to quantization, in order to increase the number of watermarked coefficients, $M_{Q\kappa,\lambda}(m,n)$ is set to 1 and the watermark coefficient $W_{\kappa,\lambda}(m,n)$ is embedded with its initial strength, which is equal to 1. In this case, the corresponding watermark strength in the DCT domain ends

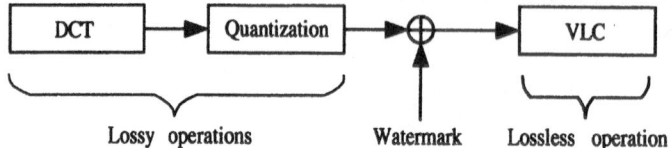

Figure 4.5.   MPEG encoding operations.

up being higher than the strength allowed by the perceptual model. However, experimentation has shown that this modification does not degrade the visual quality of the watermarked video frames. Fig. 4.6 depicts a frame from the video sequence *table tennis*, the corresponding watermarked frame and the difference between the two frames, amplified and contrast enhanced in order to make the modification produced by the watermark embedding more visible.

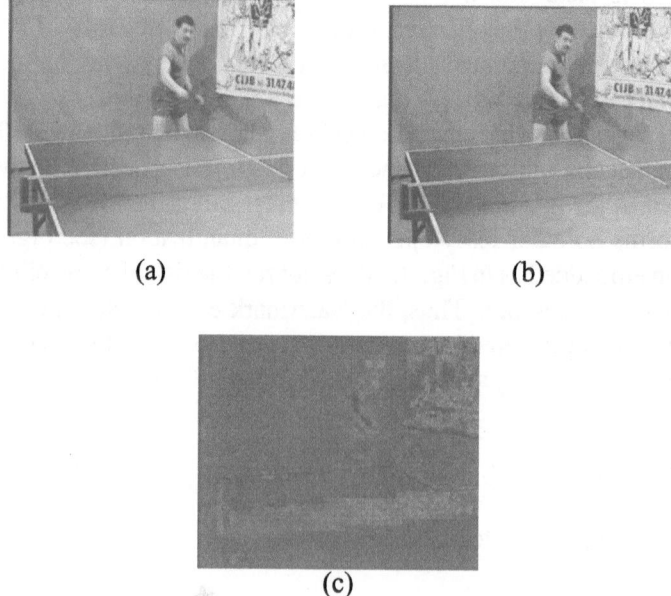

Figure 4.6.   (a) Original frame from the video sequence *table tennis*, (b) Watermarked frame, (c) Amplified difference between the original and the watermarked frame.

The absolute value of $X'_{Q\kappa,\lambda}(m,n)$ in equation (4.7) may increase, decrease or may remain unchanged in relation to $|X_{Q\kappa,\lambda}(m,n)|$, depending on the sign of the watermark coefficient $W_{\kappa,\lambda}(m,n)$ and the values of the perceptual and block classification masks. Due to the monotonicity of MPEG codebooks, when

$\left|X'_{Q\kappa,\lambda}(m,n)\right| > |X_{Q\kappa,\lambda}(m,n)|$ the codeword used for $X'_{Q\kappa,\lambda}(m,n)$ contains more bits than the corresponding codeword for $X_{Q\kappa,\lambda}(m,n)$; the inverse is true when $\left|X'_{Q\kappa,\lambda}(m,n)\right| < |X_{Q\kappa,\lambda}(m,n)|$. Since the watermark sequence has zero mean, the number of the cases where $\left|X'_{Q\kappa,\lambda}(m,n)\right| > |X_{Q\kappa,\lambda}(m,n)|$ is expected to roughly equal the number of the cases where the inverse inequality holds. Therefore, the MPEG bitstream length is not expected to be significantly altered. Experiments with watermarking of various MPEG-2 bitstreams resulted in bitstreams slightly larger (0-2%) than the original. The above result is related to the *run-level* based variable length coding of MPEG [25] and the way it is affected by the watermarking process.

As mentioned above, when the *level* (i.e. the absolute value) of a quantized DCT coefficient increases due to the addition of the watermark, some existing codewords in the bitstream are replaced with longer codewords. On the other hand, when the *level* of a quantized coefficient decreases, there is a possibility that the coefficient may be set to zero. This results in the elimination of the codeword corresponding to the zeroed coefficient from the bitstream. Therefore an increase in the *run* (the number of zero coefficients preceding a non-zero coefficient) for the next coded coefficient is observed, which results in a longer codeword. Thus, it is reasonable to assume that decreases of the *level* that do not result in setting a coefficient to zero are generally compensated by increases of the *level* in other coefficients. For the cases where quantized DCT coefficients are set to zero, the effect on the bitstream length is unpredictable and this may lead to leaving some *level* increases uncompensated. However, the coefficients that are set to zero due to the watermark embedding are very few because block classification and perceptual analysis generally do not permit small coefficients to be watermarked or to be assigned a large watermark value. This explains the slight increase in the bitstream length that was observed.

In order to ensure that the length of the watermarked bitstream will remain smaller than or equal to the original bitstream, the coefficients that increase the bitstream length may be left unwatermarked. This would, however, reduce the robustness of the detection scheme because the watermark could be inserted and therefore detected in fewer coefficients. For this reason, such a modification was avoided in our embedding scheme.

5.   Detection

The detection of the watermark is performed *without* using the original data. The original meaningful message that produces the watermark sequence W is needed in order to check if the specified watermark sequence exists in a copy of the watermarked video. Then, a correlation-based detection approach is taken, similar in some aspects to that in [21] but with the additional consideration of the quantization operations that take place during the watermark embedding.

In Subsection 5.1, a correlation-based detection method is formulated. Subsection 5.2 estimates the correlation metric for the case of DCT domain detection. Finally, in Subsection 5.3 the watermark propagation from I-frames to interframes and possible ways of detecting the watermark in interframes are studied.

5.1   Correlation-Based Detection

The detection can be formulated as the following hypothesis test:
H_0: the video sequence is not watermarked
H_1: the video sequence is watermarked
In order to determine which of the above hypotheses is true, a correla-tion-based detection scheme is applied. Variable length decoding is first performed to obtain the quantized DCT coefficients. Then, inverse quantization provides the DCT coefficients for each block. The block classification and perceptual analysis procedures are performed as described in Section 4 in order to define the set $\{X\}$ of the N DCT coefficients that are expected to be watermarked with the sequence W. Each coefficient in the set $\{X\}$ is multiplied with the corresponding watermark coefficient $W_{\kappa,\lambda}(m,n)$ producing the data set $\{X_W\}$.

The statistical characteristics (mean and variance) of the data set $\{X_W\}$ are calculated as follows:

$$mean = \frac{1}{N} \sum_{l=0}^{N-1} X_W(l) \simeq E\{X_W\} \qquad (4.9)$$

$$variance = \frac{1}{N} \sum_{l=0}^{N-1} (X_W(l) - mean)^2 \simeq E\{(X_W - mean)^2\} \quad (4.10)$$

Finally, the statistical correlation metric c for each frame is calculated as

$$c = \frac{mean \cdot \sqrt{N}}{\sqrt{variance}} \qquad (4.11)$$

The correlation metric c is compared to the threshold T_c, which is an adaptive threshold calculated for each frame of the video sequence. If the correlation metric c exceeds the threshold T_c, the examined frame is considered watermarked.

The calculation of the threshold is performed as in [21], where the threshold T_c is set to the half of the mean value of the random variable c. Such a threshold is chosen in order to minimize the total detection error (false detection under hypothesis H_0 and false detection under hypothesis H_1) assuming that c follows a normal distribution $N(0,1)$ under hypothesis H_0 and a normal distribution $N(\mu,1)$ under hypothesis H_1 (where μ denotes the mean value of the correlation metric c under hypothesis H_1). This assumption is valid, as justified by the experimental results depicted in Fig. 4.7 and Fig. 4.8. Fig. 4.7 presents the experimental probability density functions (pdfs) of the correlation metric. These were obtained by calculating the correlation metric of a single video frame for 4000 different watermarks under both hypotheses H_0 and H_1. The gaussian nature of both pdfs can be clearly observed. The gaussian distribution of the correlation metric is indeed verified by the normal probability plots depicted in Fig. 4.8. In both cases, the plots are almost linear, showing that c follows a normal distribution. This result was to be expected if N is large enough, by virtue of the Central Limit Theorem [26].

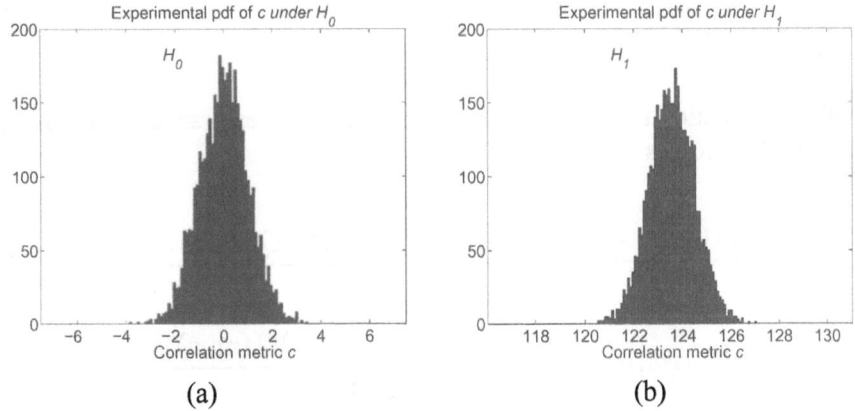

Figure 4.7. Experimentally evaluated pdfs of the correlation metric for a single video frame under (a) H_0 and (b) H_1.

5.2   Detection in the DCT Domain

In case of DCT domain detection, the set $\{X\}$ contains the DCT coefficients that were indicated by the block classification and perceptual analysis procedures. These coefficients are given by inverse quantization of the data in equation (4.7). If the above coefficients are multiplied by the watermark sequence W, the data in $\{X_W\}$ are obtained:

$$quant^{-1}\left[X'_{Q\kappa,\lambda}(m,n)\right] \cdot W_{\kappa,\lambda}(m,n) =$$

$$= \left\lfloor \frac{\left[X_{Q\kappa,\lambda}(m,n) + M_{Q\kappa,\lambda}(m,n)W_{\kappa,\lambda}(m,n)\right] \cdot Q(m,n)\,q_s}{8} \right\rfloor W_{\kappa,\lambda}(m,n)$$

$$(4.12)$$

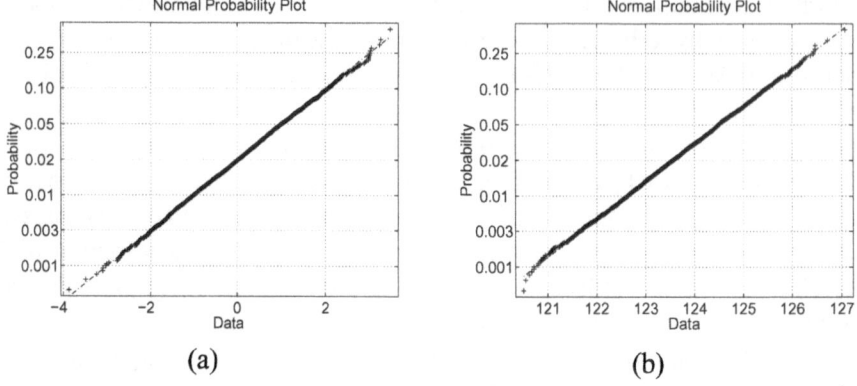

Figure 4.8. Normal probability plots for the correlation metric c using the *normplot* of MAT-LAB: (a) H_0, (b) H_1.

If the truncation noise is modeled using the zero mean random variable n_t then (4.12) becomes

$$\left(\frac{[X_{Q\kappa,\lambda}(m,n) + M_{Q\kappa,\lambda}(m,n)W_{\kappa,\lambda}(m,n)] \cdot Q(m,n)q_s}{8} + n_t\right) \cdot$$

$$\cdot W_{\kappa,\lambda}(m,n) = \frac{X_{Q\kappa,\lambda}(m,n)W_{\kappa,\lambda}(m,n)Q(m,n)q_s}{8} +$$

$$+ \frac{M_{Q\kappa,\lambda}(m,n)W_{\kappa,\lambda}(m,n)W_{\kappa,\lambda}(m,n)Q(m,n)q_s}{8} + n_tW_{\kappa,\lambda}(m,n) \quad (4.13)$$

Hereafter, indices κ, λ, m, n will be suppressed for the sake of notational simplicity. The *mean* is given by

$$mean \simeq E\left\{\frac{X_QWQq_s}{8} + \frac{M_QW^2Qq_s}{8} + n_tW\right\} = E\left\{\frac{M_QW^2Qq_s}{8}\right\}$$
$$(4.14)$$

since the first and the third term in the left hand side of (4.14) vanish. In the same manner it is trivially seen that the *variance* is given by

$$variance \simeq E\left\{\left(\frac{X_QWQq_s}{8}\right)^2\right\} +$$

$$+ E\left\{\left(\frac{M_QW^2Qq_s}{8}\right)^2\right\} - \left(E\left\{\frac{M_QW^2Qq_s}{8}\right\}\right)^2 + E\left\{(n_tW)^2\right\} \quad (4.15)$$

where the term $E\left\{(n_t W)^2\right\}$ is negligible compared with the rest of the terms in (4.15). Using (4.14), (4.15) and (4.11) we obtain

$$c \simeq \frac{E\left\{\frac{M_Q W^2 Q q_s}{8}\right\} \sqrt{N}}{\sqrt{E\left\{\left(\frac{X_Q W Q q_s}{8}\right)^2\right\} + E\left\{\left(\frac{M_Q W^2 Q q_s}{8}\right)^2\right\} - \left(E\left\{\frac{M_Q W^2 Q q_s}{8}\right\}\right)^2}}$$

(4.16)

5.3 Watermark Detection in Interframes

5.3.1 MPEG interframe coding.
The MPEG coding algorithm exploits the temporal redundancy that exists between consecutive frames in a video sequence by employing motion-compensated prediction. MPEG [1, 2] uses macroblock-based motion compensation. This means that the *target macroblock* in the picture to be encoded is matched to a set of displaced macroblocks of the same size and after the best match is found, it is used as the *prediction macroblock*. The position of the *prediction macroblock* is indicated by a motion vector that describes the horizontal and vertical displacement from *the target macroblock* to the *prediction macroblock*. The prediction error is then computed as the difference between the *target macroblock* and the *prediction macroblock*. MPEG supports *forward prediction* (see Fig. 4.9), where a past frame is used as the reference frame, and *bidirectional prediction*, where a past and a future frame are used as the reference frames. The former is used in both P- or B-frames, while the latter is used only in B-frames.

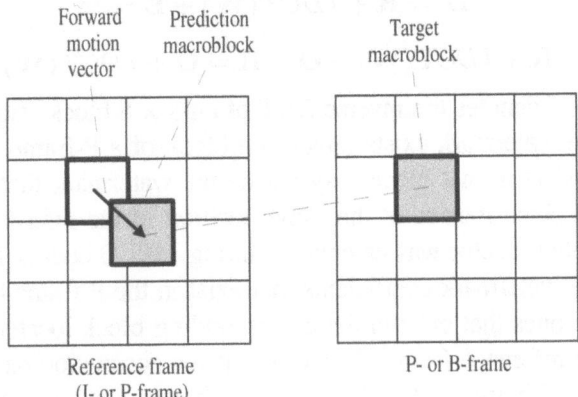

Figure 4.9. Motion estimation using forward prediction.

5.3.2   Watermark detection in interframes.

The proposed watermark embedding approach embeds watermarks only in I-frames of an MPEG video sequence. Although the MPEG coded stream contains only I-frame watermarked data, when motion compensated P- or B-frames are decoded the watermark carries over to them. The following example explains how the watermark is transferred from a block of an I-frame to a P-frame block that was coded using the motion compensation technique employed by MPEG.

Let W be the watermark matrix for an 8×8 DCT block that corresponds to the 8×8 block R of the reference frame (I-frame). Let also O and D be the original 8×8 block of a frame that is going to be coded as a P-frame and the corresponding 8×8 block that will be calculated after decoding of the P-frame respectively. Suppose that, during the motion compensation of the coding process, the motion vector for block O of the P-frame which points to block R of the reference frame is found, and that their difference (prediction error) block E is calculated. Then

$$E = O - R \qquad (4.17)$$

Note that E does not contain any watermark term, because the watermark alters only the coded I-frame data of the MPEG bitstream.

If no watermark is embedded in the I-frames, after decoding, the block D is given by

$$D = R + E = R + O - R = O \qquad (4.18)$$

Whenever the DCT data for block R of the reference frame in the MPEG bitstream are watermarked, the watermarked block D', rather than the block D, is calculated during the decoding process

$$D' = R + IDCT(W) + E =$$

$$= R + IDCT(W) + O - R = O + IDCT(W) \qquad (4.19)$$

where $IDCT(\cdot)$ denotes the inverse DCT of an 8×8 block. Equation (4.19) proves that the watermark exists intact in a block of a P-frame. But the watermark that exists in that block may not be the watermark that exists in the corresponding block located at the same position in the reference frame. If a non zero motion vector was calculated during MPEG coding for a P-frame block, then the watermark coefficients that exist in the P-frame block are different than the ones that exist in the corresponding block located at the same position in the reference frame. Therefore, if the correlation-based detection method described in this section is used, then the correlation metric value will be very low and detection will fail for the P- or B-frames of the video sequence.

A way to overcome this problem would be to use for the detection process only blocks with zero motion vectors and blocks that belong in skipped macroblocks. For such blocks, the watermark coefficients are located in the same positions as in the reference frame and the detection process can be applied in a straightforward manner. But if the motion of the scene objects from one frame to another is severe, or fast camera zoom or pan has occurred, then very few zero motion vectors will exist and the detection will almost certainly fail although watermark coefficients have been carried all over the P- or B-frame.

The above situation can be avoided by correlating with the watermark coefficients that actually exist in a block of a P- or B-frame and have been transferred there through the motion compensation process. Using the motion vectors found in the MPEG stream, for each block, these coefficients can be found and used for the correlation-based detection. In order to verify that the watermark is still detectable using this approach for the P- and B-frames, correlation-based watermark detection was applied in all frames of the 30-th Group Of Pictures (GOP) of the video sequence *table tennis* as shown in Fig. 4.10.

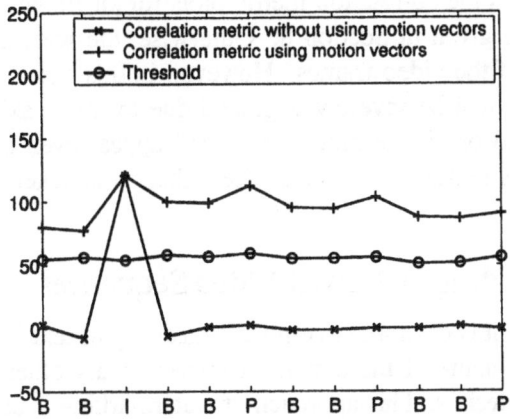

Figure 4.10.   Correlation-based detection for all frames of the 30-th GOP of the video sequence *table tennis* with and without using the motion vectors information to find the correlating watermark and the corresponding threshold.

6.   Video Watermark Detector Implementation

The proposed correlation-based DCT domain detection described in Subsection 5.1 can be implemented using two types of detectors.

The first detector (*detector-A*) detects the watermark only in I-frames during their decoding by applying the procedure described in Subsection 5.1. *Detector-A* can be used when the video sequence under examination is the original watermarked sequence. *Detector-A* can also be used in cases where the examined video sequence maintains the same GOP structure as the original watermarked sequence but is encoded at a different bit-rate using one of the techniques proposed in [27, 28]. The detection is very fast since it introduces negligible additional computational load to the decoding operation.

The second detector (*detector-B*) assumes that the GOP structure may have changed due to transcoding and frames that were previously coded as I-frames may now be coded as B- or P-frames. This detector decodes and applies DCT to each frame in order to detect the watermark using the procedure described in Subsection 5.1. The decoding operation performed by this detector may also consist of the decoding of non-MPEG compressed or uncompressed video streams, in case transcoding of the watermarked sequence to another coding format has occurred.

In cases where transcoding and I-frame skipping are performed on an MPEG video sequence, then *detector-B* will try to detect the watermark in former B- and P-frames. If object motion in the scene is slow or slow camera zoom or pan has occurred, then the watermark will be detected in B- and P-frames, as it will be depicted in the correlation metric plots for all frames of the test video sequence that are given in Section 8. Otherwise, the watermark may not be detected in any of the video frames. However, in this case, the quality of the transcoded video will be severely degraded due to frame skipping (jerkiness will be introduced or visible motion blur will appear even if interpolation is used) and for this reason it is very unlikely that an attacker will use such an attack.

7. Data Hiding in MPEG Video Sequences

In many cases, it is desirable to embed a meaningful text in the image. This text could be the name of the copyright owner or any other information the copyright owner desires. This approach of watermarking is usually referred to as data hiding [29] in related literature.

When seen from a data hiding point of view, video data is considered a communication channel which can carry information much in the same way information can be transmitted over communication channels. This brings up the issue of information capacity estimation, i.e. the estimation of the amount of information that can be imperceptibly embedded in images/video. The watermark embedding technique presented in Section 4 is equivalent to the embedding of a single bit, i.e. "1", if the existence of the watermark is verified, and "0" if it is not. The embedding of multiple bits requires the employment of a variant of the watermarking technique presented so far. This is achieved using simple error-correcting codes to correct possible bit misinterpretations.

The above issues are discussed in the sequel.

7.1   Bitstream Embedding

A bitstream B of K bits may be hidden in a video frame by modifying the embedding equation (4.1) of Section 4 as follows:

$$X'_{Q\kappa,\lambda}(m,n) = X_{Q\kappa,\lambda}(m,n)+$$

$$b_i(m,n)C_{\kappa,\lambda}(m,n)M_{Q\kappa,\lambda}(m,n)W_{\kappa,\lambda}(m,n) \qquad (4.20)$$

where b_i is the i-th data bit (-1 or 1) of the bitstream B, that will be hidden in the quantized DCT coefficient $X_{Q\kappa,\lambda}(m,n)$. Every single data bit should be embedded in more than one quantized DCT coefficient in order to extract it correctly. This requirement is imposed by the statistical approach used in the correlation-based detector described in Subsection 5.1, a variance of which is used for the bit extraction process that will be described in Subsection 7.3.

The set of coefficients that each bit will be embedded in, may be determined in a number of ways. A simple approach is to embed each bit b_i into all coefficients (actually only into the ones that the block classification and perceptual analysis allow) of a number of consecutive blocks of a video frame (see Fig. 4.11a). This approach will generally lead to a correct extraction of the data bit b_i. However, if the blocks that b_i is going to be embedded in are consecutive and happen to lie in an area of the image that the block classification and perceptual analysis indicate most coefficients cannot be watermarked, then this bit may be embedded in very few coefficients or even no coefficients at all and the danger of incorrect extraction will arise. Therefore, in order to achieve more reliable data extraction, a randomly selected bit b_i out of the K bits of the bitstream may be embedded in each block of a video frame (see Fig. 4.11b). The same random selection process is used during the extraction in order to extract each bit from the blocks it was embedded.

In order to further improve the extraction results, a randomly selected bit b_i out of the K bits of the bitstream may be embedded in each one of the DCT coefficients of the video frame the perceptual analysis allows. If the embedding of the bitstream is done in the way described above, every bit is embedded in coefficients with very close statistical properties and therefore, the probability of correct extraction for each one of the K bits of bitstream B is equalized. Fig. 4.11c depicts an example for embedding the bits b_i of a bitstream with $K = 70$ in the coefficients of an 8×8 DCT block using the above method. This method was used for the experiments on bitstream embedding and detection that will be presented in Subsection 8.3.

7.2   Error Correcting Codes

In conjunction with the data embedding techniques mentioned previously, Error Correcting Codes (ECCs) [30] can be used so as to further improve the

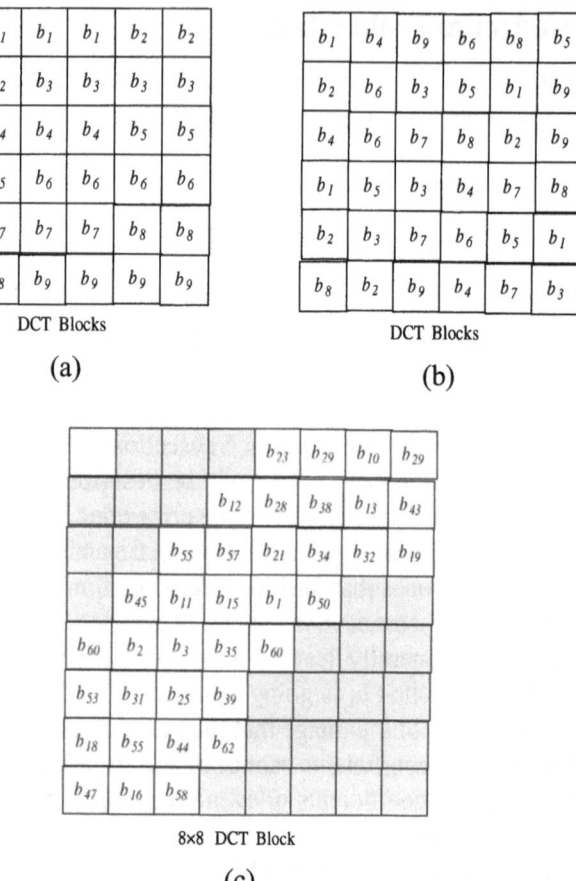

Figure 4.11. (a) Bit embedding in consecutive blocks, (b) Bit embedding in randomly selected blocks, (c) Random bit embedding in an 8×8 DCT block for a bitstream of $K = 70$. This block is classified as a *textured block* and the grey filled rectangles represent the DCT coefficients for which the block classification and perceptual analysis do not allow watermark embedding.

reliability of the information extraction process. Error correcting codes can detect and correct bit errors that may occur in the extracted bitstream. In order to be able to correct such errors, adequate error control bits should be added in the embedded bitstream. Therefore, in the proposed scheme, out of the K bits that are embedded in the image, only M constitute useful information bits and the rest $K - M$ bits serve error correction purposes.

Many ECCs, such as Hamming codes [31], BCH codes [32, 33] or turbo codes [34, 35], have already been used for improved data hiding performance. However, there is a tradeoff between the number of bits that can be imperceptibly embedded and the probability of correct extraction of those bits. In order to study the effect of using ECCs for data hiding we applied the bitstream embedding method depicted in Fig. 4.11c in conjunction with a simple Hamming ECC. This code adds three error control bits C_1, C_2, C_3 for every four information bits I_1, I_2, I_3, I_4. The control bits are computed from the information bits in the following way

$$C_1 = I_1 \oplus I_2 \oplus I_4$$
$$C_2 = I_1 \oplus I_3 \oplus I_4$$
$$C_3 = I_2 \oplus I_3 \oplus I_4$$

where $\oplus$ denotes the XOR operator.

The embedding bitstream takes the form $C_1C_2I_1C_3I_2I_3I_4$ for every four information bits. If a single error occurs while extracting the four information bits from the video frame, the error can be detected and corrected [31].

7.3   Bitstream Extraction

The bitstream extraction process is performed in the DCT domain using a process similar to the one described in Section 5. Specifically, for each set of the DCT coefficients $X'(l)$ that are expected to contain each embedded bit b_i, correlation with the correlating watermark sequence W is initially performed and then the *mean* is calculated:

$$mean = \sum_{l=0}^{N_{b_i}} X'(l)W(l)$$

where N_{b_i} is the number of DCT coefficients that are expected to contain each embedded bit b_i.

As in equation (4.14), by omitting the terms that contribute in the calculation of *mean* but are assumed to have zero mean we obtain

$$mean \simeq E\left\{ \frac{b_i M_Q W^2 Q q_s}{8} \right\} \tag{4.21}$$

But

$$M_Q, Q, q_s, W^2 > 0 \Rightarrow E\{M_Q W^2 Q q_s\} > 0 \tag{4.22}$$

From equations (4.21) and (4.22), it is clearly seen that the value of the embedded bit b_i determines the sign of the *mean*. Therefore, the value of each extracted bit b_i^E is given by

$$b_i^E = sign\,(mean) \simeq sign\left(E\left\{ \frac{b_i M_Q W^2 Q q_s}{8} \right\} \right) \tag{4.23}$$

where the $sign(\cdot)$ is equal to 1 when its argument is positive and -1 when its argument is negative.

After the K embedded bits are extracted, error correction is applied. For each sequence of seven bits $C_1 C_2 I_1 C_3 I_2 I_3 I_4$, the control bits C_1', C_2', C_3' are calculated from the information bits I_1, I_2, I_3, I_4 as in the embedding process. If C_1', C_2', C_3' are the same as C_1, C_2, C_3 then no correction of the information bits is needed. If error correction is needed, the bitstrings $C_3 C_2 C_1$ and $C_3' C_2' C_1'$ are added using binary addition. The resulting sum indicates the position of the incorrectly extracted bit and its value is flipped. In this way, all M information bits can be correctly extracted under the condition that a single error has occurred in each quadruplet of information bits $I_1 I_2 I_3 I_4$.

Using the techniques described above, the embedding and extraction of bitstreams of various lengths was tested on I-frames of MPEG-1/2 video sequences coded at various bitrates. The results are presented in Subsection 8.3

8.   Experimental Evaluation

The video sequences used for most of the experiments were the MPEG-1 video sequence *mountain*, which is part of a movie, and the MPEG-2 video *spokesman*, which is part of a TV broadcast. The former is an MPEG-1 system stream and the latter a MPEG-2 program stream i.e. both are multiplexed streams containing video and audio. These were produced from PAL VHS sources using a hardware MPEG-1/2 encoder. The reason for using such test video sequences instead of most commonly used sequences like *table tennis* or *foreman* was that the latter are short video-only sequences that are not multiplexed with audio streams, as is the case in practice. Thus, in order to reliably test the overall performance of the system, most experiments were conducted using the multiplexed sequences *mountain* and *spokesman*, although the system also supports video-only MPEG-1/2 streams. A number of experiments were also conducted for the MPEG-2 video-only sequence *table tennis*. In general, the embedding and detection scheme supports constant and variable bitrate main profile MPEG-2 program streams and MPEG-1 system streams.

8.1   Speed Performance of Watermark Embedding and Detection Scheme

A software simulation of the proposed embedding algorithm was implemented and executed using a Pentium III 800 MHz processor. The total execution time of the embedding scheme for the 22 sec MPEG-2 (5 Mbit/sec, PAL resolution) video sequence *spokesman* is 72% of the real-time duration of the video sequence. Execution time is allocated to the three major operations performed for embedding: file operations (read, write headers and packets), partial decoding and partial encoding and watermarking as shown in Fig. 4.12. In Fig. 4.12 the embedding time is also compared to the decoding time (without saving each decoded frame to a file) using the MPEG Software Simulation Group

(MSSG) decoder software. Clearly, the embedding time is significantly shorter than the decoding and re-encoding time that would be needed if the watermark embedding were performed in the spatial domain. Fig. 4.12 also presents the time required for detection using the *detector-A* described in Section 6. Detection time (I-frame decoding and detection) is only 23% of the real-time duration of the video sequence, thus enabling the detector to be incorporated in real-time decoders/players.

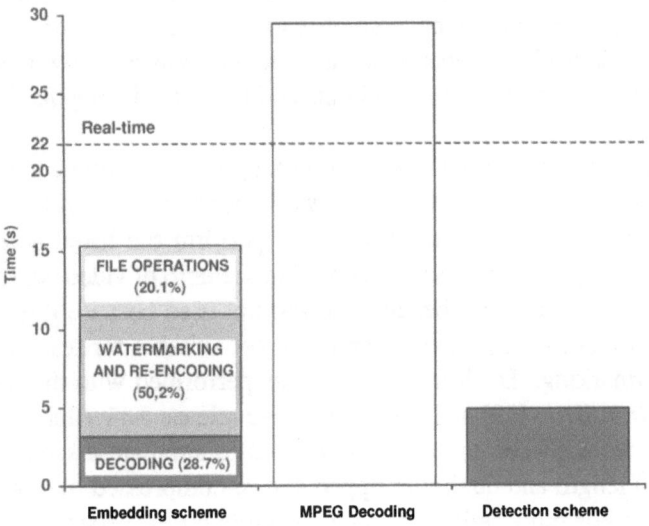

Figure 4.12. Speed performance of the embedding and detection schemes.

8.2 Experiments on DCT Domain Detection and its Robustness to Attacks

The watermarking algorithm selects the set of AC coefficients of an I-frame that can be altered without degrading the visual quality of the video frames. Because of the visual quality constraint, only a portion of the total number of DCT coefficients is actually altered. However, the experiments demonstrate that the detection is very effective and accurate even if the watermarked sequence is manipulated in a number of ways.

In Figs 4.13 and 4.14 the correlation metric is evaluated for 795 consecutive frames of the MPEG-1 video sequence *mountain* and 576 frames of the MPEG-2 video sequence *spokesman* using *detector-B*. As seen, the correlator output exceeds the adaptively calculated threshold for all I-frames. The first 263 frames of the *spokesman* sequence are surrounded by a thick black rectangle around them. Since the latter cannot be watermarked, lower correlation metric values are observed. However, the correlation metric for all I-frames included in these frames is still above the threshold.

Figs 4.15 and 4.16 depict the correlation metric plot for 30 different correlating watermark sequences, which were produced from 30 different seeds, for all frames of the MPEG-1 video sequence *mountain* and the MPEG-2 video sequence *spokesman* respectively. The 16th seed corresponds to the correlating watermark sequence that is embedded in the video.

In our experiments various attacks were directed on the watermarked MPEG-2 video *spokesman* with the purpose of rendering the watermark undetectable. The watermarked video underwent the following attacks: low-pass filtering, cropping 64% of the frame area, and transcoding using the popular DivX codec [36].

For all the attacks except DivX transcoding, *Adobe Premiere* [37] was used. *Adobe Premiere* is a video editing software widely used by video professionals. Built-in filters were used in order to perform the listed attacks on the watermarked sequence to simulate a possible scenario of video editing and processing. The *Stirmark* [38] benchmark was not used because it is targeted on still image watermarking systems and it is not suitable for testing attacks on video watermarking. DivX transcoding was performed with the freeware application *VirtualDub* [39] and the DivX codec release 3.11 Alpha. The coding parameters were chosen in such a way so as to simulate a scenario where a file of DVD length and quality is ripped and re-compressed in order to fit into a single CD. For more reliability both available codec versions, low and fast motion, were used.

Snapshots of an attacked frame of the video and the correlation metric plot for all frames of the video for the corresponding attack are demonstrated in Figs 4.17 and 4.18. In all cases the watermark is shown to be maintained in all I-frames. Fig. 4.19 illustrates the correlator output for each frame type, averaged over the entire sequence, for the GOP structure of the original watermarked and attacked watermarked MPEG-2 video sequences *spokesman*. It should be noted, that the picture types shown on the horizontal axis of each diagram represent the picture coding types of the frames in the original watermarked stream. In the case of DivX transcoding attacks, most I-frames have been re-encoded in another coding type but the watermark is still detectable.

8.3   Retrieval of Data in Data Hiding Applications

A software simulation of the bitstream embedding and extraction method described in Section 7 was implemented. In order to test the performance of the extraction process, we embedded bitstreams having lengths ranging from $K = 105$ to $K = 2065$ bits. The corresponding information bits range from

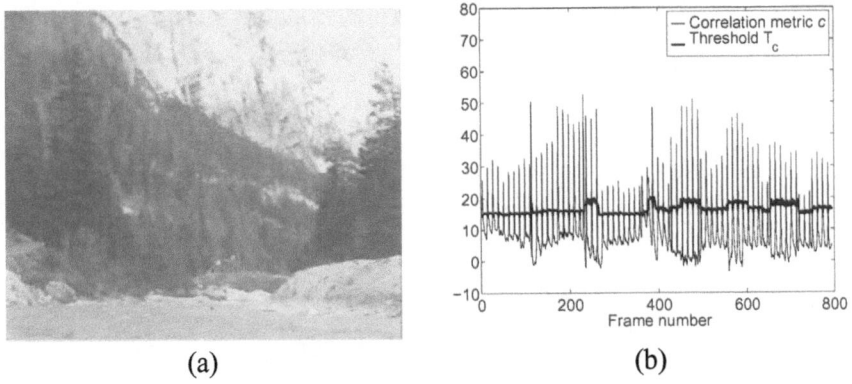

Figure 4.13. (a) Snapshot from the MPEG-1 video *mountain* used in the experiments, (b) Correlation metric plot.

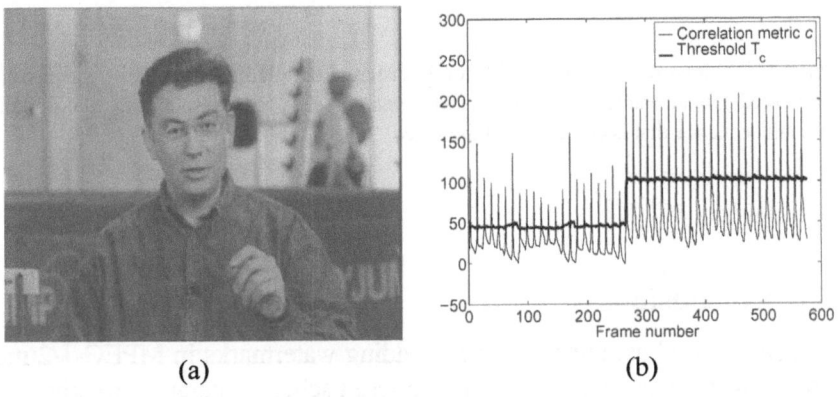

Figure 4.14. (a) Snapshot from the MPEG-2 video *spokesman* used in the experiments, (b) Correlation metric plot.

$N = 60$ to $N = 1180$ respectively. Fig. 4.20 illustrates the extraction process performance for various lengths of embedded bitstreams on a single MPEG-2 coded I-frame of the video sequence *table tennis* coded at seven different bitrates. As the bitrate increases and/or as the embedded bitstream length decreases, the extraction performance improves. The extraction performance is measured as the ratio of the number of successfully decoded information bits over the total number of information bits. It is worth noting that if the video sequence bitrate exceeds 4Mbits/sec, for most of the tested embedded bitstream lengths all bits were correctly extracted.

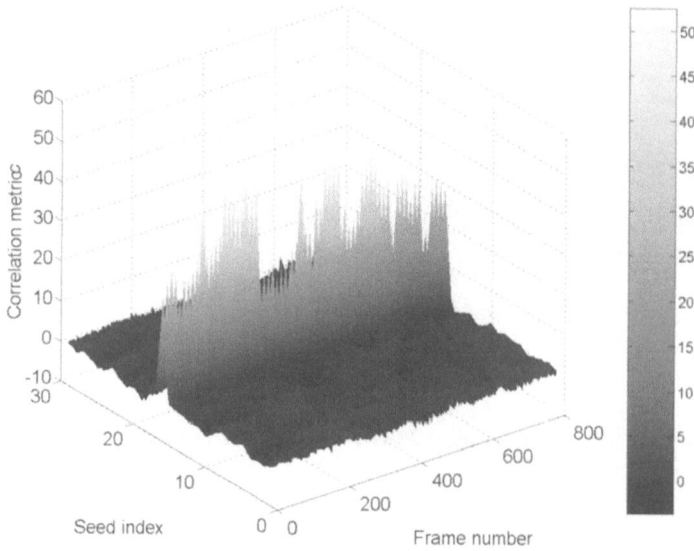

Figure 4.15.   3D-correlation metric plot for 30 different correlating watermark sequences (30 different seeds) for all frames of MPEG-1 video *mountain* (seed 16 corresponds to the correlating watermark sequence that is embedded in the video).

9.   Conclusions

A novel and robust method for embedding watermarks in MPEG-1/2 multiplexed streams was presented. The proposed scheme operates directly in the compressed domain and is able to embed copyright information without causing visible degradation to the quality of the video. Experimental evaluation showed that the embedded watermarks are able to withstand a variety of attacks. Apart from being very effective and reliable, the detection procedure used in the proposed scheme is very fast due to the fact that it actually introduces negligible additional computational load to the decoding operation. This enables the proposed system to be useful not only for copyright protection but also as a component of real-time decoders/players that are used for applications such as broadcast monitoring.

Acknowledgments

This work was supported by the European IST Project ASPIS. The assistance of the European COST254 and COST276 Actions is also gratefully acknowledged.

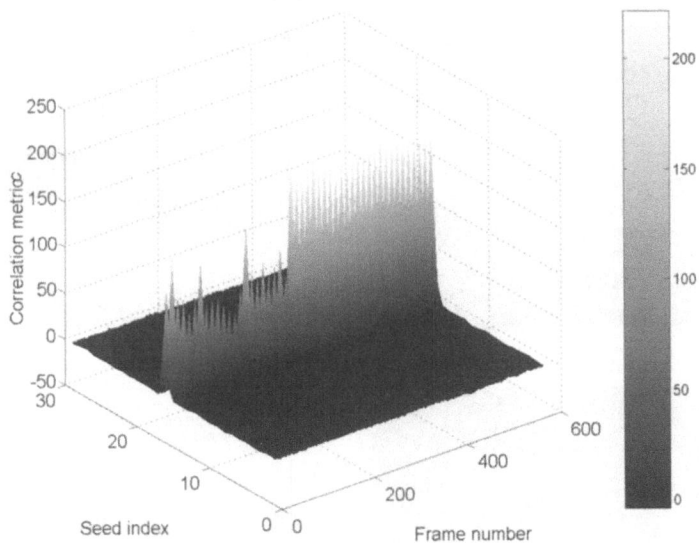

Figure 4.16. 3D-correlation metric plot for 30 different correlating watermark sequences (30 different seeds) for all frames of MPEG-2 video *spokesman* (seed 16 corresponds to the correlating watermark sequence that is embedded in the video).

References

[1] ISO/IEC, "13818-2 Information technology - Generic coding of moving pictures and associated audio: Video."

[2] B. Haskell, A. Puri, and A. Netravali, *Digital Video: An introduction to MPEG-2.* Dordrecht, The Netherlands: Kluwer Academic Publishers, 1997.

[3] I. J. Cox, J. Kilian, T. Leighton, and T. Shamoon, "Secure spread spectrum watermarking for multimedia," *IEEE Trans. Image Processing*, vol. 6, pp. 1673–1687, Dec. 1997.

[4] M. Barni, F. Bartolini, V. Cappelini, and A. Piva, "A DCT-domain system for robust image watermarking," *Signal Processing*, vol. 66, pp. 357–372, May 1998.

[5] D. Tzovaras, N. Karagiannis, and M. G. Strintzis, "Robust image watermarking in the subband or discrete cosine transform domain," in *9th European Signal Processing Conf. (EUSIPCO'98)*, (Rhodes, Greece), pp. 2285–2288, Sept. 1998.

[6] D. Simitopoulos, N. V. Boulgouris, A. Leontaris, and M. G. Strintzis, "Scalable detection of perceptual watermarks in JPEG2000 images,"

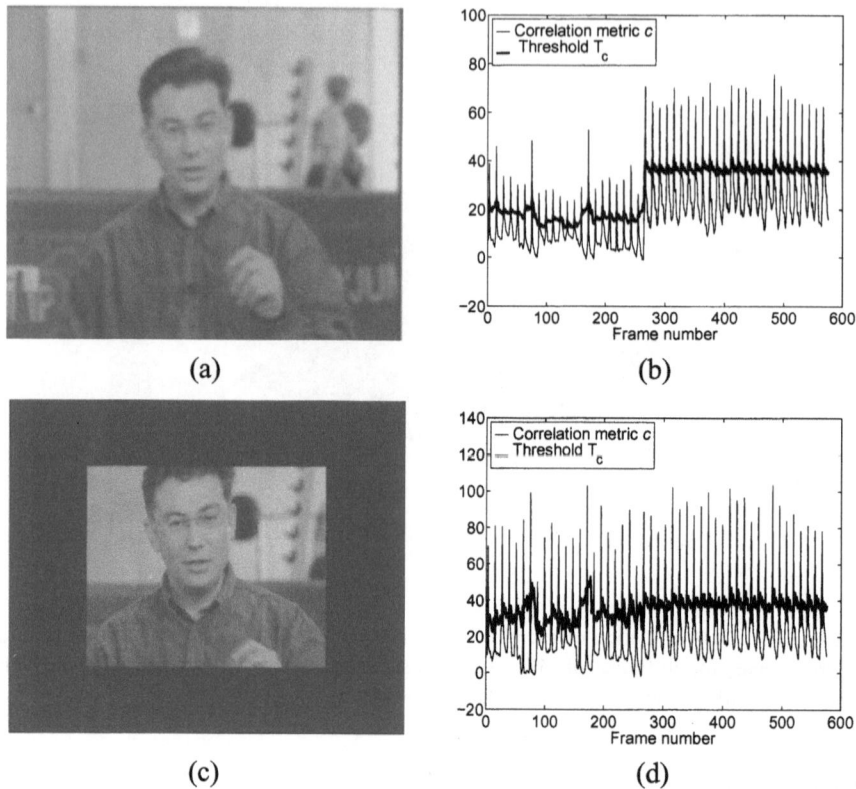

(a)   (b)

(c)   (d)

Figure 4.17.   Snapshots of performed attacks and corresponding correlation metric values for the MPEG-2 video *spokesman*: Case *I*: Blurring attack, (a) Corresponding snapshot, (b) Corresponding correlation metric value. Case *II*: Cropping 64% of the frame area, (c) Corresponding snapshot, (d) Corresponding correlation metric value.

in *Communications and Multimedia Security*, (Darmstadt, Germany), pp. 93–102, May 2001.

[7] M. D. Swanson, B. Zhu, A. H. Tewfik, and L. Boney, "Robust audio water-marking using perceptual masking," *Signal Processing*, vol. 66, pp. 337–355, May 1998.

[8] M. Maes, T. Kalker, J. P. M. G. Linnartz, J. Talstra, F. G. Depovere, and J. Haitsma, "Digital watermarking for DVD video copy protection," *IEEE Signal Processing Magazine*, vol. 17, pp. 47–57, Sept. 2000.

[9] R. B. Wolfgang, C. I. Podilchuk, and E. J. Delp, "Perceptual watermarks for digital images and video," *Proc. IEEE*, vol. 87, pp. 1108–1126, July 1999.

[10] D. Simitopoulos, S. A. Tsaftaris, N. V. Boulgouris, and M. G. Strintzis, "Digital watermarking of MPEG-1 and MPEG-2 multiplexed streams for

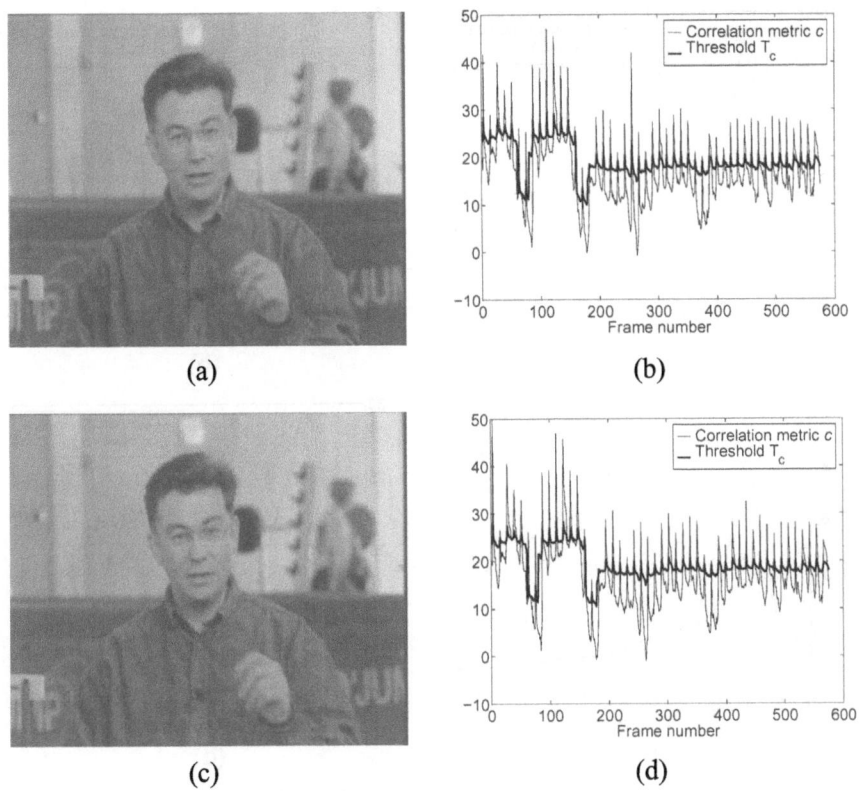

Figure 4.18. Snapshots of performed attacks and corresponding correlation metric values for the MPEG-2 video *spokesman*. Case *III*: DivX transcoding using low motion codec, bitrate at 910Kbit/s and keyframes every 10 sec, (a) Corresponding snapshot, (b) Corresponding correlation metric value. Case *IV*: DivX transcoding using fast motion codec, bitrate at 910Kbit/s and keyframes every 10 sec, (c) Corresponding snapshot, (d) Corresponding correlation metric value.

copyright protection," in *Second USF International Workshop on Digital and Computational Video (DCV'01)*, (Tampa, FL, USA), pp. 140–147, Feb. 2001.

[11] N. Nikolaidis and I. Pitas, "Robust image watermarking in the spatial domain," *Signal Processing*, vol. 66, pp. 385–403, May 1998.

[12] F. Hartung and B. Girod, "Watermarking of uncompressed and compressed video," *Signal Processing*, vol. 66, pp. 283–301, May 1998.

[13] C. Y. Lin, M. Wu, J. A. Bloom, I. J. Cox, M. L. Miller, and Y. M. Lui, "Rotation, scale and translation resilient public watermarking for images," in *Proc. of SPIE in Security and Watermarking of Multimedia Contents II*, vol. 3971, (San Jose, CA, USA), pp. 90–98, Jan. 2000.

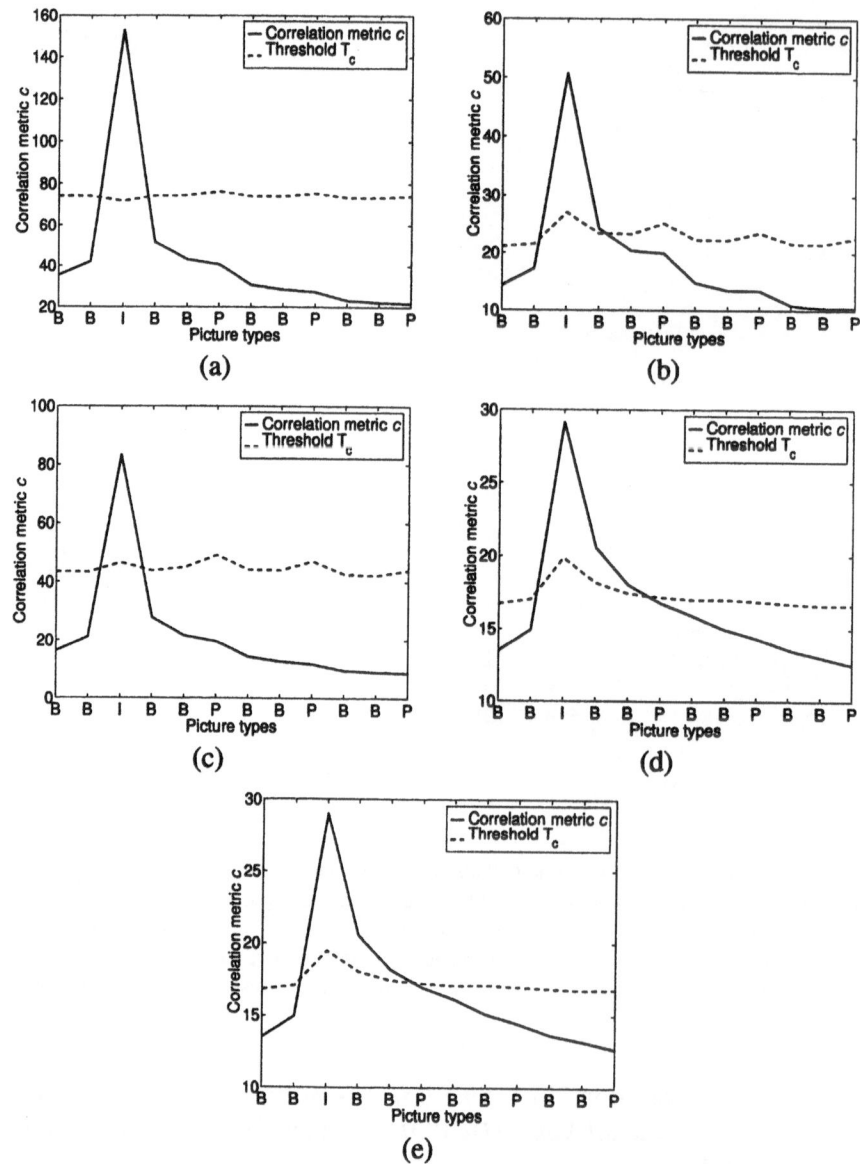

Figure 4.19. Plots of the mean value of the correlation metric for each one of the frames contained in a GOP for all frames of the original watermarked and attacked watermarked MPEG-2 video sequences *spokesman*: (a) Original watermarked, (b) After Blurring, (c) After cropping 64% of the frame area, (d) After DivX transcoding using low motion codec at 910 kbit/s and keyframes every 10 sec, (e) After DivX transcoding using fast motion codec at 910 kbit/s and keyframes every 10 sec.

[14] J. O'Ruanaidh and T. Pun, "Rotation, scale and translation invariant digital image watermarking," in *Proc. International Conference on Image*

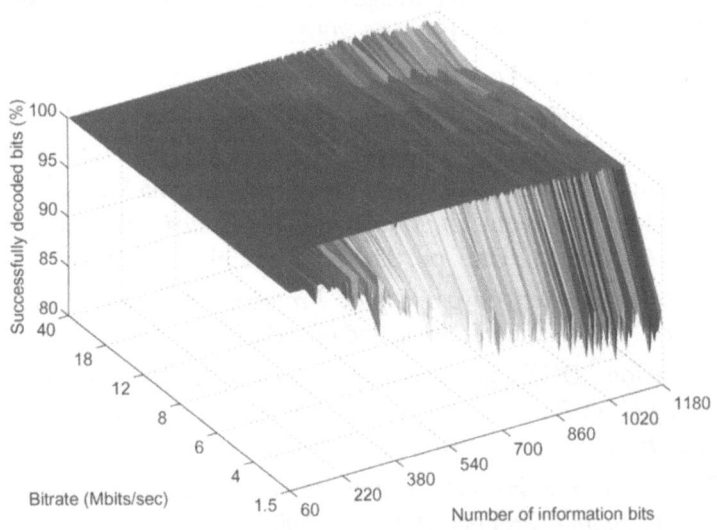

Figure 4.20. Hidden bitstream extraction performance for various lengths of embedded bitstreams in a single MPEG-2 coded I-frame of the video sequence *table tennis* coded at seven different bitrates.

Processing, vol. 1, (Santa Barbara, CA, USA), pp. 536–539, Oct. 1997.

[15] J. R. Hernandez, M. Amado, and F. Perez-Gonzalez, "DCT-domain watermarking techniques for still images: detector performance analysis and a new structure," *IEEE Trans. Image Processing*, vol. 9, pp. 55–68, Jan. 2000.

[16] H. Inoue, A. Miyazaki, and T. Katsura, "An image watermarking method based on the wavelet transform," in *Proc. International Conference on Image Processing*, vol. 1, (Kobe, Japan), pp. 296–300, Oct. 1999.

[17] P. C. Su, H. J. Wang, and C. C. J. Kuo, "An integrated approach to image watermarking and JPEG2000 compression," *Journal of VLSI Signal Processing*, vol. 27, pp. 35—53, Feb. 2001.

[18] G. C. Langelaar, R. L. Lagendijk, and J. Biemond, "Real-time labeling of MPEG-2 compressed video," *J. Visual Commun. Image Representation*, vol. 9, pp. 256–270, Dec. 1998.

[19] T. Y. Chung, M. S. Hong, Y. N. Oh, D. H. Shin, and S. H. Park, "Digital watermarking for copyright protection of MPEG-2 compressed video," *IEEE Trans. Consumer Electronics*, vol. 44, pp. 895–901, Aug. 1998.

[20] G. C. Langelaar and R. L. Lagendijk, "Optimal differential energy watermarking of DCT encoded images and video," *IEEE Trans. Image Processing*, vol. 10, pp. 148–158, Jan. 2001.

[21] W. Zeng and B. Liu, "A statistical watermark detection technique without using original images for resolving rightful ownerships of digital images," *IEEE Trans. Image Processing*, vol. 8, pp. 1534–1548, Nov. 1999.

[22] K. Nahrstedt and L. Qiao, "Non-invertible watermarking methods for MPEG video and audio," in *Multimedia and Security Workshop at ACM multimedia*, (Bristol, UK), pp. 93–98, Sept. 1998.

[23] B. Schneier, *Applied Cryptography: Protocols, Algorithms, and Source Code in C*. John Wiley & Sons, 2nd ed., 1995.

[24] A. B. Watson, "DCT quantization matrices visually optimized for individual images," in *Proc. SPIE Conf. Human Vision, Visual Processing and Digital Display IV*, vol. 1913, pp. 202–216, Feb. 1993.

[25] K. R. Rao and J. J. Hwang, *Techniques and standards for Image, Video and Audio coding*. Prentice Hall PTR, 1996.

[26] A. Papoulis, *Probability Random Variables and Stochastic Processes*. New York, NY: McGraw-Hill, 3rd ed., 1991.

[27] A. Eleftheriadis and D. Anastasiou, "Constrained and general dynamic rate shaping of compressed digital video," in *Proc. International Conference on Image Processing*, vol. 3, (Washington, DC, USA), pp. 396–399, Oct. 1995.

[28] R. J. Safranek, C. R. Kalmanek, and R. Garg, "Methods for matching compressed video to ATM networks," in *Proc. International Conference on Image Processing*, vol. 1, (Washington, DC, USA), pp. 13–16, Oct. 1995.

[29] D. Mukherjee, J. J. Chae, and S. K. Mitra, "A source and channel-coding framework for vector-based data hiding in video," *IEEE Trans. Circuits and Systems for Video Technology*, vol. 10, pp. 630–645, June 2000.

[30] S. Lin and D. J. Costello, *Error Control Coding: Fundamentals and Applications*. Englewood Cliffs, NJ: Prentice-Hall, 1983.

[31] J. Proakis, *Digital communications*. New York, NY: McGraw-Hill, 3rd ed., 1995.

[32] J. R. Hernández, J. M. Rodríguez, and F. Pérez-González, "Improving the performance of spatial watermarking of images using channel coding," *Signal Processing*, vol. 80, pp. 1261–1279, July 2000.

[33] S. Baudry, P. Nguyen, and H. Maitre, "Channel coding in video watermarking: Use of soft decoding to improve the watermark retrieval," in *Proc. International Conference on Image Processing*, (Vancouver, Canada), Sept. 2000.

[34] C. Berrou and A. Glavieux, "Near optimum error correcting coding and decoding: turbo codes," *IEEE Trans. Communications*, vol. 44, pp. 1261–1271, Oct. 1996.

[35] S. Pereira, S. Voloshynovskiy, and T. Pun, "Effective channel coding for DCT watermarks," in *Proc. International Conference on Image Processing*, (Vancouver, Canada), Sept. 2000.

[36] DivX.com, "The official site of divx video." http://www.divx.com/, Jan. 2003.

[37] "Adobe premiere." http://www.adobe.com/products/premiere/main.html, Jan. 2003.

[38] "Stirmark benchmark 4.0." http://www.cl.cam.ac.uk/~fapp2/watermarking/stirmark, Jan. 2003.

[39] "Virtualdub homepage." http://www.virtualdub.org/, Jan. 2003.

Chapter 5

ROBUST WATERMARKING OF VIDEO FOR COPYRIGHT PROTECTION

Mauro Barni, Franco Bartolini, Roberto Caldelli, Vito Cappellini, Alessia De Rosa, and Alessandro Piva

University of Florence, Via di Santa Marta 3

50139 Florence, Italy

barni@dii.unisi.it, {barto,caldelli,piva}@lci.det.unifi.it, {cappellini,derosa}@lci.die.unifi.it

Abstract   The future development of networked multimedia services is conditioned by the achievement of efficient methods to grant that the Intellectual Property Rights (IPR) are well respected and the assets properly managed. In the framework of digital watermarking for the copyright protection of multimedia data, the specific case of video watermarking will be considered. In particular three different approaches for embedding a code in the video domain will be deeply described, and a particular attention will be focused on watermarking of MPEG-4 objects. Specific characteristics, advantages and drawbacks of the different systems will be highlighted.

Keywords: Copyright protection, video watermarking, raw-frame-based watermarking, object-based watermarking, MPEG-4.

1.   Introduction

The advent of Internet, satellite and cable communications, and mobile telephony has allowed to develop new services for the users based on the distribution of multimedia data. The telecommunications industry is investing to deliver audio, image and video data in electronic form to an ever increasing number of its customers, and broadcast televisions, major corporations and photo archives are converting their contents from analogue to digital form to secure for themselves a part of a new market. Initially the problems of bandwidth have strongly restricted the spreading of true multimedia services, so that only audio files and still images have been highly distributed. Recently, the bottleneck represented by the limited bandwidth has been overcome thanks both to the improvements

J.F. Tasič et al. (eds.), Intelligent Integrated Media Communication Techniques, 153-194.
© 2003 *Kluwer Academic Publishers.*

of transmission technology, and to the use of different and powerful communication media, so, nowadays, it is possible to transmit also video data over the Internet, where the users will be soon able to enjoy services like video-on-demand and interactive television.

Nevertheless a new problem has raised: the protection of the Intellectual Property Rights (IPR) of the multimedia data distributed in an open-networked environment. As a matter of fact, the future development of networked multimedia services is conditioned by the achievement of efficient methods to safeguard data owners against non-authorised copying and redistribution of the delivered material, to grant that the IPR are well respected and the assets properly managed.

Copyright protection of multimedia data has been initially accomplished by means of cryptography algorithms to provide control over data access and to make data unreadable to non-authorised users. However, encryption systems do not completely solve the problem, because once encryption is removed there is no more control on the dissemination of data; a possible solution envisages the use of a rising technology like *digital watermarking*. Digital watermarking grants multimedia works protection and allows their distribution to be tracked. In this way the number of permitted copies is not limited, but the possibility to control the path the original work has been disseminated through really exists.

A digital watermark is a signal permanently embedded into digital data that can be detected or extracted later by means of computing operations to make an assertion about the data. The watermark is hidden into the host data in such a way that it is inseparable from them and can resist to many operations not degrading the host document. By means of watermarking the work is still accessible, but permanently marked. This code can be used to bring different sorts of informative data: it may indicate the content creator, the copyright owner, the authorized distributor, the legitimate purchaser or something else. Watermarking must supply the technical tools to establish a distributive channel, which it is possible to make goods flow through, in a secure and checkable way.

In this chapter the specific case of video watermarking will be analyzed and peculiar attention will be paid to the particular aspects concerning it. Most of the general features and basic requirements, a watermarking system must satisfy, may be considered as valid independently from the type of medium at hand, anyway for a more detailed analysis see [1].

This chapter is organized as follows: in Section 2 some particular aspects regarding specifically video watermarking are addressed; then in the subsequent sections three watermarking systems developed in the last few years by the LCI (Laboratorio Comunicazioni e Immagini) of the Department of Electronics and Telecommunications of the University of Florence will be outlined. In particular, in Section 3 a frame-based method dealing with raw video [2] is described; in Section 4 some general considerations about the MPEG-4 standard characteristics are outlined before entering Section 5 where another approach, this one working on video-objects in the compressed domain is reported [3] and at last in Section 6 the third technique [4], still acting on video-objects, but in the raw-video domain is described. In Section 7 some conclusions are drawn.

2. How Can Video be Watermarked

As it can be observed in literature, many different techniques oriented to video watermarking have been developed. These algorithms can be basically grouped in two separate categories on the basis of which kind of video they deal with: some of them work directly in the MPEG-2/MPEG-4 coded domain embedding the watermark in the encoded video-stream [5, 6], on the contrary, other systems need a non-coded video [7, 8, 9, 10, 11] to be able to insert a mark in an appropriate way. Both of them seem to show advantages and drawbacks as well, either from the point of view of robustness against intentional and non-intentional attacks, or from the point of view of computational complexity. The use of a certain methodology instead of another one, often depends on the sort of application at hand. In the next subsection some specific issues regarding video watermarking will be considered.

2.1 Video Watermarking: Specific Issues

Most of the requirements that has been indicated in literature with reference to image watermarking [1], fit well also for video. Also the idea of robustness, as the capacity of an algorithm to resist to any kind of manipulations and modifications the watermarked document undergoes, is still valid, but it has to be pointed out that new, different and specific needs must be taken into account in this case. In fact, unlike audio and still images, video contains a wide quantity of data presenting a high redundancy, therefore it can sustain some processing, as frame swapping, frame elimination and so on, that do not cause a significant quality degradation but can determine the failure of the watermark retrieval, especially if the used algorithm needs to know perfectly the timing of the image sequence. In the light of these considerations, some of the most significant demands video watermarking must answer to, are exploited in the sequel.

First of all, from the point of view of video applications, particularly in video delivering, assets are usually distributed towards a huge amount of users. If everyone needs to be identified through a watermark without ambiguity, for tracking purposes, the embedding technique has to be able to create many different watermarks, without leading to uncertainty. Moreover if watermarking process has to be applied on-the-fly when a given video sequence of a specified bit-rate has been chosen by an end-user, the casting software has to grant high computational performance, not to add further delay to usual downloading time. Directly connected to the needs just exposed for the transmission, one more requirement arises, it concerns with the necessity to respect the bit-rate established for that sequence. Stated in another way, the modifications introduced in the video sequence during the embedding procedure, must not increase the amount of the original data and, consequently, the bandwidth occupied by the sequence itself.

After these general considerations, regarding video delivery, let us see other basic aspects to be carefully highlighted mainly for visibility and robustness issues, a brief list is itemized below.

- A video sequence is composed by many frames, there is not just a still picture and the amount of data to be protected is huge with respect to the still image case. It could be interesting to wonder if it is better to insert in each frame the same watermark without taking into account the existing differences among them, or to exploit these features to properly hide the code in the data.

- Video is accessed by a generic user in a different way with respect to still images; heavy artifacts deriving from watermark embedding can be easily recognized in a still image but the same is not true in a real-time sequence in which the HVS (Human Visual System) is not able to deeply check the changing scene.

- Is it necessary to protect a video by watermarking all the frames belonging to the sequence or is it enough to mark only a given number of frames, thus resulting in lower decreasing of video quality? For example one frame every six, at a frame rate of 25 Hz, it would mean that about a quarter of a second of the video is not protected, this would not be, in general, a significant part of the whole sequence. Anyway, in some applications and for very strategic sequences, it might be important to provide a high level of safeguarding because also a short-timed piece of video could be very crucial.

- Video, more than still images, has to be considered in relation to the standards (H263, MPEG-1, MPEG-2, MPEG-4, MPEG-7 and in the next future MPEG-21) that define their storing and transmission format; it is important to analyze what happens when a conversion from a format to another is applied and in particular when the resulting bit-rate of the coded sequence is reduced, for instance, to satisfy transmission and/or distribution requirements. It is evident that in this case a choice has to be made in order to adopt the best watermarking technique either working in raw-video or compressed-video domain, according to the application at hand.

- Watermarking robustness aspects are slightly different with respect to still images; resistance against a geometric attack like rotation could not be so important as instead resistance against scaling would. New kinds of manipulations like frame sequence exchange or elimination should be carefully considered, in fact they do not result in a strong video quality degradation, but the watermark detection might dramatically fail. A flipping operation continues to be an attack to be taken into account because most of the sequences maintain their quality and, above all, their substantial meaning if left-right reversed (people go towards left-side than right-side); on the contrary a flopping process, which turns video upsidedown, drastically damages quality and, except very particular circumstances, it can be classified as a requirement not to be satisfied.

3. A Raw Video Frame-Based Watermarking Algorithm

The proposed approach consists in applying a tested and well-performing watermarking technique [12], originally devised for still images, to non-coded video, by considering it as a collection of single frames. In the following two subsections the still image algorithm will be roughly described and then the main aspects of its adaptation to the case of video watermarking explained.

3.1 Watermarking of Still Images

The watermark casting is carried out, as described in [12], extracting the brightness of the to-be-marked frame, computing its full-frame DFT (Discrete Fourier Transform) and then taking the magnitude of the coefficients. The watermark is embedded by slightly modifying the magnitude of some DFT coefficients, belonging to a specific middle frequency region of the transformed domain, successively the IDFT (Inverse DFT) is performed to obtain the watermarked image. Moreover to better preserve the original quality of the single frame, a particular masking operation, exploiting knowledge of the characteristics of the HVS (Human Visual System), is performed. In the detection step the luminance of the image to be checked for watermark presence is extracted and the magnitude of its DFT is considered again; only who knows the private code strings, used during the coding phase, is able to generate the exact code to look for. An optimum criterion to verify if the mark is present in the image is derived, based on statistical decision theory. In practice the response of a maximum likelihood function is compared to a threshold set to impose a given probability of false allarm. This kind of watermark is detectable, unperceivable and presents a good robustness to the usual image processing as linear/non-linear filtering, sharpening, JPEG compression and so on; furthermore resistance to geometric transformations as scaling, rotation, cropping, etc., is well-granted thanks to the insertion of a template during the coding step. With this method the original unmarked image is not needed to perform the detection phase, i.e. the technique can be defined as blind, besides, the algorithm permits the insertion/detection of multiple marks, to possibly manage multiple ownership situations.

3.2   Watermarking of Video Sequences

Here an application of the watermarking algorithm, outlined in the previous subsection, to non-coded video sequences, where each frame is processed in a distinct and different way [13], is presented. In fact, thanks to the used watermarking technique, the inserted code is frame-dependent and though the private key is always the same, the mark really introduced in the image is diverse every time. Doing so we will embed correlated watermarks between correlated frames and uncorrelated watermarks between uncorrelated frames, allowing changes, due to code insertion, to adapt gracefully to the video content. Practically a set of still images is available and only those frames which have to be watermarked, are passed to the marker, the others are let unaltered. Obviously dealing with raw video allows to achieve video-coding format independence and moreover to be able to choose how many and which are the frames to be marked. In particular for this application it has been decided to watermark the first frame of each GOP (Group Of Pictures), which was composed of 12 frames, leaving the other ones uncorrupted. Assuming a frame rate equal to 25 frame/sec, this approach grants that at least 2 frames per second are marked and this would seem to be a considering part with respect to video length. Anyway if a superior protection is needed a higher number of frames can be marked, at most the entire GOP. Moreover by-passing most of frames without changes (referring to this application case 11/12 that is 91.6%) results in a good preserving of the whole video quality. Other experiments have been carried out by marking one frame every six, that is two in a GOP, or frames located in intermediate positions of GOP, for example the ninth or the eleventh one; these changes have led to similar conclusions as in the above considered case.

During the detection phase all the video is checked for the mark presence, not only the first frame of each GOP, in such a way a synchronization of the stream is not required and the knowledge of the exact position within the sequence, is not needed, as it would seem to be for other algorithms [10]. When the checked code is found, that is the detector response has got a peak over the established threshold, it means that the video contains at least a watermarked frame; the process could go on with its search into the remaining part of video, but it would not be necessary if a more accurate validation is not specifically required. Therefore the revealing process might be stopped, after a sufficient number of positive check of watermark presence is reached, thus resulting in a saving of computational time, obviously this is true only if the watermark has been detected, in the case of no-detection the whole video has to be definitely checked.

3.3 Robustness Evaluation

The possibility to decide to watermark one or more frames in a GOP (at most all of them), together with the choice of considering each frame as a still image [9] yield some important advantages from a robustness point of view. First a trade-off between time spent for marking and the degree of robustness needed for the sequence can be achieved, in other words the lower the number of watermarked frames in the GOP, the faster the coding phase, but, conversely, just a minor part of the video stream will be watermarked, thus causing a robustness decrease; obviously if a superior security has to be obtained a higher amount of frames might be considered. Moreover, if some attacks like frames exchange or frames dropping/replacing, which do not result in a strong video quality degradation, are applied to the watermarked sequence, it will be always possible to reveal the watermark; in the first case its position will be different with respect to the code insertion phase, and in the second one the mark may be found in the remaining watermarked frames belonging to the same GOP or to the successive ones. Experiments carried out in this direction have confirmed these assumptions. Furthermore it has been verified that MPEG-2 coding/decoding operations, at various and lower bit-rates, do not harm to the correct watermark detection. Thanks to the good robustness the watermarking algorithm had already shown, in the case of still images, against usual image processing as linear/non-linear filtering, noise addition, JPEG compression, and geometric transformations as rotation, scaling, cropping, also for video applications these longed-for characteristics are held.

3.4 Experimental Results

In this section some experimental results to point out how the proposed approach really works and its robustness, particularly against geometric manipulations and MPEG-2 coding/decoding, at various bit-rates, are shown.

The results obtained with a video sequence, named *Quintana* (*Quintana* has kindly been provided by RAI, the Italian National Broadcasting Company), in which a medieval banquet is represented, are proposed. The sequence is composed by 250 frames (4 : 2 : 2 PAL format) of size 720×576 pixels and has been first watermarked and then MPEG-2 coded at 6 Mbit/sec and 25 frame/sec.

In Figure 5.1(a) and Figure 5.1(b) the frame 36 of the watermarked video and the detector response have been depicted respectively. The whole video is checked for the mark presence; the watermarking code is revealed when the detector response is higher than the established threshold; in Figure 5.1(b) the difference between the response and the threshold is visualized (i.e. the threshold is normalized to 0), so when a watermarked frame is detected a positive spike is obtained. Going through the 250 frames a set of periodical peaks (1 over 12 in this test case) is displayed.

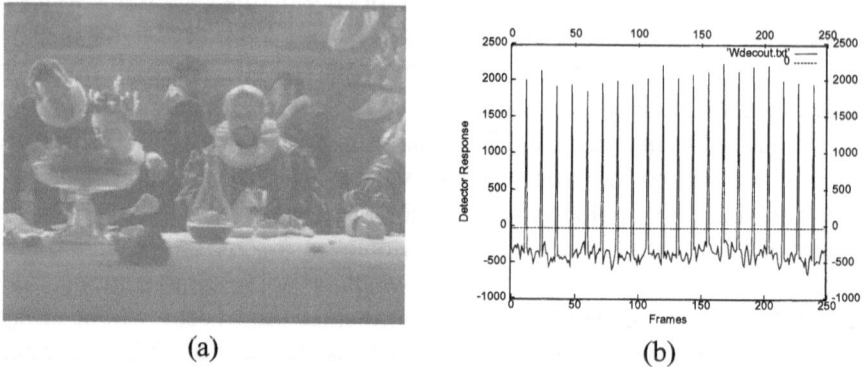

(a)   (b)

Figure 5.1.   Test video sequence: (a) Frame 36 of the watermarked sequence MPEG-2 coded at 6 Mbit/sec, (b) Detector response.

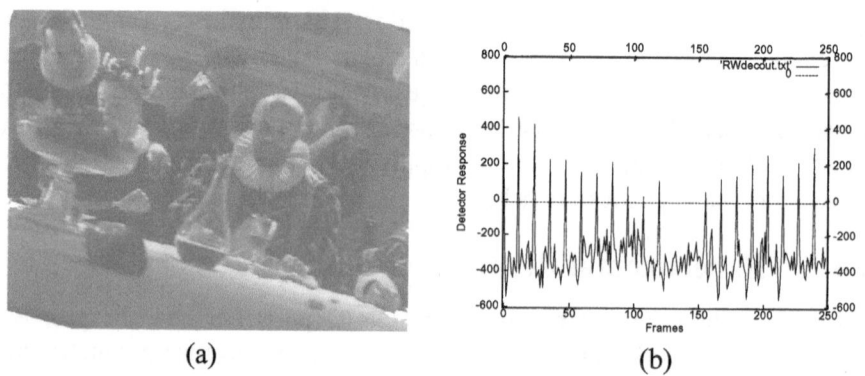

(a)   (b)

Figure 5.2.   Test video sequence: (a) Frame 36 of the watermarked sequence that has been rotated by a 15 degrees, and scaled to size 1000 × 900 and finally cropped to the original size, (b) Detector response.

In Figure 5.2(a) the same frame of the test video is pictured. Now the sequence has undergone a geometrical composite attack as a 15 degrees rotation, then an asymmetric scaling to 1000 × 900 pixels and finally a cropping to the original size. Like the above case, the respective detector response is proposed on the right; because of the transformations occurred the spikes, though still well distinguishable, have a height lower than before and moreover two peaks in the middle of the sequence are missed, but, as explained in Section 2.3, this does not invalidate the extraction process at all.

Finally, as third robustness test, another kind of attack is considered: an MPEG-2 coding where bit-rate is reduced to half, in this case from 6 Mbit/sec

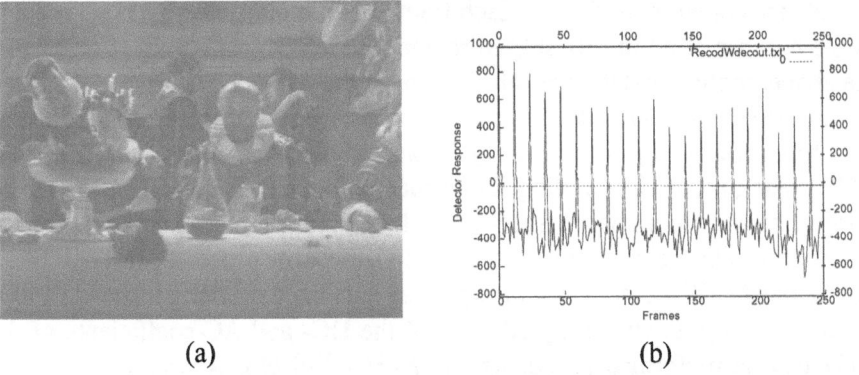

(a)   (b)

Figure 5.3. Test video sequence: (a) Frame 36 of the watermarked sequence after an MPEG-2 coding at lower bit rate (3 Mbit/sec), (b) Detector response.

to 3 Mbit/sec (in Figure 5.3(a) frame 36 is depicted again). As it can be seen in Figure 5.3(b) the detector is always able to exactly reveal the mark, there are no missed frames, but the height of the peaks is lower with respect to the first case.

4. The MPEG-4 Standard

In this section some basic aspects of MPEG-4 standard will be introduced before presenting, in the next two sections, two algorithms that deal with it. In particular in Subsection 4.1 a brief explanation of the structure of MPEG-4 video coding will be presented and in Subsection 4.2 some aspects of the standard specifically dedicated to IPMP (Intellectual Property Management and Protection) reported.

4.1 MPEG-4 Video Coding

The recent finalisation of MPEG-4 (a good overview can be found in [14]) will make this standard very attractive for a large range of old and novel applications, such as video editing, internet video distribution, wireless video communications. One of the key point of the MPEG-4 video coding standard is the possibility to access and manipulate objects within a video sequence directly in the compressed domain. Thus object watermarking will be reasonably achieved in such a way that, while a video object is transferred from a sequence to another (object manipulation), it is still possible to correctly access the copyright data of the object itself.

The structure of MPEG-4 coding is not essentially different from previous video standards such as MPEG-1 and MPEG-2, in that block-based motion compensation and motion-compensated hybrid DPCM/transform coding techniques are used, the main difference is that coding is content-based, i.e. single objects are coded individually. Each frame of an input sequence is segmented into a number of arbitrarily shaped regions, Video Object Planes (VOPs), and the shape, motion and texture information of the VOPs belonging to the same Video Object (VO) are coded into a separate Video Object Layer (VOL). The first VOP of a Group Of Video object planes (GOV) is coded intra-frame (I-VOP coding mode) by splitting it into macroblocks (MB). Each MB contains luminance and chrominance 8×8 pixel blocks $f[y][x]$ (e.g. in the 4:2:0 format 4 luminance and 2 chrominance 8×8 blocks), these are processed through DCT, producing the DCT blocks $F[v][u]$, and then quantized resulting in the $QF[v][u]$ blocks. Finally an efficient prediction of the DC- and AC-coefficients of the DCT can be performed producing the $PQF[v][u]$ blocks which are zig-zag scanned and run-level entropy encoded to get the bit-stream. Each subsequent VOP in the GOV is coded using inter-frame VOP prediction (P-VOP or B-VOP, Video Object Planes related to P and B frames respectively), i.e. it is motion compensated, and the residual prediction error signal is split into MBs, and then to blocks which are compressed in the same way as I-VOP blocks.

4.2   MPEG-4 IPMP

The problem of intellectual property management and protection (IPMP) [15, 16] was deeply felt in the sphere of MPEG-4 standard and peculiar attention was dedicated to this issue; the interest was basically to achieve some effective tools for copyright safeguarding, without limiting MPEG-4 characteristics. In fact, MPEG-4's target applications are very disparate and also involve different kinds of media; hence, the intellectual property management and protection systems to be adopted had to be so general to permit a large integrability with all the diverse applications the standard supported. Despite these considerations, the first proposed solution was to conceive an MPEG-4 IPMP including facilities such as encryption and watermarking within the MPEG-4 standard itself. It was soon realized that this structure was in contrast with the aforementioned requirements, that is the need not to bind MPEG-4 application potentialities and to adequately answer to the widely different IPMP possibilities. So the solution was to achieve the standardization of a generic interface to private and non-normative IPMP systems.

This choice allows to optimize the domain-specific IPMP tools for the individual application. This interface is designed to be a simple extension of basic MPEG-4 systems constructs, and it consists of IPMP-Descriptors (IPMP-Ds) and IPMP-Elementary Streams (IPMP-ES), as evidenced in Figure 5.4. The latter ones are like any other MPEG-4 elementary stream while the first ones are extensions to the MPEG-4 object descriptors. Both provide a communication mechanism between IPMP systems and the MPEG-4 terminal; MPEG-4 objects requiring management and protection have IPMP-Ds associated with them which indicate which IPMP systems are to be used and give information to these systems about how to manage and protect the content. In Figure 5.4 some possible IPMP control points (*hooks*) are indicated with a cross; these represent points, in the MPEG-4 terminal, where IPMP control might be desired and where a proprietary DRM (Digital Rights Management) system could be attached.

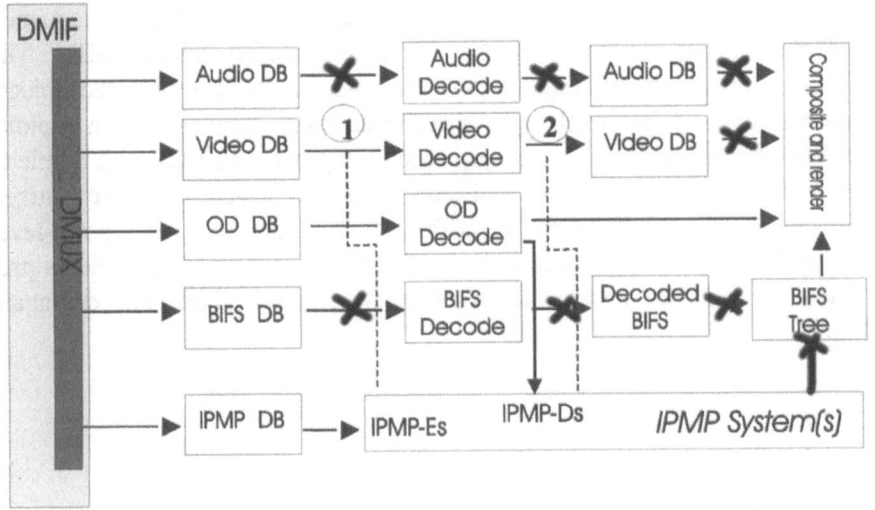

Figure 5.4. IPMP Framework in the ISO/IEC 14496 Terminal Architecture. DMIF: Delivery Multimedia Integration Framework, DB: Data Buffer, BIFS: Binary Format for Scenes, OD: Object Descriptor.

Anyway, which has to be the watermarking approach to use, has not been standardized and which must be the peculiar features to be supplied, is not fixed, and only generic indications are to be followed. From a primary analysis, many of the existing watermarking techniques could be adopted to work in this scenario and their integration with the MPEG-4 standard does not appear to be critical.

In Figure 5.4, the labels "1" and "2" indicate the possible location of the watermarking systems to be presented in Section 5 and in Section 6, respectively, within the MPEG-4 decoding interface. In fact, the former deals with compressed video, while the latter works in the raw video domain. They could use the information carried by the IPMP elementary streams and descriptors to make protected content available to the terminal; generally IPMP elementary streams bring time-variant information, for example about keys to be used, while IPMP descriptors are usually adopted to convey information about methods to associate the IPMP system to a particular elementary stream.

5.   A Compressed Domain Object-Based Watermarking Algorithm

The algorithm, that is going to be described, embeds a watermark in each video object of an MPEG-4 coded video bit-stream by imposing specific relationships among quantized DCT coefficients in a way similar to those presented in [17] and [18]. In particular a relationship is imposed, if not naturally occurring, between some predefined pairs of quantized DCT middle frequency coefficients in the luminance blocks of pseudo-randomly selected MBs. The adaptation of such a technique is here presented to the case of video objects watermarking. An innovative approach to watermark recovery which exploits the data obtained from the whole sequence to improve reliability, and is able to give a measure of the confidence of reading, is also presented. The quantized coefficients are recovered from the MPEG-4 bit-stream by reversing run-level entropy coding, zig-zag scanning and intra DC and AC DCT coefficients prediction; they are then modified to embed the watermark and then encoded again (Figure 5.5).

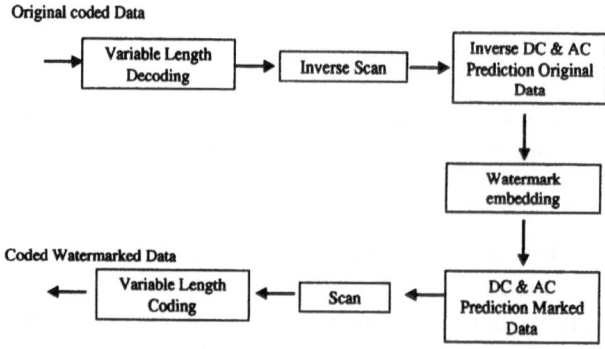

Figure 5.5.   Texture decoding process on an MPEG-4 decoder, modified to hide the watermark.

Watermarks are equally embedded into intra and inter MBs. A masking method is also adopted to limit visual artifacts in the watermarked VOPs and to improve, at the same time, the robustness of the system. Watermark recovery does not require the original video and is done in the compressed domain.

5.1 Watermark Embedding

The watermarking algorithm proposed here hides a bit of the copyright code in every luminance block belonging to MBs selected on a pseudo-random basis. If the MB is skipped the bit is also skipped. The watermarking code is repeated over the whole VOP (i.e. after the last bit of the code has been embedded, the process considers the first bit again, and so on); on every VOP of the same VOL the code is embedded again starting from the first bit. Every bit is thus embedded more than once during a sequence of VOPs, but, due to MBs skipping in the coded bit-stream, some bits are embedded less frequently than others. In Figure 5.6 the watermark embedding scheme is shown. The watermark is

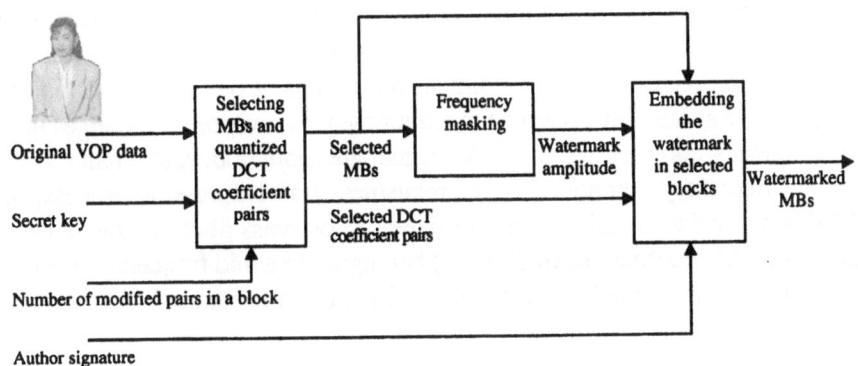

Figure 5.6. Diagram of the watermark embedding procedure.

embedded in the video through the following steps:

1 select MBs and the DCT quantized coefficients pairs to be modified;

2 for each block belonging to a selected MB:

    (a) compute the frequency mask;

    (b) use the mask to weight the watermark amplitude;

    (c) modify, according to the algorithm rule, selected pairs of DCT coefficients creating the watermarked block.

At the start of each VOP, a pseudo-random binary sequence is generated, based on a secret key and on the characteristics (number of MBs) of the VOP itself, for choosing those MBs where the watermarking code has to be embedded and those coefficients pairs that should be modified to embed the watermark. If the chosen MB is not skipped, one bit of the watermarking code is embedded into it by imposing a particular relationships between the values of selected pairs of coefficients that belong to each luminance block existing on it, otherwise the bit is skipped too. The use of the pseudorandom sequence permits to improve the robustness of the watermark, by preventing the possibility for an attacker to alter the watermarking code having identified the positions where the watermark was embedded. This makes the watermarking algorithm private [19], i.e. only allowed people (who knows the secret key) can read the watermark. Moreover it is widely know that applying an identical watermark in each frame of a video, leads to problems of maintaining statistical invisibility. To prevent statistical attacks (e.g. the averaging attack) pseudo random sequence generations is also VOP dependent (i.e. the locations where the watermark is embedded change from VOP to VOP according to the dimension of the VOP itself); this is done by seeding the pseudo random generator not with the secret key alone but with the sum of the secret key and the dimension in MBs of the VOP to be marked.

To achieve a trade-off between the requirement of invisibility (changes in the low-frequency components of a video signal are more noticeable than those in the high-frequency components) and robustness (in compression process, like MPEG4, the video signal can be considered as low-pass filtered), the values of quantized DCT coefficients ($QF(u, v)$) belonging to a mid frequency range are considered for watermark embedding (as depicted in Figure 5.7).

0	1	2	3	4	5	6	7
8	9	10	11	12	13	14	15
16	17	18	19	20	21	22	23
24	25	26	27	28	29	30	31
32	33	34	35	36	37	38	39
40	41	42	43	44	45	46	47
48	49	50	51	52	53	54	55
56	57	58	59	60	61	62	63

Figure 5.7.   Sketch of the mid frequency quantized DCT coefficients that are watermarked.

The watermark is carried by the signal (watermarking feature) which is the difference between the magnitudes of some selected pairs of the quantized DCT coefficients belonging to the mid-frequency region sketched in Figure 5.7, i.e.:

$$W(u_1, v_1, u_2, v_2) = |QF(u_1, v_1)| - |QF(u_2, v_2)| \qquad (5.1)$$

where (u_1, v_1) and (u_2, v_2) are the coordinates of the two coefficients of one of these pairs. It is expected that $W(u_1, v_1, u_2, v_2)$ is a non-stationary random process having zero mean, and a moderate variance if the coefficients composing each pair are sufficiently near one each other. If $\{QF(u_1, v_1), QF(u_2, v_2)\}$ is a randomly selected pair, the corresponding watermarked pair should be outlined as $\{QF'(u_1, v_1), QF'(u_2, v_2)\}$. By supposing the bit to be embedded being a 1, two cases can hold:

- both coefficients of the pair are non zero;

- at least one of the coefficients of the pair is zero.

In the first case the watermark is inserted with maximum strength: the sign of the watermarked coefficients is not changed with respect to the original sign, while the respective magnitudes become:

$$|QF'(u_1, v_1)| = (|QF(u_1, v_1)| + |QF(u_2, v_2)| + A_F(1 + n))/2 \quad (5.2)$$

where $/$ is an integer division and A_F is the watermark strength, and:

$$|QF'(u_2, v_2)| = \begin{cases} |QF'(u_1, v_1)| - A_F(1 + n) & \text{if } |QF'(u_2, v_2)| > 0, \\ 0 & \text{otherwise.} \end{cases}$$
$$(5.3)$$

Instead, when one or both coefficients of the pair are zero, it is more difficult to maintain the watermark perceptually invisible, because the masking effect between DCT frequency components is absent and cannot be exploited. In this case the coefficients of the pair are changed less heavily, in a way not to disturb the retrieval phase:

$$QF'(u_{1,2}, v_{1,2}) = \begin{cases} QF(u_{1,2}, v_{1,2}) & \text{if } |QF(u_1, v_1)| > |QF(u_2, v_2)|, \\ 0 & \text{otherwise.} \end{cases}$$
$$(5.4)$$

When a bit 1 has been inserted, for the coefficients pair (u_1, v_1) and (u_2, v_2) it results $W'(u_1, v_1, u_2, v_2) \geq 0$, where $W'(u_1, v_1, u_2, v_2) = |QF'(u_1, v_1)| - |QF'(u_2, v_2)|$. A similar algorithm is adopted for embedding a bit 0: in this case, the roles applied to the coefficients pair (u_1, v_1) and (u_2, v_2) are exchanged, thus resulting $W'(u_1, v_1, u_2, v_2) \leq 0$. The parameter n takes into account that the quantization step (QP) of a given coefficient can change during the coding process, in order to keep the bit-rate as constant as possible, from block to block. Since we work on quantized levels, the same modification applied to coefficients quantized with large or small quantization step, can produce very different visible effects: on that basis, n is increased when the quantization step decreases. Furthermore, to achieve a reasonable trade-off between invisibility and robustness, it is necessary that the larger is the number (C_{num}) of pairs to be marked in a block, the smaller is the value of n. The adaptation rule of n was obtained experimentally and resulted in:

$$n = \begin{cases} \frac{15 - 4c_{num}}{QP} - 1 & \text{if } (15 - 4c_{num}) > QP \\ 0 & \text{otherwise.} \end{cases} \quad (5.5)$$

Finally a masking method is also used to improve the invisibility and the robustness of the watermark embedded into each VOP . The implemented system is based on the model proposed in [20] and it works by changing the strength of the watermark (A_F) according to the smoothness and the edgeness characteristics of each block to be marked.

It can obviously happens that a pair of coefficients is naturally high, i.e. $W(u_1, v_1, u_2, v_2) \geq A_F(1 + n)$ in which case any modification is not needed for embedding the bit 1 (and the same happens for bit 0, in naturally low pairs).

5.2   Watermark Recovery

Watermark retrieval is done in two steps as can be seen in Figure 5.8. The

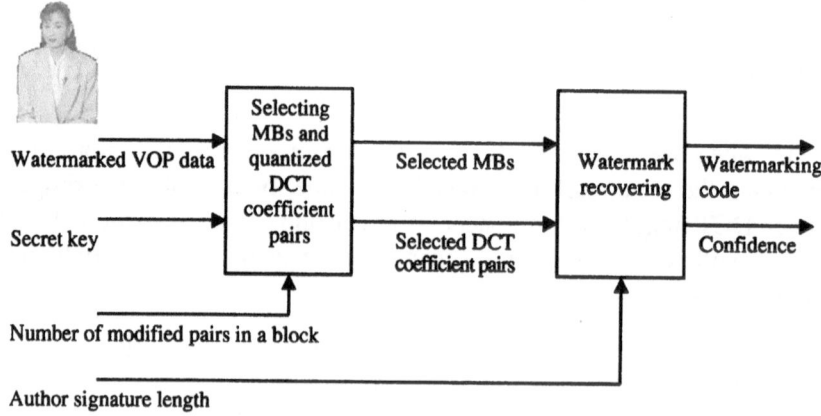

Figure 5.8.   Diagram of the watermark recovering procedure.

first step is analogous to the one used in the embedding process (see Subsection 5.1) and requires the knowledge of the parameters used in the embedding phase (i.e. the secret key and the number of DCT coefficient pairs that was modified in each selected block) to correctly identify MBs and coefficients pairs where the watermark was effectively hidden. In the second step the relationships between the coefficients of the selected pairs are analyzed. The knowledge of the watermarking code length is needed to compute the repetition step of the watermarking code in each VOP (i.e. how many times the copyright code was embedded in the considered VOP).

For reading the jth bit of the watermarking code, an accumulator is considered Acc_j, where the values of $W'(u_1, v_1, u_2, v_2)$ corresponding to all the pairs of coefficients where the bit itself was inserted, are summed up. Let us call ψ_j the set of these pairs:

$$Acc_j = \sum_{pairs \in \psi_j} W'(u_1, v_1, u_2, v_2). \tag{5.6}$$

Such a sum is then compared to a threshold T_D, to decide for the value of the embedded bit:

$$bit_j = \begin{cases} 1 & if \quad Acc_j > 0, \\ indeterminate & if \quad -T_D \leq Acc_j \leq T_D, \\ 0 & if \quad Acc_j < 0. \end{cases} \tag{5.7}$$

The value of T_D is fixed to 0 in order to minimize the overall error probability; in fact it is expected for Acc_j to be positive when the embedded bit is a 1 and negative in the opposite case. On the other hand choosing $T_D = 0$ leads to read a watermarking code also when no code was actually embedded. To obviate this problem, the accuracy (confidence) of the reading is also provided for each recovered bit. A hypothesis testing problem is solved for this purpose, in which only one of the following situations is possible:

- $Hp.A$: the VOL is not marked,

- $Hp.B$: the VOL is marked.

It can be easily verified that for each bit:

$$E[Acc|_{Hp.A}] \approx 0 \tag{5.8}$$

and that a good estimate of $\sigma^2_{Acc|_{Hp.A}}$ is given by:

$$\sigma^2_{Acc|_{Hp.A}} \cong 2\left(A^2 - B^2\right) \tag{5.9}$$

where

$$A^2 = \sum_{i=1}^{N_p} QF_{i_1}^2 \cong \sum_{i=1}^{N_p} E[QF_{i_1}^2] \tag{5.10}$$

and

$$B^2 = \sum_{i=1}^{N_p} |QF_{i_1}||QF_{i_2}| \cong \sum_{i=1}^{N_p} E[|QF_{i_1}||QF_{i_2}|]. \tag{5.11}$$

are estimated over the values of the DCT coefficients belonging to the non-watermarked blocks (i.e. the blocks not selected by the random key). The confidence value for each bit is thus defined as:

$$C_j = \frac{Acc_j}{\sigma_{Acc|_{Hp.A}}}. \tag{5.12}$$

The smaller this value, the higher the probability that the sequence is not watermarked. Small values will also be obtained when the sequence is watermarked with another key: in this case the absolute value of the accumulator will be in general larger than for $Hp.A$, because some coefficient pairs where some bits are actually embedded can be selected, thus contributing to increase it. As a global measure of the possibility that a video has been watermarked with the used key, the sum of the confidence values resulting for all the bits can be used. Furthermore, the availability of a confidence measure for each bit, would allow the use of soft decoding [21] techniques for decreasing the Bit Error Rate (BER). (Anyway the use of error correcting codes has not been considered in this chapter.)

5.3   Experimental Results

This section presents part of the experiments that were performed to prove the effectiveness of the above described algorithm. In the embedding phase, the test software was implemented so as to receive as input, an MPEG-4 VOL, a secret key, the number of pairs that should be modified in a block, a binary watermarking code, and the maximum strength A_{MAX} of the watermarking signal A_F [20]; then the software parses the VOL and writes it to a new file. In the recovery phase the implemented software takes a watermarked VOL, the watermarking code length, a secret key and the number of pairs that were modified in a block), as input parameters, and computes the watermarking code and the confidence of each bit read.

Two kinds of tests were conducted with this system on different video sequences for proving on one side the invisibility of the embedded watermark and on the other side the robustness against all processing which does not seriously degrade the quality of the video. Among the tested video sequences the results regarding "News" and "Stefan" are here presented. The standard video sequences "News" and "Stefan" are coded using binary alpha planes and consist of 300 frames in CIF format with a frame rate of 25 frame/sec. "News" is composed of four VOs (see Figure 5.9(a)): the anchor-man on the left is labeled as VO0, the woman on the right as VO1, the monitor in the center as VO2 and the background as VO3. The video sequence "Stefan", instead, is composed of two VOs (see Figure 5.10(a)): the tennis player (VO1) and the background (VO0).

5.3.1   Visibility experiments.   To evaluate the quality of the watermarked videos, a series of tests has been performed in which the original video and a video in which each VO is watermarked are displayed in sequence to a viewer. The order in which the original and the watermarked videos were displayed (i.e. as original-watermarked or watermarked-original) was randomly selected. The viewer was asked to select which of the sequences has better

(a)

(b)

Figure 5.9. Frames from "News" video sequence: (a) original and (b) watermarked.

quality. The embedded watermark appears perceptually undetectable and so each video was selected approximately 50% of the time. Two original frames of the two videos are shown in Figures 5.9(a) and 5.10(a). The corresponding watermarked frames are shown in Figures 5.9(b) and 5.10(b). As can be seen the watermarked and the original frame for each video appear identical.

5.3.2   Robustness against bit-rate decreasing.   It is widely known that if each frame of a video is watermarked the complete removal of the watermark by an attacker requires enormous computing power and memory resources. Hence, only attacks which can be done by an average consumer (i.e. attacks requiring a limited computing power) are considered. The robustness against bit-rate decreasing is very important because this kind of attack can be either of intentional or "incidental" nature. In fact, in most application involving storage and transmission of digital video, it is required to reduce the bit rate in order to improve the coding efficiency. In the proposed test each watermarked video sequence is decoded and encoded again by decreasing the bit-rate, and

(a)

(b)

Figure 5.10.   Frames from "Stefan" video sequence: (a) original and (b) watermarked.

watermark detection is attempted. This test is performed, in each video, for a set of different watermarking code lengths.

In Figures 5.11(a), 5.11(b), 5.11(c) and 5.11(d) the total number of correctly read bits is plotted against the bit-rate for the four VOs of the "News" sequence. It appears that in most cases free error decoding is achieved also when the bit-rate is reduced to half the original (i.e. very low coding quality). A particular observation is worth for VO3 (the background), where only 15 bits can be reliably embedded due to the scarcity of motion in the scene, and thus to the low number of inter-coded MBs available in the bit-stream for watermark embedding.

In Figures 5.12(a) and 5.12(b) the results for the "Stefan" sequence are presented. In this case up to 30 bits can be reliably embedded, thanks to the larger number of MBs available for embedding.

5.3.3   Robustness against frame dropping.   Frame drops may arise either intentionally or not. The encoding process may in fact result in frame

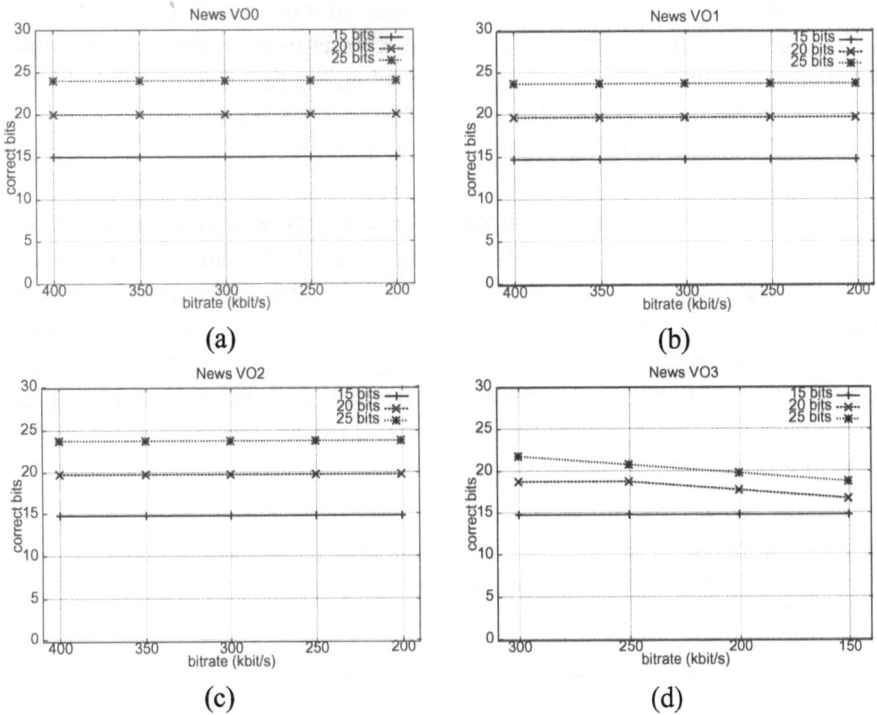

Figure 5.11. Plot of the number of correctly read bits for each VO of the "News" sequence, at decreasing bit rate and for three different values of the code length (15, 20 and 25 bits).

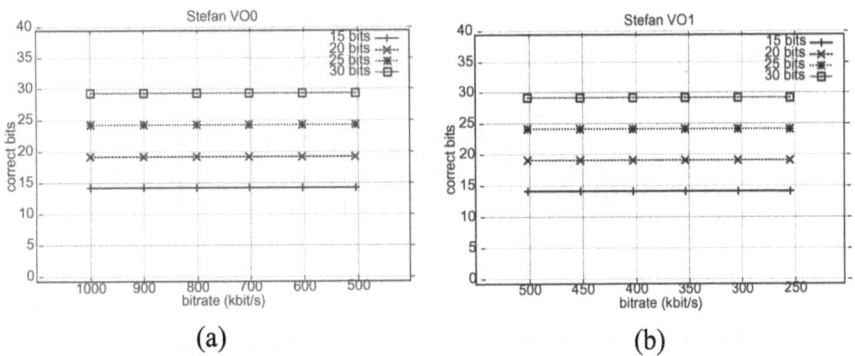

Figure 5.12. Plots of the number of correctly read bits for (a) VO0 and (b) V01 of the "Stefan" sequence, at decreasing bit rate and for four different values of the code length (15, 20, 25 and 30 bits).

skipping; similarly, in videos with very low motion components (i.e. high interframe correlation) frame cutting can be done without significantly degrading the quality. The proposed watermarking algorithm inserts the watermark without modifying zero coefficients; for that reason the strength of the embedded watermark is higher in intra coded VOPs than in inter coded VOPs. Changes in the GOV structure (i.e frame cutting) are then critical; in this way it is obvious that one of the worst cases is obtained when the first frame of the video is dropped. In that case, every intra coded VOP become the last inter coded VOP in the GOV structure (i.e. coarsely quantized and heavily affected by motion compensation errors).

In the proposed test each watermarked video sequence is decoded and encoded again (at the same bit-rate) after cutting the first frame, and watermark detection is attempted. In the Tables from 5.1 to 5.4, the percentage bit error rate $BER_\%$ (i.e. the percentage of wrong bits in the recovered watermarking code) for each video object is shown against A_{MAX} for two different watermarking code lengths. Although such an attack is strongly effective, at least 15 bits can still reliably be hidden inside almost all the video objects. The most critical video object is again the background of News.

Table 5.1. $BER_\%$ for the two video objects of Stefan when 15 bits are embedded, after cutting the first frame of the video and re-encoding it to the bit rate indicated.

	$A_{MAX} = 1.0$	$A_{MAX} = 3.0$
VO0 (1000 kbit/s)	0	0
VO1 (500 kbit/s)	0	0

Table 5.2. $BER_\%$ for the two video objects of Stefan when 30 bits are embedded, after cutting the first frame of the video and re-encoding it to the bit rate indicated.

	$A_{MAX} = 1.0$	$A_{MAX} = 3.0$
VO0 (1000 kbit/s)	6.7	0
VO1 (500 kbit/s)	0	0

Table 5.3. $BER_\%$ for the four video objects of News when 15 bits are embedded, after cutting the first frame of the video and re-encoding it to the bit rate indicated.

	$A_{MAX} = 1.0$	$A_{MAX} = 2.0$
VO0 (400 kbit/s)	0	0
VO1 (400 kbit/s)	6.7	0
VO2 (400 kbit/s)	0	0
VO3 (300 kbit/s)	13.3	6.7

Table 5.4. BER$_\%$ for the four video objects of News when 20 bits are embedded, after cutting the first frame of the video and re-encoding it to the bit rate indicated.

	$A_{MAX} = 1.0$	$A_{MAX} = 2.0$
VO0 (400 kbit/s)	0	0
VO1 (400 kbit/s)	15	5
VO2 (400 kbit/s)	0	0
VO3 (300 kbit/s)	30	20

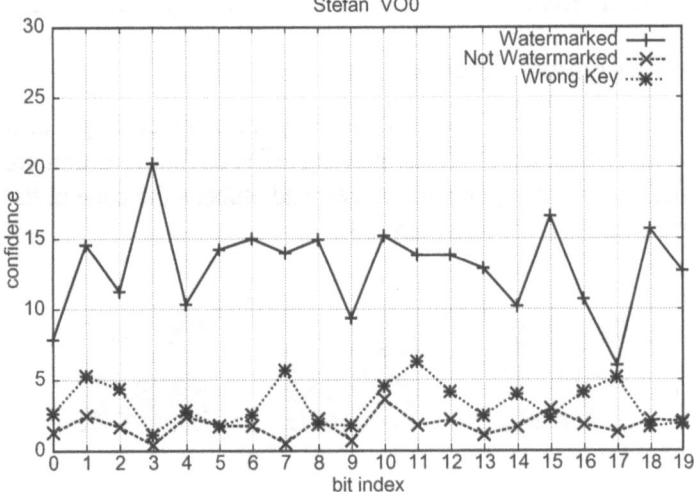

Figure 5.13. Plot of the confidence measure obtained for each of the 20 bits embedded into the "Stefan" sequence.

5.3.4   The confidence measure.

Regarding the confidence measure defined by equation (5.12), the values obtained for VO 0 of the "Stefan" sequence are plotted for each bit in Figure 5.13: the three cases of reading the watermark from a watermarked copy using the correct key (label "Watermarked"), reading the watermark from a non watermarked copy (label "Not Watermarked"), and reading the watermark from a watermarked copy by using the wrong key (label "Wrong Key") are considered. It is evident that the confidence values are always higher when the correct key is used on a watermarked sequence than in the other two cases. For some of the bits (e.g. bit 0 and bit 17) the value of the confidence is quite low also when the correct key is used: this is due to the fact that these particular bits have been repeated only a few times in the sequence, and thus their reading is quite unreliable. These are also the two bits that are lost when the first frame of the sequence is cut (see Table 5.2). Anyway the average of the confidence values of all the bits results to be 13.05 for the "Watermarked" case, 1.83 for the "Not Watermarked" case, and 3.41 for the "Wrong Key" case: it can, thus, be assumed to be a good parameter for deciding if the sequence is really watermarked with that key or not.

6.   A Raw Video Object-Based Watermarking Algorithm

In this section an object-based watermarking method that works on non-coded video will be outlined. The algorithm, though not fully integrated with MPEG-4 coding and decoding systems, permits to directly interact with video objects and to embed in each of them a specific code; doing so it allows to separately and differently deal with each object belonging to the sequence.

6.1   Watermark Embedding

The proposed algorithm operates frame by frame by casting a different watermark in each video object. Watermarking relies on the algorithm presented in [22], originally developed for still images, and embeds the code in the Discrete Wavelet Transform (DWT) domain [23].

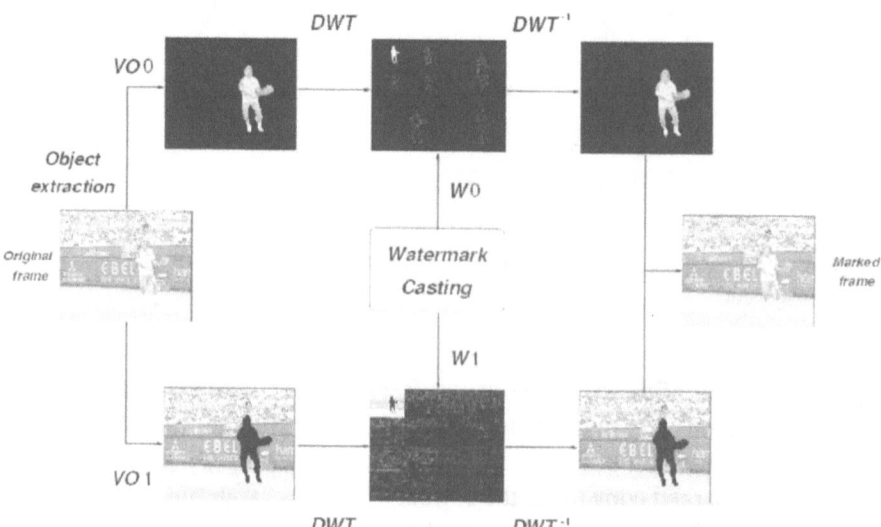

Figure 5.14. Watermark casting process. The objects composing the original frame, on the left, are separated; next, each object is watermarked, and finally the embedded objects are merged together, obtaining the watermarked frame, on the right.

According to Figure 5.14, the objects contained in each frame are extracted, obtaining a different image for each one (video objects are located on a black background), and in each image a different code is embedded by means of the system presented in [22] and resumed in the following. The procedure, described in Figure 5.14, is then applied to all the frames of the sequence.

The image to be watermarked is first decomposed in four levels through DWT: let us call I_j^θ the sub-band at resolution level j (where $j = 0, 1, 2$) and having orientation θ (where $\theta = LL, LH, HL, HH$). The watermark, consisting of a pseudo-random binary sequence, is inserted by modifying the wavelet coefficients belonging to the three detail bands at level 0, i.e. I_0^{LH}, I_0^{HL} and I_0^{HH}. Before adding it to the DWT values, each binary value is multiplied by a weighting parameter which is obtained by a noise sensitivity function. In this way the maximum tolerable level of disturbation (i.e. watermark coefficient) is added to each DWT coefficient. The construction of the sensitivity function is mainly based on the analysis of the degree of image activity in the neighborhood of the pixel to be modified (for more details see [22]).

Let us consider in detail how the watermark casting takes place. The mark is converted in a code sequence $x_i \in \{+1, -1\}$, with $i = 0, \ldots 3MN - 1$, where $2M \times 2N$ is the image size, and the three DWT sub-bands I_0^{LH}, I_0^{HL} and I_0^{HH} are modified according to equation (5.13):

$$\tilde{I}_0^{LH}(i, j) = I_0^{LH}(i, j) + \alpha w^{LH}(i, j) x_0(i, j)$$
$$\tilde{I}_0^{HL}(i, j) = I_0^{HL}(i, j) + \alpha w^{HL}(i, j) x_1(i, j) \qquad (5.13)$$
$$\tilde{I}_0^{HH}(i, j) = I_0^{HH}(i, j) + \alpha w^{HH}(i, j) x_2(i, j)$$

The code sequence is split in three sub-sequences each one for each DWT sub-band $x_k(i, j) = x_{kMN+iN+j}$ with $k = 0, 1, 2$; α is a parameter accounting for watermark strength (usually its value is in the range 1.0, 2.5) and $w(i, j)$ is a weighting function taking into account the local sensitivity of the image to noise [22]. Of course $\tilde{I}_0^{LH}$, $\tilde{I}_0^{HL}$ and $\tilde{I}_0^{HH}$ are the DWT modified coefficients.

To properly adapt this method to the video-objects which generally are small with respect to the rest of the image (for example see the tennis-player in Figure 5.14), the sensitivity function is firstly computed on the whole frame and then the visual mask related to each video-object is extracted from it and used during the watermark embedding. In fact if the mask was calculated directly on the video-object, the high image activity along the border of the object, related to the transition to the black background, would determine the insertion of a strong mark that would be perceivable once the original background is replaced instead of the black one.

The inverse DWT is then computed obtaining the watermarked video-objects. The watermarked video objects are then merged together in order to rebuild the frame containing the copyright information concerning each object present in the scene. When all the frames have been marked, the sequence can be compressed, obtaining the watermarked MPEG-4 coded bit-stream.

The proposed system is invariant to the conversion to a different compression standard, because it works in the raw-video domain and it is able to detect a watermark also in very small regions of an image, this is very useful because video-objects are often not so big with respect to the frame size.

6.2   Watermark Detection

In this section the process of watermark detection is analyzed. The watermarked MPEG-4 coded video bit-stream is decoded obtaining a sequence of frames. Once again, the objects present in the scene are extracted frame by frame, obtaining a different image for each object. The DWT of each image is then computed and the code corresponding to each object is detected by means of the correlation product between the watermark x_i and the DWT marked coefficients $\tilde{I}_0^\theta$, as evidenced in equation (5.14):

$$\rho = \frac{1}{3MN} \sum_{i=0}^{N-1} \sum_{j=0}^{M-1} \tilde{I}_0^{LH}(i,j)x_0(i,j) + \tilde{I}_0^{HL}(i,j)x_1(i,j) + \tilde{I}_0^{HH}(i,j)x_2(i,j)$$

(5.14)

The value of the correlation is compared to a threshold T_ρ, which depends on the variance σ_ρ^2 of the DWT coefficients of the watermarked image, as stated in equation (5.15), to decide upon presence of watermark. When the correlation value is higher than the threshold, the tested code is considered to be present in the examined video-object:

$$T_\rho = 3.97\sqrt{2\sigma_\rho^2}$$

(5.15)

where

$$\sigma_\rho^2 = \frac{1}{(3MN)^2} \cdot \sum_{i=0}^{N-1} \sum_{j=0}^{M-1} E[\tilde{I}_0^{LH}(i,j)^2] + E[\tilde{I}_0^{HL}(i,j)^2] + E[\tilde{I}_0^{HH}(i,j)^2]$$

(5.16)

The threshold has been set in such a way to minimize the missed detection probability given a false alarm probability of 10^{-8} [22]. In particular, the value of this threshold can be estimated *a-posteriori* without the need of knowing data concerning the original frame. Since the detection process is computed frame by frame, the watermark embedded into a video object can be revealed also if the VO is transferred from a sequence to another. However, since the DWT is not invariant to translation, if the video object is placed in a different position of the new scene, the synchronization between the watermark and the VO is lost; to cope with this situation, the watermark detector would need to compute the correlation for all shifts of the frame. To reduce the computational complexity, the properties of the 2-D Discrete Fourier Transform can be exploited.

By multiplying the DFT of the image DWT coefficients and the DFT of the watermark, and by computing the IDFT of the product, a peak, indicating the original position of the VO, can be identified. In Section 6.3 this aspect will be discussed more in detail and experiments regarding this particular MPEG-4 application will be proposed in Subsection 6.4.2.

6.3   Detection on Translated Objects

In this watermarking technique, as debated in Section 6.1, the watermark is embedded by using the DWT (Discrete Wavelet Transform) and it is well-known that this kind of transform is not invariant to shift operations. Thus there is a problem which is common with all correlation based detection: the synchronization [24]. In fact manipulations like cropping and translation determine the loss of the absolute position where the code is located, thus resulting in a failure of the detection itself.

In the approach adopted in this work, the watermark is inserted in the DWT domain by modifying the coefficients belonging to the three bands LH, HL and HH, at level 0, as explained in Section 6.1, and Equation (5.14) has to be rewritten as specified in (5.17) to take into account all possible shifts:

$$\rho(h, k) =$$

$$\frac{1}{3MN}\{ \sum_{i,j=0}^{N-1,M-1} \tilde{I}_0^{LH}(i-h, j-k)x_0(i,j)+$$

$$\sum_{i,j=0}^{N-1,M-1} \tilde{I}_0^{HL}(i-h, j-k)x_1(i,j)+ \qquad (5.17)$$

$$\sum_{i,j=0}^{N-1,M-1} \tilde{I}_0^{HH}(i-h, j-k)x_2(i,j)\}$$

where indices h and k take account for all possible shifts. The indices $i - h$ and $j - k$ have to be intended as cyclic wrap around with modulo respectively N and M.

This operation is very heavy from a computational point of view and it depends on the image size; to avoid this hurdle and drastically reduce the complexity, a two-dimensional DFT (Discrete Fourier Transform) and its inverse IDFT (Inverse Discrete Fourier Transform) can be used [25]. In fact the detection is boiled down to computing as stated in equation (5.18).

$$\rho(h, k) =$$

$$\frac{1}{3MN}\{IDFT[DFT[\tilde{I}_0^{LH}] \otimes DFT^*[x_0]]+$$

$$IDFT[DFT[\tilde{I}_0^{HL}] \otimes DFT^*[x_1]]+ \tag{5.18}$$

$$IDFT[DFT[\tilde{I}_0^{HH}] \otimes DFT^*[x_2]]\}$$

where the DFT and IDFT are respectively the Discrete Fourier Transform and Inverse DFT; the symbol $\otimes$ states for the point-to-point product of the two arguments. Doing so all the correlation values for all the possible shifts are obtained; by taking the maximum of these ones the correlation peak (CP), related to the right matching between the watermarked image coefficients and the watermark, can be obtained (see equation (5.19))

$$CP = MAX_{h,k}[\rho(h, k)] \tag{5.19}$$

and then compared with the decision threshold, calculated as already mentioned in equation (5.15). It can be easily demonstrated that the threshold does not depend on the particular shift. Moreover the exact translation values of the image plane will be indicated by the values of h and k corresponding to the maximum of $\rho(h, k)$:

$$SHIFT_x = arg_k\{MAX_{h,k}[\rho(h, k)]\}$$
$$SHIFT_y = arg_h\{MAX_{h,k}[\rho(h, k)]\} \tag{5.20}$$

The parameters $SHIFT_x$ and $SHIFT_y$ will provide a spatial measure of the shift undergone by the video object along x and y directions. The whole approach will be applied and tested in detail in Subsection 6.4.2 for the MPEG-4 video objects shifting.

6.4   Experimental Results

The algorithm has been tested on a large set of experiments: in the following some of the most significant results are presented.

6.4.1   Detection on MPEG-4 sequences.

The proposed watermarking algorithm has been tested on different video sequences. Here after, the results concerning the CIF sequence *Stefan* are shown. The two objects each frame is composed of, namely the background (Video Object 0) and the tennis-player (Video Object 1), are reported in Figure 5.15.

A different random sequence has been introduced into each object, as illustrated in Figure 5.14. After watermarking, the video sequence has been compressed obtaining an MPEG-4 coded video bit-stream, with a target bit rate

Figure 5.15. Video objects belonging to the *Stefan* sequence: (a) VO 0, and (b) VO 1.

of 1 Mbit/sec for Video Object Layer 0 (VOL0) and 500 Kb/sec per Video Object Layer 1 (VOL1), in both cases a GOP (Group Of Pictures) of 12 frames has been considered. The video stream has been next decompressed, and in each frame the two objects were separated obtaining two different image sequences, where the detection process was applied frame by frame. Detection results, related to one of the intra frame, are shown in Figure 5.16 (a) for VO 0, and (b) for VO 1: the response of the watermark detector to 200 marks randomly generated shows that the response to the right embedded watermark (i.e. no. 60 for VO 0 and no. 70 for VO 1) is much larger than the response to the others, and it is higher than the detection threshold.

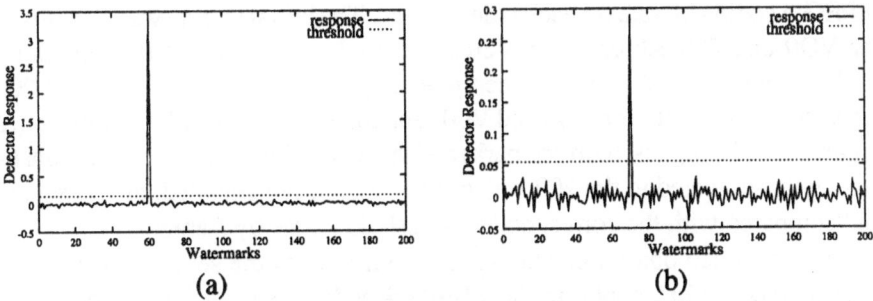

Figure 5.16. Detector response related to the video objects belonging to the *Stefan* sequence: VO 0 watermark no. 60 (a) and VO 1 watermark no. 70 (b).

It has to be carefully noted that the correlation peak for VO 1 (the tennis-player), though distinctly over the threshold has got a value, about 0.3, which is decidedly lower with respect to the VO 0, which is about 3.5; this is mainly due to the small size of the tennis-player that, roughly speaking, can contain a reduced amount of watermark energy.

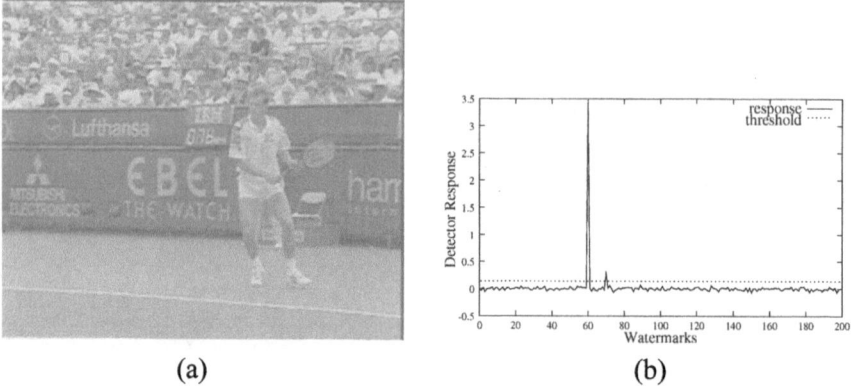

Figure 5.17.   Detector response related to the whole frame of the *Stefan* sequence.

Moreover, the detection process was also applied to the whole frame, after MPEG-4 decoding, without selecting the objects: as demonstrated in Figure 5.17, the two watermarks embedded in the two objects are easily detected; let us note again that the correlation peak of the tennis player (VO 1) is lower than the response of the background (VO 0), due to the fact that a low watermark energy can be embedded into the tennis player. The correct detection of the two objects indicates that the system is robust to video-format conversion, since the two watermarks are revealed in the frame, indicating the presence of the two objects, even if, clearly, in this case it is no longer possible to associate each watermark to the corresponding video object.

Another interesting test has been performed by gradually decreasing the MPEG-4 coding bit rate for each video object. The initial bit rate was 1 Mbit/sec for VO0 and 500 Kbit/sec for VO1, as already evidenced at the beginning of this section, and then it has been gradually reduced firstly to 700 and 200 Kbit/sec, respectively for VO0 and VO1, and finally to 600 and 100 Kbit/sec. In Figure 5.18, the highest (correct mark) and the second highest detector responses among those related to the 200 watermarks randomly generated, are presented and compared with the correspondent threshold frame by frame.

Only the video objects belonging to even frames of the sequence have been watermarked before applying the MPEG-4 coding, so, during the detection phase, it should happen that if a watermark is present, it will be present just in even frames, but looking at Figure 5.18, in particular on the left side where the graphs of the VO0 are plotted, it can be noticed that the detector response is higher than the threshold also for odd frames, especially from number 24 up to number 39, where no watermark has been embedded. This effect is due to temporal correlation exploited by the MPEG-4 encoder that, through the motion estimation and compensation, predicts the following pictures by using the previous ones where the watermark is contained, thus it results that the

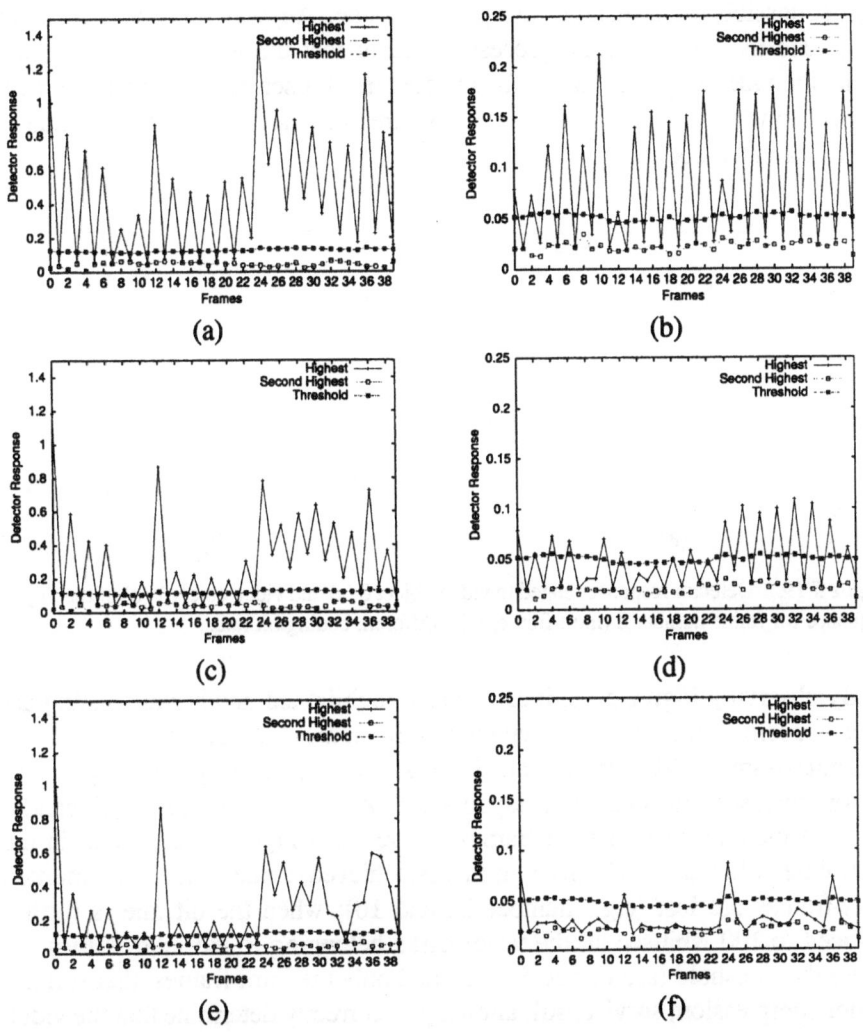

Figure 5.18. Detector response related to video object 0 (left side) and video object 1 (right side) of the *Stefan* sequence for 40 frames (0 − 39); different bit rates have been considered respectively for each video object: 1000 (a) and 500 (b), 700 (c) and 200 (d), and 600 (e) and 100 (f).

watermark is dragged along the sequence. The same dragging effect can be also recognized for the VO1 (right side of Figure 5.18), but, in this case, the detector response is always under the threshold, though it is always the highest one compared to the second highest response that is visualized in the graphs. This can be explained taking into account that the VO1 (tennis player) presents much more motion components with respect to the background (VO0) that is practically still from a frame to another, so in this case, the temporal correlation is lower and the MPEG-4 encoder cannot transfer much information from a frame, possibly watermarked, to a predicted one. Anyway this effect does not invalidate the watermarking process at all, in fact the watermark is not lost along the video sequence, it is only detected in a higher number of frames with respect to what it was expected, but these are not false alarms, they are due to MPEG-4 video coding.

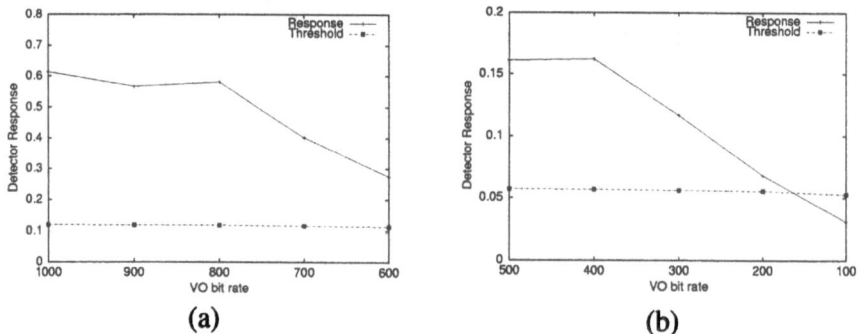

Figure 5.19.   Detector response related to video object 0 (a) and video object 1 (b) of the *Stefan* sequence for frame number 6 (intra-frame) with different coding bit rate.

By observing Figure 5.18, it can be also highlighted another interesting aspect, concerning the different behavior of intra-frames (numbered as 0, 12, 24, 36) and inter-frames. When the bit rate decreases (or equivalently the compression factor increases), the detection response for inter-frames reduces significantly, whereas the response for intra-frames still remains high. In particular, in the VO1 case, when a 200 Kbit/sec bit rate is achieved, some of the watermarked inter-frames are lost (e.g. number 14 and 16); when the bit rate is further reduced to 100 Kbit/sec all the watermarks belonging to the inter-frames are under the threshold (see Figure 5.18(f)) and only the intra-frames, that suffer a minor compression, survive, still allowing to correctly determine that the video object is watermarked. The same happens to the VO0, though in this case, the watermarks belonging to the inter-frames resist also when the bit rate is decreased from 1000 Kbit/sec to 600 Kbit/sec.

In Figure 5.19, the specific situation of the detector response regarding the frame number 6 of the sequence compared to the threshold is presented in detail per each video object. It can be easily seen that the when the coding bit rate decreases, the gap between the detector response and the threshold becomes low; in particular, for the VO1 when the coding bit rate is 100 Kbit/sec, the watermark is not revealed anymore and the video sequence becomes unprotected. Anyway it has to be pointed out that with this coding factor the video quality is really degraded.

Another CIF sequence, *News*, containing a higher number of video objects, is worth to be considered. The four different video objects of this sequence,

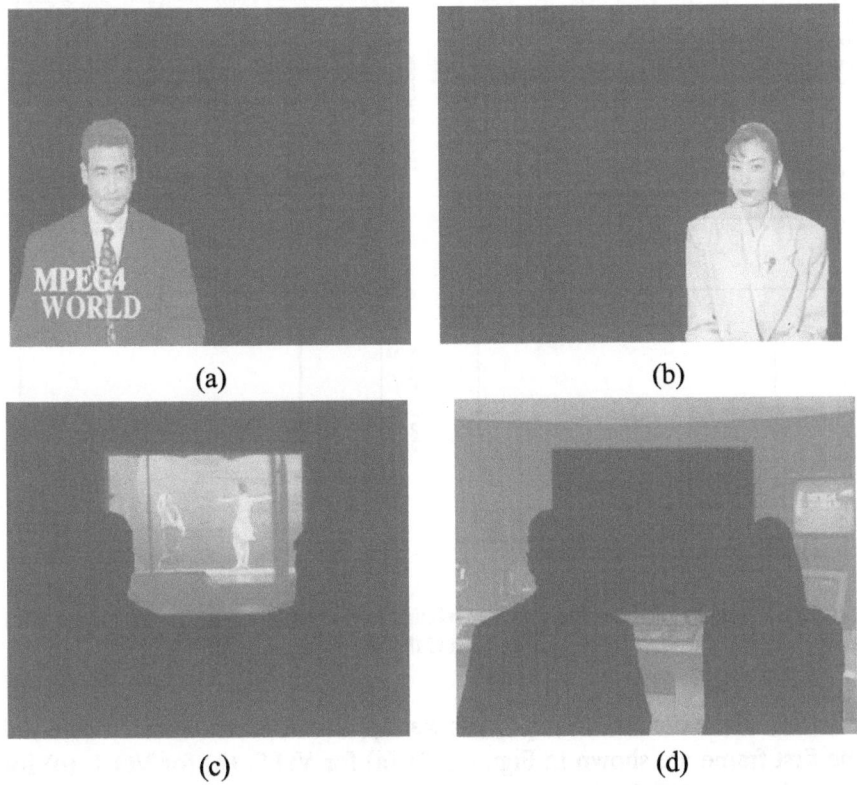

(a)   (b)

(c)   (d)

Figure 5.20. Video objects belonging to the *News* sequence: (a) The male speaker VO 0, (b) The female speaker VO 1, (c) The screen with the ballet VO 2 and (d) The background VO 3.

the male speaker (Video Object 0), the female speaker (Video Object 1), the screen with the ballet (Video Object 2) and the background (Video Object 3), are illustrated in Figure 5.20.

Contrary to the former situation, where a relatively smaller object, as the tennis-player, was distinguished from the background occupying the major part of the frame, four objects are now considered featuring a more equally distributed size.

As for *Stefan*, a different random sequence has been inserted into each object, according to the approach explained in Figure 5.14. After the watermark casting, the sequence was compressed obtaining an MPEG-4 coded video bitstream; the resulted target bit rate has been of 500 Kb/s per each Video Object Layer (VOL), globally 2 Mbit/sec, and the GOP was set to 12 as before. Indeed, in this situation, it has been noticed that the MPEG-4 coder is able to compress the video sequence more efficiently, resulting in an effective global bit rate of around 1 Mbit/sec. The video stream was next decompressed, and in each frame the four objects were separated obtaining four resulting image

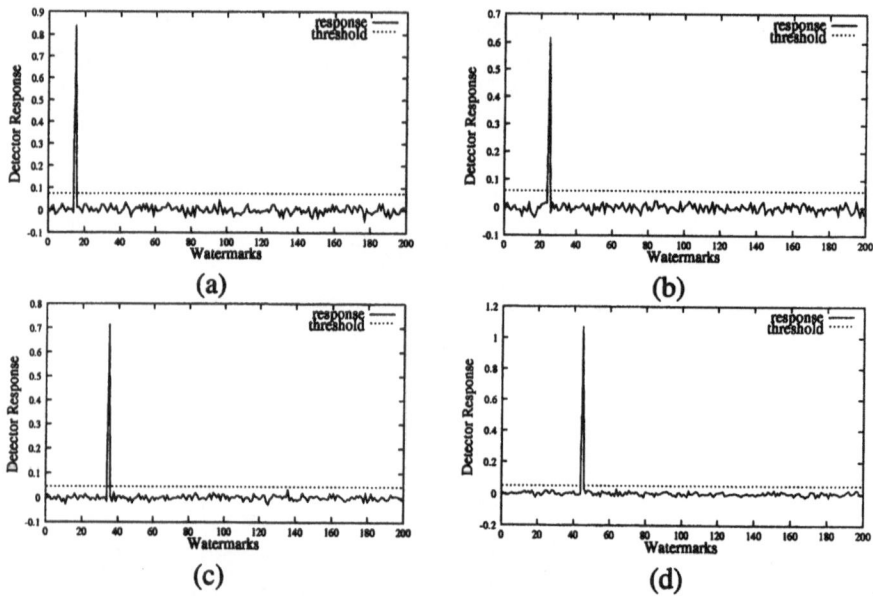

Figure 5.21.   Watermark detection response relating to the VO 0 watermark 15 (a), the VO 1 watermark 25 (b), the VO 2 watermark 35 (c) and the VO 3 watermark 45 (d).

sequences, where the detection process was applied. Detection results referred to the first frame are shown in Figure 5.21 (a) for VO 0, (b) for VO 1, (c) for VO 2 and (d) for VO 3: the response of the watermark detector to 200 marks randomly generated shows that the response to the right embedded watermark (i.e. no. 15 for VO 0, no. 25 for VO 1, no. 35 for VO 2, and no. 45 for VO 3) is much larger than the response to the others, and it is higher than the detection threshold.

Referring to what has been previously highlighted about the height of the correlation peaks, in this case it can be noted that each spike has a value close to the others, around 0.7, only the background VO 3 has got a value around 1.1; in fact looking at Figure 5.20(d) the amount of black pixels present is noticeably less than for the other video objects and so the watermark energy to put on it can be superior.

The detection process has been also applied to the whole frame, after MPEG-4 decoding, without exploiting the information regarding the single objects: as demonstrated in Figure 5.22 (b), the four watermarks embedded in the four objects are correctly detected. Let us note again that the correlation peaks of the objects are very similar, due to the fact that watermark energy embedded into each one is more or less the same, because their sizes are not so different. As for the *Stefan* sequence, the correct detection of the four objects shows that the system is robust to video-format conversion, since the watermarks are revealed in the frame, even if, it is no longer possible to know any information about

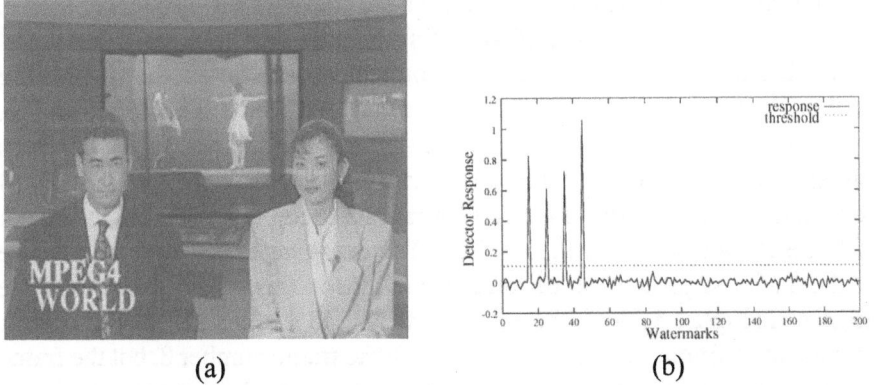

Figure 5.22. Watermark detection relative to the whole *News* frame: (a) frame of the sequence and (b) detector response.

the objects present in the scene and the detection phase must be applied to the whole frame.

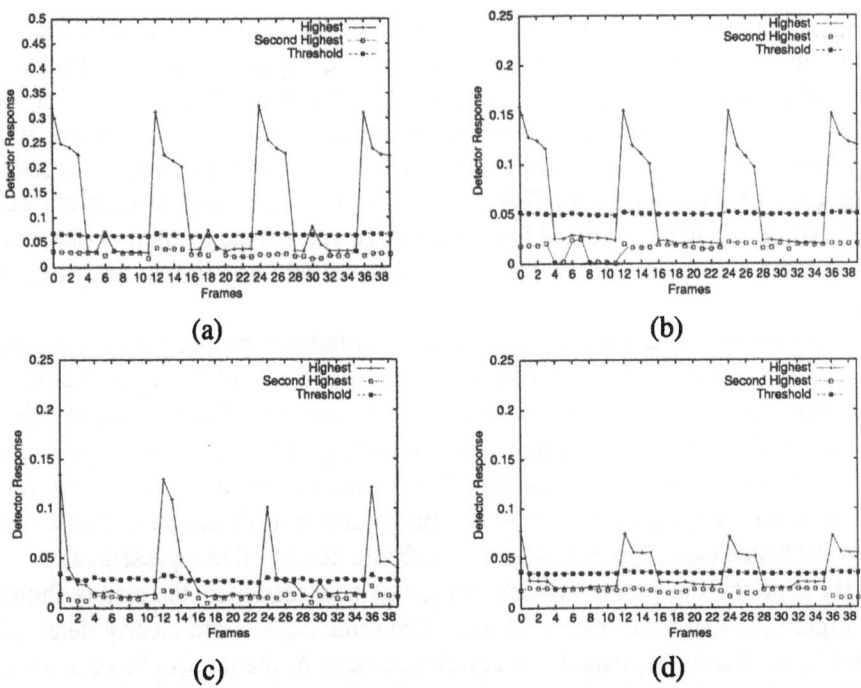

Figure 5.23. Detector response related to video object 0 (a), video object 1 (b), video object 2 (c) and video object 3 (d) of the *News* sequence for 40 frames (0 − 39); GOP equal to 4 frames and the watermark is embedded once every six frames.

According to the approach adopted for the previous test sequence, also in this case experiments on MPEG-4 coding bit rate reduction have been performed leading to results similar to the previous case. So another interesting test has been verified: the GOP has been reduced from 12 to 4 frames (i.e. frames number $0, 4, 8, 12, ..., 36$ are intra-frame) and the watermark has been embedded once every six frames (i.e. frames number $0, 6, 12, 18, 24, 30, 36$ are watermarked), besides the target bit rate has been left at 500 Kbit/sec per VOL; results are shown in Figure 5.23. The dragging effect explained before for the *Stefan* sequence can be found again also in this case (see for example the graph in Figure 5.23(a) where the trend for VO0 is reported); the watermark from the intra-frame number 0 is dragged till the frame number 3, but the frame number 4, which is an intra-frame, but not watermarked, does not contain the watermark because it is not time-correlated to the previous ones and does not suffer from the dragging effect. When the number of the watermarked frame is not a common multiple of the GOP value (e.g. frames number 6, 18 and 30), the watermarked frame is an inter-frame and there is not propagation of the watermark to the successive frames. As it happened for the *Stefan* sequence, the watermark inserted in the inter-frame is weaker than that introduced in the intra-ones and even it becomes undetectable like for VO1 and VO3.

6.4.2   Detection on MPEG-4 modified sequences.

In this section experimental results, specifically devoted to investigating if the proposed watermarking technology is really feasible for MPEG-4 applications and if the adopted approach effectively allows to directly interact with video objects, will be presented. These tests have been considered to study the main characteristic the MPEG-4 standard offers, that is the possibility to deal with objects. Video objects can be extracted from a sequence and moved to another one, thus composing a completely new video sequence without any link to the previous one.

The first experimental test that has been carried out, properly considers this aspect. In fact the video object 0 (the male speaker), belonging to the *News* sequence, watermarked as before with the mark number 15 and the video object 1 (the tennis player), belonging to the *Stefan* sequence, watermarked with the mark number 70, have been transferred over a new background named *Flowers* which has been watermarked with the mark number 100: this results in a very new sequence (see Figure 5.24(a) where frame number 0 is represented).

By applying the watermark detector to this new sequence, the results shown in Figure 5.24(b) have been obtained. The three objects are clearly detected, even if, as already explained, the correlation peak of the tennis player is lower with respect to the peak of the speaker and to the peak of the background.

In this first test the video objects coming from *News* and from *Stefan* sequences have been placed at their original location with respect to the original

Figure 5.24. Watermark detection relative to the new generated sequence composed of the VO 0 of *News* watermark no. 15, the VO 1 of *Stefan* watermark no. 70 and the background of *Flowers* watermark no. 100: (a) Frame number 0 of the sequence and (b) The detector response.

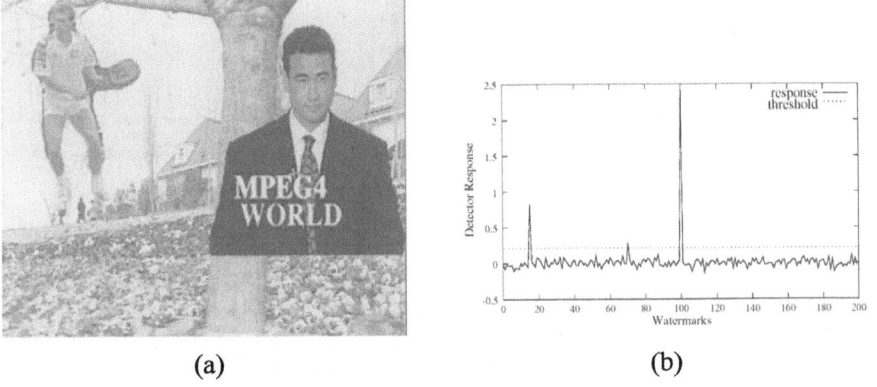

Figure 5.25. Watermark detection relative to the new generated sequence composed by the VO 0 of *News*, the VO 1 of *Stefan* and the background of *Flowers*, the speaker and the tennis player having in this case been translated with respect to their original positions: (a) Frame number 0 of the sequence and (b) The detector response.

sequences, but if they were translated at different locations of the new frame, as visualized in Figure 5.25(a) (all the VOPs have been rolled of the same amount), the watermarks that are present on the whole frame can still be detected (see Figure 5.25(b)). As explained in Section 6.3, to correctly extract the watermarks of the tennis-player and of the speaker, the approach based on the 2-D DFT has been used to recognize the exact matching between the mark and the translated video-object. It can be noted that the values of the correlation peaks for the watermark no.15 and no.70 are obviously the same in both Figures 5.24(b) and 5.25(b) but the peak for the background is different because diverse regions are covered by the two video-objects in foreground and so different watermarked coefficients take part to the correlation computation.

6.4.3   Detection on MPEG-2 trans-coded sequences.

In this section another kind of experimental tests is presented. This time the robustness of the proposed approach against MPEG-2 coding with different bit rates has been checked. Each of the four video objects of the *News* sequence has been watermarked as before, embedding the watermark once every six frames and then the watermarked frames have been MPEG-2 coded with different bit rates (GOP is 12 frames again). In Figure 5.26 the detector response for each video object with diverse bit rates in a decreasing order (starting with 1500 Kbit/sec (a) down to 600 Kbit/sec (d)) are illustrated.

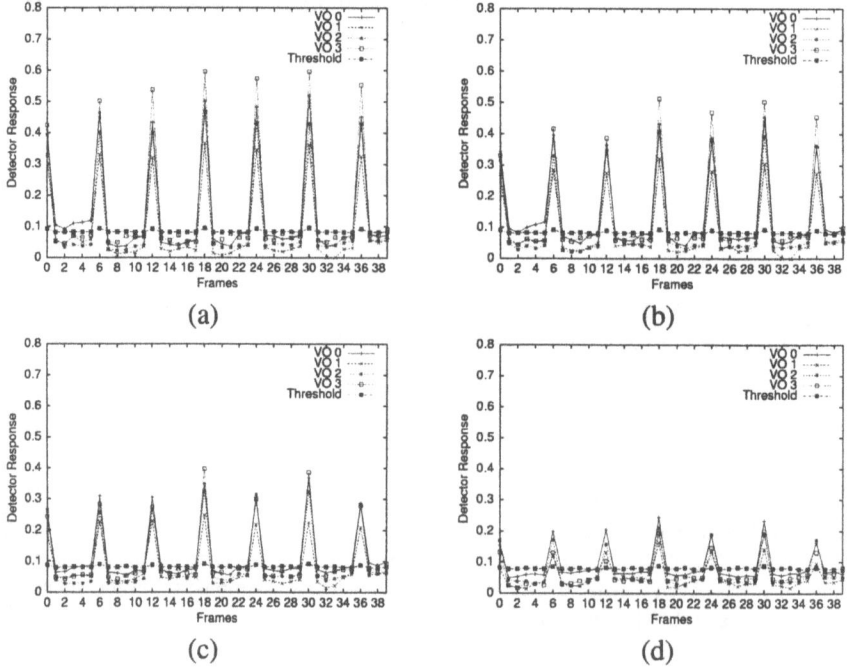

Figure 5.26. Detector response related to each of the four video objects of the *News* sequence for 40 frames (0 – 39), with different MPEG-2 coding bit rate: 1500 Kbit/sec (a), 1200 Kbit/sec (b), 900 Kbit/sec (c) and 600 Kbit/sec (d). GOP equal to 12 frames and the watermark is embedded once every six frames.

In this case, looking at the graphs presented in Figure 5.26, it can be realized that the adopted approach is robust also against MPEG-2 coding, in fact all the four video objects are detected in the watermarked frame. By decreasing the MPEG-2 coding bit rate, the detector responses get lower, though till to 600 Kbit/sec (Figure 5.26(d)) they are always over the threshold; when the compression is pushed to 300 Kbit/sec the watermark is not revealable anymore in all the video sequence. It is interesting to notice that for the frames number 6, 18, 30, though they are inter-frames, the watermark is well detected as for the intra-frames; this can be explained because these frames are predicted, so they receive part of watermark information from previous watermarked frames, and are also watermarked themselves, so a sort of *folding effect* happens and the detector response is sometimes higher with respect to that for the intra-frames.

7. Conclusions

The future development of networked multimedia services is conditioned by the achievement of efficient methods to safeguard data owners against non-authorised copying and redistribution of the delivered material, to grant that the IPR are well respected and the assets properly managed.

In the framework of digital watermarking for the copyright protection of multimedia data, the specific case of video watermarking has been analyzed and specific issues regarding it detailed. In particular three different approaches for embedding a code in the video domain have been deeply described: two of them work in the non-coded video domain, while the other one inserts the mark directly in the coded domain. Advantages and drawbacks of the different systems, either from the point of view of robustness, or from the point of view of computational complexity, have been highlighted. A particular attention has been focused on watermarking of MPEG-4 video sequences, since this standard seems to be very attractive for a large range of novel applications, such as internet video distribution and wireless video communications.

In the present chapter three algorithms have been presented. The first and the third ones, that were presented in Sections 3 and 6 respectively, deal with raw video and offer the chance to directly interact with frames, so this allows to better exploit the video sequence features, for example for watermark masking purposes, and to grant a superior robustness against usual attacks; specifically the third system (Section 6) also grants the advantage to work with video-objects according to MPEG-4 indications. On the contrary, the second technique (Section 5) takes into consideration codified video and, because it operates straightly with video streams, does not need coding-decoding operations to succeed in embedding and extracting the watermarks, thus resulting in a reduced computational complexity.

Acknowledgments

The research described in this chapter has been performed in the framework of the COST 254 Action and COST 276 Action, of the Project "Image Sequence Analysis and Processing Techniques for Indexed Video Watermarking and Quality Evaluation", Grant no. 9909117848_004, funded by the Italian

Ministry of Research (MURST), of the EC funded IST-1999-21031 TRADEX Project, and during a cooperation with the Italian National Broadcasting Company (RAI).

References

[1] M. Barni, F. Bartolini, and A. Piva, "Digital watermarking of visual data: State of the art and new trends," in *Proceedings of European Signal Processing Conference (EUSIPCO 2000)*, (Tampere, Finland), pp. 1657–1664, Sept. 5-8 2000.

[2] R. Caldelli, M. Barni, F. Bartolini, and A. Piva, "A robust frame-based technique for video watermarking," in *Proceedings of EUSIPCO 2000*, (Tampere, Finland), pp. 1037–1040, 4-8 Sept. 2000.

[3] M. Barni, F. Bartolini, V. Cappellini, and N. Checcacci, "Object watermarking for MPEG-4 video streams copyright protection," in *Security and Watermarking of Multimedia Contents II, Wong, Delp, Editors, Proceedings of SPIE Vol. 3671*, (S. Jose', CA), pp. 465–476, Jan. 2000.

[4] R. Caldelli, M. Barni, F. Bartolini, V. Cappellini, and A. Piva, "Digital video watermarking for MPEG-4 applications," in *International Conference on Media Futures*, (Florence, Italy), pp. 249–252, 8-9 May 2001.

[5] F. Hartung and B. Girod, "Digital watermarking of MPEG-2 coded video in the bitstream domain," in *Proc. Internat. Conf. on Acoustic, Speech, & Signal Processing (ICASSP'97)*, vol. 4, (Munich, Germany), pp. 2621–2624, Apr. 1997.

[6] G. C. Langelaar, R. L. Lagendijk, and J. Biemond, "Watermarking by DCT coefficient removal: A statistical approach to optimal parameter settings," in *Security and Watermarking of Multimedia Contents* (P. W. Wong and E. J. Delp, eds.), Proc. SPIE 3657, (San Jose, California), pp. 2–13, Jan. 1999.

[7] C. Hsu and J. Wu, "Digital watermarking for video," in *Proc. IEEE Intern. Conf. on Digital Signal Processing*, vol. 1, (Santorini, Greece), pp. 217–220, Jul. 1997.

[8] M. D. Swanson, M. Kobayashi, and A. H. Tewfik, "Multimedia data-embedding and watermarking technologies," *Proc. of the IEEE*, vol. 86, pp. 1064–1087, Jun. 1998.

[9] T. Kalker, G. Depovere, J. Haitsma, and M. Maes, "A video watermarking system for broadcast monitoring," in *Security and Watermarking of Multimedia Contents* (P. W. Wong and E. J. Delp, eds.), Proc. SPIE 3657, (San Jose, California), pp. 103–112, Jan. 1999.

[10] F. Hartung and B. Girod, "Digital watermarking of raw and compressed video," in *Digital Compression Technologies and Systems for Video Communications, SPIE Proceedings Series*, vol. 2952, pp. 205–213, Oct. 1996.

[11] F. Deguillaume, G. Csurca, J. O'Ruanaidh, and T. Pun, "Robust 3D DFT video watermarking," in *Security and Watermarking of Multimedia Contents, Proc. SPIE Vol. 3657*, (S.Jose, CA), pp. 113–124, Jan. 1999.

[12] A. Piva, M. Barni, F. Bartolini, V. Cappellini, A. D. Rosa, and M. Orlandi, "Improving DFT watermarking robustness through optimum detection and synchronisation," in *Proc. of ACM Workshop on Multimedia and Security '99*, (Orlando, Florida, USA), pp. 65–69, Oct. 1999.

[13] M. Holliman, W. Macy, and M. M. Yeung, "Robust frame-dependent video watermarking," in *Security and Watermarking of Multimedia Contents II* (P. W. Wong and E. J. Delp, eds.), vol. 3971 of *Proc. SPIE*, (San Jose', California), pp. 186–197, Jan. 2000.

[14] T. Sikora, "The MPEG-4 video standard verification model," *IEEE Trans. on Circuits and Systems for Video Technology*, vol. 7, Feb. 1997.

[15] J. Lacy, N. Rump, and P. Kudumakis, "MPEG-4 Intellectual Property Management and Protection (IPMP) Overview and applications," in *MPEG doc. ISO/IEC JTC1/SC29/WG11/N2614*, Dec. 1998.

[16] F. Hartung and F. Ramme, "Digital rights management and watermarking of multimedia content for M-commerce applications," *IEEE Communications Magazine*, vol. 38, pp. 78–84, Nov. 2000.

[17] E. Koch and J. Zhao, "Towards robust and hidden image copyright labeling," in *Proc. NSIP'97, IEEE Workshop on Nonlinear Signal and Image Processing*, (Neos, Marmaras, Halkidiki, Greece), pp. 452–455, Jun. 1995.

[18] C. Hsu and J. Wu, "Hidden signatures in images," in *Proc. IEEE Internat. Conf. Image Processing '96*, (Lausanne, Switzerland), pp. 223–226, Sept. 16-19 1996.

[19] A. Piva, M. Barni, F. Bartolini, and V. Cappellini, "Application-driven requirements for digital watermarking technology," in *Europ. Multimedia Microproc. System and Electronic Commerce Conf. and Exhibition (EMMSEC98)*, (Bordeaux, France), pp. 513–520, Septt. 1998.

[20] J. Dittman, M. Stabenau, and R. Steinmetz, "Robust MPEG video watermarking technologies," in *Proceedings of the 6th ACM international conference on Multimedia*, (Bristol, UK), pp. 71–80, Sept. 1998.

[21] S. Lin and D. J. C. Jr, *Error Control Coding: Fundamentals and Applications*. Englewood Cliffs, NJ: Prentice Hall, 3rd ed., 1983.

[22] M. Barni, F. Bartolini, and A. Piva, "Improved wavelet-based watermarking trough pixel-wise masking," *IEEE Trans. on Image Processing*, vol. 10, pp. 783–791, May 2001.

[23] M. Vetterli and J. Kovacevic, *Wavelets and Subband Coding*. Englewood Cliffs, NJ: Prentice Hall, 1995.

[24] T. Kalker and A. J. E. M. Janssen, "Analysis of watermark detection using SPOMF," in *Proc. IEEE Internat. Conf. Image processing '99*, vol. 1, (Kobe, Japan), pp. 316–319, Oct. 24-28 1999.

[25] M. Maes, T. Kalker, J. Haitsma, and G. Depovere, "Exploiting shift invariance to obtain a high payload in digital image watermarking," in *Proc. IEEE Internat. Conf. Multimedia Computing and Systems (ICMCS'99)*, vol. I, (Florence, Italy), pp. 7–12, Jun. 1999.

III

VIDEO AND IMAGE CODING

Chapter 6

ERROR-RESILIENT CODING FOR MULTIMEDIA COMMUNICATIONS

Nikolaos V. Boulgouris, Nikolaos Thomos, and Michael G. Strintzis
Department of Electrical and Computer Engineering
Aristotle University of Thessaloniki, Greece
and
Informatics and Telematics Institute
Thermi-Thessaloniki, Greece
{nblg,nthomos,strintzi}@iti.gr

Abstract   Error-resilient techniques are proposed for the efficient transmission of still images and video over unreliable channels. The proposed techniques are applied in conjunction with multiresolution decomposition of images or video frames. The resulting coders produce scalable bitstreams which can deliver very good quality over a variety of different bandwidths. The error-resilient structure of the stream endows the proposed systems with the capability to localize and discard the corrupted portion of the transmitted information and thus, attain superior reconstruction quality. When combined with forward error correction, the resulting streams are shown to yield very good performance in terms of error resilience and reconstruction quality.

Keywords: Wireless transmission, error-resilient coding, wavelets, scalable video, layered coding, forward error correction.

1.   Introduction

The transmission of multimedia over today's heterogeneous and often unreliable networks has necessitated provision of protection of information against possible channel failures [1, 2]. Even though, in theory, under certain circumstances source and channel coding can be studied independently (according to the so-called "Shannon's separation principle" [3]), channel coding strategies which take into consideration the structure of the underlying source coder produce significantly better performance [1]. Error-resilient and error-correction

J.F. Tasič et al. (eds.), Intelligent Integrated Media Communication Techniques, 197-220.
© 2003 *Kluwer Academic Publishers.*

techniques are thus becoming increasingly popular. This chapter describes methods for the reliable transmission of images and video over unreliable channels.

Robust transmission of images/video can be achieved using the following techniques:

- Error-resilient coding.

- Forward Error Correction.

- Retransmission using Automatic Repeat Request (ARQ) protocols.

Forward Error Correction involves the transmission of redundant bits that can be used to correct errors. *Error-resilient coding* aims to limit the impact of errors not corrected by Forward Error Correction. Finally, *ARQ protocols* are used to retransmit corrupted information.

Since in many transmission applications an ARQ mechanism may not be available, in the present work we focus our attention on the first two of the above classes of techniques. Although these techniques can be applied to streams generated using existing image/video coding standards, in this chapter we propose novel layered coding schemes which are much more suitable for robust transmission of images/video over unreliable networks. The combination of error-resilient coding and forward error correction will be seen to result in very efficient transmission of multimedia over unreliable channels.

Apart from being very efficient in terms of error-resilience, the proposed coding techniques have the valuable additional convenience of producing scalable bitstreams. The transmission of appropriate portions of the same coded streams allows the transmission of the same image/video sequence over a variety of channels having different bandwidths.

A block diagram of the multimedia communication system proposed in this chapter is depicted in Fig. 6.1. Techniques are proposed for the layered coding of images/video, as well as Forward Error Correction (FEC) techniques for the appropriate protection of layers. The layers are formed in the wavelet domain whereas FEC is based on Rate Compatible Punctured Convolutional codes [4].

This chapter is arranged as follows. Section 2 describes a general framework for the coding and organization of visual information in layers so that the resulting streams are endowed with error-resilience capabilities. Section 3 investigates the efficient transmission of images and video over noisy channels using Forward Error Correction. Section 4 describes the error handling process carried out by the decoder when a corrupted stream is processed. Experimental evaluation of the proposed schemes for the transmission of images and video over noisy channels is carried out in Section 5. Conclusions are drawn in Section 6. Finally, Rate Compatible Punctured Convolutional codes and List-Viterbi decoding are briefly described in Appendices A and B respectively.

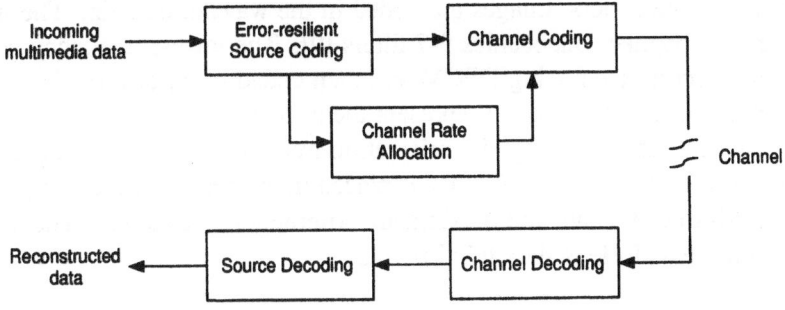

Figure 6.1. Generic multimedia communication system based on layered coding.

2. Error-Resilient Layered Coding of Images and Video

The efficiency of most state-of-the-art coders for images and video is heavily dependent on the deployment of Adaptive Arithmetic codes [5] or other Variable Length Codes (VLC). Despite their indisputable efficiency, such codes produce bitstreams that are very sensitive to bit errors. For this reason, even slight modifications of a compressed image/video stream usually derail the decoding process and render the stream undecodable.

In this section a layered coding methodology for efficient compression of images and video is proposed. Information is coded in such a way so that the resulting stream is error-resilient and thus, very suitable for transmission over unreliable channels. This is achieved using a multiresolutional (wavelet) decomposition for spatial decorrelation of images and motion compensation residuals, and by dividing their wavelet representation into blocks that are independently coded. The block-based approach for the coding of wavelet coefficients resembles the one employed in the recently finalized JPEG2000 image coding standard [6] and attains significant advantages in terms of error-resilience when compared to zerotree-based [7, 8] approaches.

In the specific case of video transmission over unreliable channels, additional problems are caused by the use of predictive coding which causes error propagation from one frame to the other via the motion compensation process. Therefore, in addition to individual block coding of residual frames, further error resilience is achieved by temporal decoupling of information, i.e. by individual coding of each motion compensated residual frame. The layered systems for the error-resilient coding of images and video are described in the sequel.

2.1 Layered Image Coding

The block diagram of the proposed layered image coding scheme is shown in Fig. 6.2. In order to maintain high coding performance as compared to state-of-the-art image coders, images are coded in the wavelet domain. The filters used are the popular Daubechies 9-7 filters [9]. The lowest frequency coefficients are decorrelated using DPCM and then coded using arithmetic coding. The number of bits required to represent the maximum absolute coefficient in each subband and in each spatial orientation tree, originating from the lowest frequency subband (see Fig. 6.4(a)), is placed in the header of the compressed stream. All tree and subband maxima are arithmetically encoded. The header of the compressed file is shown in Fig. 6.3.

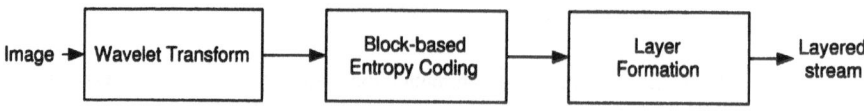

Figure 6.2.   Proposed layered coding scheme.

Dimensions	Levels	DC coefficients	Tree Maxima	Subband Maxima

Figure 6.3.   Header of the compressed file.

The subbands in the two highest resolution levels of the image are further split into blocks of dimensions 32 x 32 pixels, see Fig. 6.4(b). The transmission of information for each subband/block is done in a bitplane-wise manner starting from the most significant bit to the least significant bit. For each subband/block, first the coefficients whose most significant bit (MSB) lie in the bitplane currently coded are identified by comparing them to a threshold $T = 2^n$ where n is the index of the bitplane that is being coded. If a coefficient becomes significant (i.e. its MSB lies in the current bitplane) then its sign is also coded. This process is often called *significance identification* and the layer containing this information is denoted by $\mathcal{L}_{nk}^s$ [7] where k is the block index and the superscript s stands for "significance". Similarly, the *refinement* layer containing the nth bitplane of coefficients found significant in previous passes is denoted by $\mathcal{L}_{nk}^r$, where the superscript r stands for "refinement".

To achieve a more robust bitstream, appropriate for transmission over noisy channels, the arithmetic coder is initialized before the transmission of each layer. A context modelling strategy like the one in [10] was implemented for the coding of insignificant coefficients. The refinement layers, as well as the signs of the coefficients, are entropy coded using a single adaptive arithmetic model.

Information is coded about a coefficient if and only if the maximum coefficient in the tree is greater than or equal to the current threshold and the maximum coefficient in the subband is greater than or equal to the current threshold. The deployment of the above rules reduces drastically the number of coefficients whose significance is tested during the coding of a significance identification layer. However, in order to further reduce the number of symbols that have to be coded during the layer coding stage, a single bit is initially coded for each block to indicate if all coefficients in the block are insignificant.

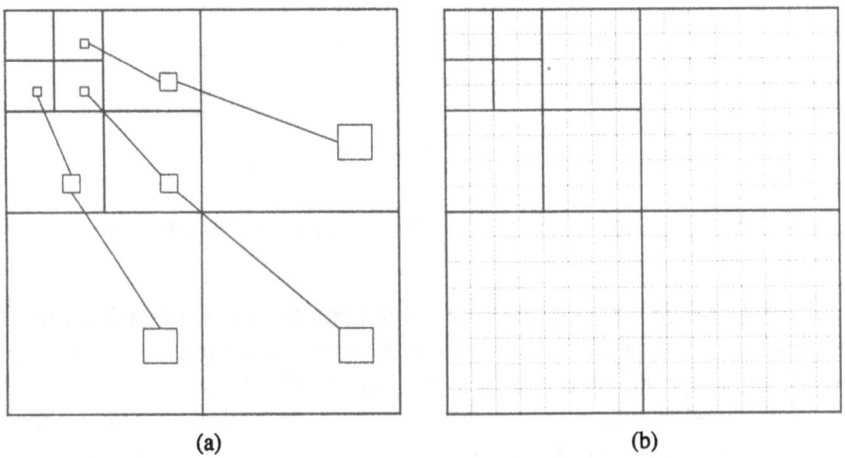

(a)   (b)

Figure 6.4.   The efficiency of the proposed scheme is based on the combination of a tree and a block approach.

The bitplanes of subbands are transmitted in a predefined order. The reason is that since the 9/7 biorthogonal filter bank is used for the transformation, all layers of the transformed images have approximately equal importance regardless of the resolution level in which they belong. The transmission order is depicted in Fig. 6.5. For each bitplane and for all subbands, the *significance identification* layers are transmitted first, followed by the *refinement* layers.

2.2   Layered Video Coding

Two main classes of scalable wavelet video coders are distinguished in recent literature. Coders in the first class apply motion compensation first and then encode the resulting motion compensation difference. Coders in the second class apply direct 3D wavelet transform on the data [11, 12]. Coders applying 3D transforms in all frames of a Group of Pictures (GOP) are not very suitable for transmission over noisy channels since uncorrectable errors can easily propagate in all frames of a GOP producing a very undesirable effect on the quality of the reconstructed frames. For this reason the coder presented in this

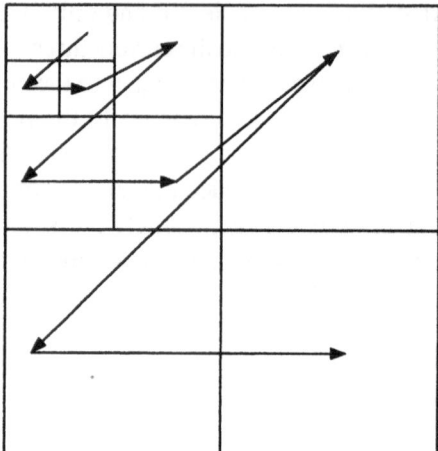

Figure 6.5.   Scanning order of bitplanes when the 9-7 biorthogonal filter bank is used.

chapter is based on the application of motion compensation of each frame and the subsequent independent coding of motion compensation residuals. Additionally, the proposed coder employs data partitioning and periodic arithmetic coder initialization which in conjunction with error correction codes endows the resulting scheme with the capability to prevent and correct bit errors.

The proposed error-resilient video coding scheme produces a *base* layer, which provides a basic level of quality, and a sequence of *enhancement* layers $\mathcal{L}$. The enhancement layers aim to incrementally improve the quality of the sequence transmitted at the base rate. In practical video coding applications, the transmitted enhancement rate depends on the available bandwidth. Due to the embedded nature of the bitstream produced using the proposed coder, each low rate is a prefix of a higher rate and thus, all rates can be extracted from the same coded file.

In the proposed encoding scheme the first frame in each GOP is intra-coded. The correlation between consecutive frames is subsequently removed using overlapped block motion compensation (OBMC) [13]. The original frames are used as reference frames for the calculation of motion vectors so that all available information is exploited during the estimation of the motion vector fields. The motion field is losslessly coded using the techniques in [14].

Using the previously estimated motion vector fields, the procedure for the generation of the base layer stream continues as follows: initially the first inter-frame is compensated using the decoded I-frame as reference. The prediction error is wavelet decomposed and quantized (see Q in Fig. 6.6) using embedded bitplane transmission. Then it is dequantized (Q^{-1} in Fig. 6.6), and, after the inverse wavelet transfrom, the reconstructed error frame is added to the

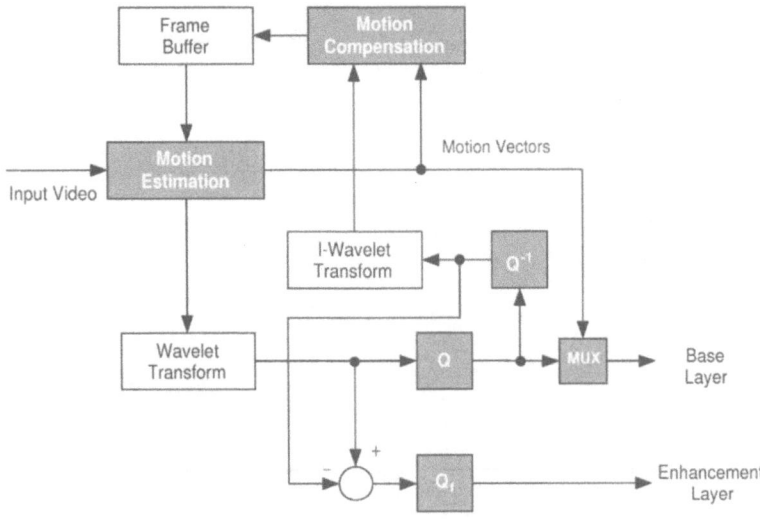

Figure 6.6. Block diagram of the source coder.

compensated image. The resulting reconstructed frame (rather than the original) will serve as the reference frame for the compensation of the next inter-frame. This method has been adopted to ensure that the motion compensation process performed at the encoder can be identically replicated at the decoder even if only the base layer is transmitted. Otherwise, errors would accumulate in the decoded video sequence causing a distortion usually termed drift [15].

The enhancement layers consist of bitplanes of wavelet coefficients which have *not* been transmitted during the transmission of the base layer. This operation can be seen as requantizing wavelet coefficients with a finer quantizer and is indicated as Q_f in Fig. 6.6. Hereafter, the bitstream corresponding to a coded bitplane of a block in a wavelet decomposed frame will be called an *ELementary Layer* (ELL). Thus, an enhancement layer is a succession of ELLs corresponding to all frames in a GOP. Such layers are schematically described in Fig. 6.7. Two kinds of ELLs may be distinguised: significance identification ELLs ($\mathcal{L}^s$) and refinement ELLs ($\mathcal{L}^r$) much in the same way as significance and refinement coding is differentiated in zerotree compression of still images.

The arithmetic coders employed by the proposed coder are initialized in the beginning of the coding of each layer so that layers corresponding to different frames are decoupled and uncorrectable bit errors in a layer of a specific frame do not affect the decoding of a layer of another frame. This technique prevents propagation of errors among frames in a GOP and makes the bitstream output by the proposed coder more resilient.

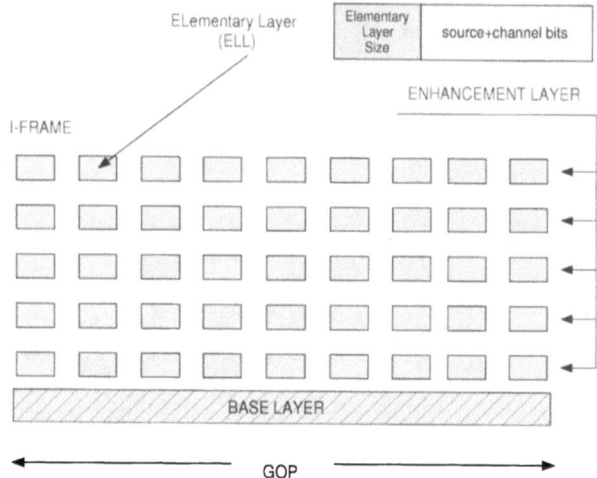

Figure 6.7.   Bitstream organization in a Group of Pictures. Bitplanes of wavelet coefficients for all frames are transmitted after the base layer.

The previously described error resilience tools that are used by the proposed system are based on data partitioning and do not lower the performance of the proposed scheme during error-free transmission. However, further protection of the coded bitstream can be achieved using Forward Error Correction (FEC). This will be discussed in the ensuing sections.

3.   Protection of Compressed Streams Using Forward Error Correction

The layered image/video streams produced as described in the previous section are coded using channel coding [4]. Since each bitplane of a block is coded without using information from other blocks, protection can be individually applied to each such block. A schematic description of the system used for the generation of robust streams is shown in Fig. 6.8.

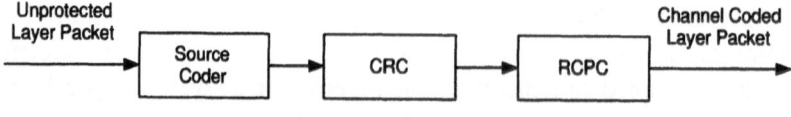

Figure 6.8.   Cascade of operations for the efficient protection of layers.

Specifically for images, header information, i.e. DC coefficients, tree and subband maxima are considered very important information and are highly protected. An error in the header would be catastrophic and would render the rest of the stream useless. For the same reason, the base layer of video sequences are highly protected.

Layers $\mathcal{L}_{nk}^s$ and $\mathcal{L}_{nk}^r$ which correspond to significance identification and refinement coding are channel coded. The basic structure for the application of forward error correction is depicted in Fig. 6.8. Each layer is independently protected by employing a field in its header which indicates the size of the set of source bits used for the coding of that layer. Another field in the header specifies the matrix with which the Rate Compatible Punctured Convolutional (RCPC) codes are punctured. This is very useful in cases where an entire layer has to be discarded (due to uncorrectable errors) since the length of the Source+Channel rate of the layer can be deduced at the decoder side and thus, the corrupted layer can be discarded without preventing subsequent layers (that do not depend on the discarded one) from being decoded correctly (see Fig. 6.9).

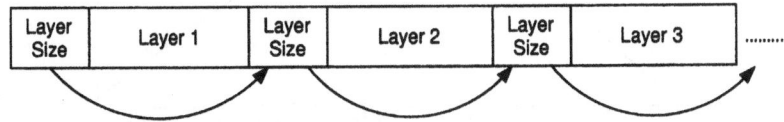

Figure 6.9. Bitstream structure. The beginning of each layer is a highly protected header indicating the size of the layer. If an uncorrected error occurs in a layer, the corrupted layer can be discarded and the decoding process can proceed with the next uncorrupted layer.

For its efficient protection, each layer $\mathcal{L}$ is partitioned into $N_p(\mathcal{L})$ packets of equal size (apart form the last packet which may be shorter) and protected using the coder shown in Fig 6.8. This is shown in Fig. 6.10. Note that the (non-constant) size of the last packet in a layer can be implicitly calculated from the size of the layer and the puncturing matrix identifier (both are stored in the layer header). Thus, no other side information is needed for its coding and decoding.

4. Error Handling

A significant feature of a robust coder is its ability to detect and confine errors not corrected by the channel code. In our coder, due to the bitstream generation and organization strategy described in the previous sections, errors not corrected by the channel code, affect usually *only* the packet in which the error occurred and occasionally a few subsequent packets.

For the detection of errors, Cyclic Redundancy Codes (CRC) [16] are employed in conjunction with the RCPC codes [4] described in detail in Appendix

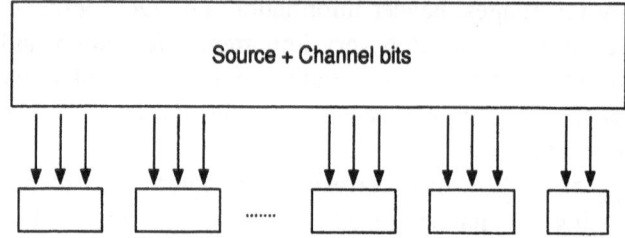

Figure 6.10. Organization of information in a robust layer. The layer is divided into packets of equal size (apart from the last packet) and each packet is protected separately.

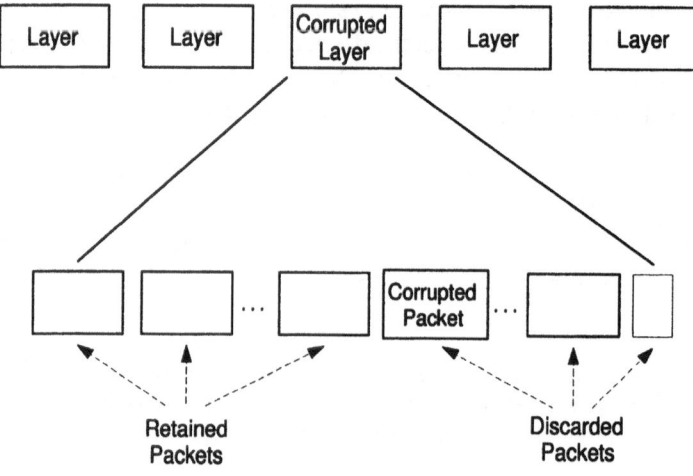

Figure 6.11. Packet disposal as performed by the proposed coder in case of uncorrectable errors.

A. For the efficient correction of errors, the serial List-Viterbi algorithm (LVA) [17] was used with a list of 100 paths. When the LVA is used, the optimal path in the Viterbi decoding is chosen among the paths that follow the constraints imposed by the CRC [18] (see Appendix B).

Alternately, in the highly unlikely case in which an uncorrectable error is not detected by the CRC check, the detection of errors can be performed using another mechanism. Since each layer is decoded independently and the size of the layer is a priori known to the decoder, the occurrence of an uncorrectable error can be easily detected due to loss in the arithmetic code resynchronization. This means that the arithmetic decoder attempts to decode symbols beyond the anticipated end of the layer, disclosing the existence of an uncorrectable error.

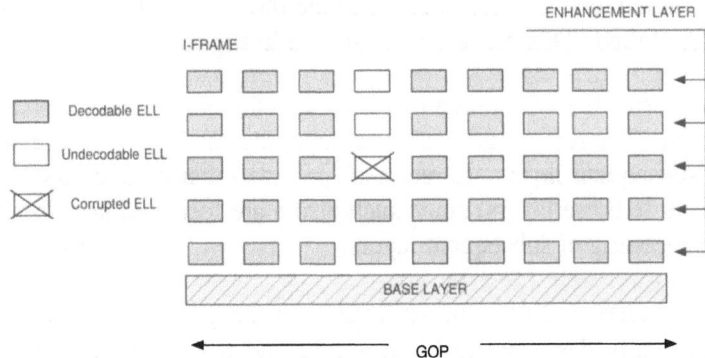

Figure 6.12. Error handling when an uncorrectable error occurs in an enhancement layer of transmitted video streams.

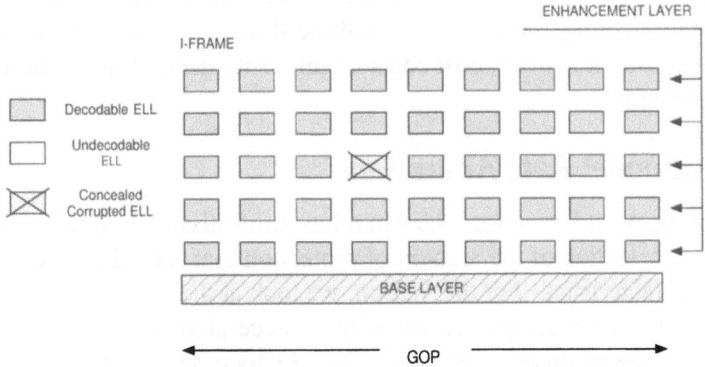

Figure 6.13. Error handling when an uncorrected error occurs in a refinement layer of transmitted video streams.

The reader who is interested in gaining additional insight to issues concerning the resynchronization properties of arithmetic codes is referred to [19].

The detection of an uncorrectable error during decoding triggers the following actions.

- If the error is in the image header or in the base layer of the video, the decoding stops. However, this likelihood is almost entirely negligible since header and base layer data are strongly protected.

- If the error is in layer $\mathcal{L}_{nk}^s$, then this layer is retained up to the first corrupted packet and all subsequent layers $\mathcal{L}_{jm}^s$, $\mathcal{L}_{jm}^r$, $j < n$, $k = m$ for the same block are discarded since the information they contain can not be exploited. This process is illustrated in Fig. 6.11.

- If the error is in $\mathcal{L}^r_{nk}$, then this layer is retained up to the first corrupted packet. The rest of the packets comprising the layer are discarded, but all subsequent layers are retained (provided that no uncorrectable error occurs in those layers) since such errors are localized and do not affect the decoding of subsequent layers.

The aforementioned rules in application to the general video coding case are schematically described in Figures 6.12 and 6.13. Apart from errors in the base layer, all other errors can be localized and the corrupted portion of the bitstream can be discarded. The ability of our robust coding methodology to discard corrupted portions of the bitstream in order to confine errors and achieve the best possible reconstruction quality endows the proposed scheme with the capability of achieving superior performance in comparison to non error-resilient coding methods. This will be shown in the experimental results section where a family of robust coders, built using the techniques described so far, are evaluated.

5.   Channel Rate Allocation

One of the most important issues in robust transmission of information is the appropriate selection of the amount of protection that should be used given the channel characteristics. For this reason, we look at the image/video streams as an initial very important part followed by a succession of layers (see Fig.6.14) that improve the reconstruction quality but are not essential for the decoding of the stream. In case of image transmission the integrity of the image header is highly important whereas for video the base layer is required to be transmitted errorlessly. Such very important data should be highly protected.

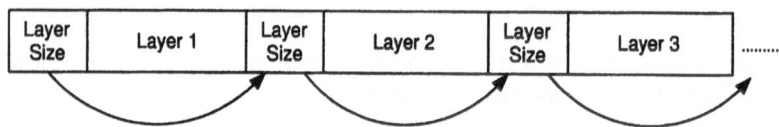

Figure 6.14.   Bitstream organization. The zeroth layer is strongly protected ensuring error-free reception. The subsequent layers are protected according to the rate allocation algorithm.

In order to allocate source and channel bits we first note that each additional portion of the bitstream that is made available to the decoder reduces the distortion between the original and the reconstructed image. Thus, the problem can be described as that of maximization of the distortion decrease D achieved when bitplanes from $Q(k)$ to $N(k)$ for each block k are transmitted

$$D = \sum_{k=1}^{M} D_k = \sum_{k=1}^{M} \sum_{i=Q(k)}^{N(k)} D(i, k) \qquad (6.1)$$

where D_k is the distortion decrease for block k, M is the number of blocks of wavelet coefficients (some blocks may be as large as an entire subband), $N(k)$ is the number of non-zero bitplanes in the kth block and $D(n, k)$, denotes the distortion reduction achieved by the transmission of bitplanes $n, \ldots, N(k)$ of the kth block. Finally, if $Q(k)$ is the bitplane at which transmission stops for block k, the average distortion decrease caused by significance layers for the kth block is:

$$D(n, k) = (1 - P(n, k)) \sum_{m=n}^{N(k)} D_{mk} \qquad (6.2)$$

where D_{nk} denotes the individual distortion decrease caused by layers $\mathcal{L}_{nk}$ and $1 - P(n, k)$ denotes the probability that *only* layers $\mathcal{L}_{mk}, m = n, \ldots, N(k)$ are correctly decoded. Since the decoding of a layer is possible only if *all* previous (more significant) layers have been decoded correctly, this probability is equal to

$$1 - P(n, k) = P_{n-1}(r_{n-1,k}) \prod_{m=n}^{N(k)} (1 - P_{mk}(r_{mk})) \qquad (6.3)$$

where $P_{nk}(r_{nk})$ denotes the *individual* probability that a significant layer is not decoded correctly (i.e. supposing all layers it depends on are correctly decoded) and r_{nk} denotes the channel code rate used for its coding.

For the formulation of an efficient rate-allocation algorithm, each layer $\mathcal{L}$ is divided into $N_p(\mathcal{L})$ constant-length packets and each packet is individually protected. The probability that a layer will be discarded is equal to the probability that at least one packet in this layer will be plagued by uncorrectable errors. If p is the probability that a packet is corrupted, then the probability of l corrupted packets among the $N_p(\mathcal{L})$ packets that comprise a layer $\mathcal{L}$ coded using channel code rate r is

$$P(l, N_p(\mathcal{L}), r) = \binom{N_p(\mathcal{L})}{l} p^l (1 - p)^{N_p(\mathcal{L}) - l} \qquad (6.4)$$

and, therefore, the probability of a layer error (of the existence of at least one packet in the layer in error) is given by the expression

$$P(r) = \sum_{l=1}^{N_p(\mathcal{L})} P(l, N_p(\mathcal{L}), r) \qquad (6.5)$$

Since the probability of the occurrence an of uncorrectable packet depends on the code used, this probability is experimentally evaluated for the set of channel codes used.

Provided the channel condition is known, the error probability P can be easily calculated for each layer, and optimal selection of the code rates r is possible by maximization of (6.1), subject to a rate constraint, using exhaustive search or dynamic programming techniques.

6. Multimedia Transmission over Noisy Channels

The proposed schemes were experimentally evaluated for the transmission of images and video over Binary Symmetric Channels (BSC). Direct application of the proposed scheme to the case of fading channels is possible if an interleaver is applied to spread out the protected bits before transmitting them over the communication channel. In this case the fading channel can be treated as a BSC. However, for fading channels which would require impractical interleavers in order to be treated as BSCs, Reed-Solomon codes [16] should be used instead of the RCPC codes that are used in the present work.

Comparison was based on the Peak-Signal-to-Noise-Ratio (PSNR) of the reconstructed images/frames for two channel conditions. Specifically, two BSCs were simulated with Bit Error Rates (BER) equal to 0.01 and 0.001 respectively. The PSNR was used as a measure of the reconstruction quality (in dB),

$$PSNR = 10 log_{10} \frac{255^2}{MSE}$$

where MSE is the Mean Squared Error between the original and the reconstructed images/frames. For each test, a thousand independent trials were simulated on a computer and the results were averaged.

The CRC codes used were taken from [20]. The family of RCPC codes that was used is based on a rate $1/4$, memory 6 mother code given by the generator tap matrix in Appendix A.

Four puncturing matrices were employed with rates $\{8/9, 8/10, 8/11, 8/12\}$. In most practical applications, for $BER \leq 10^{-2}$, puncturing with the above matrices is sufficient. Extending the set of available matrices would yield vanishingly negligible gain since the more appropriate protection would be outbalanced by the increase in the cost for the transmission of matrix indices.

6.1   Application to Image Transmission

The algorithms compared to the present Error Resilient Wavelet Image Coder (ERWIC) were those proposed by Sherwood [21] and Man [22]. The results are reported in Tables 6.1 and 6.2. Ten thousand MSE values were averaged and converted to PSNR for calculating the entries in the tables. As seen, for low

BERs ($\approx$ 0.001) the performance of all coders appears to be roughly equivalent. For higher BERs ($\approx$ 0.01) the performance of the coder proposed here is clearly superior to that in [21] and competitive to that in [23]. This is also demonstrated in Fig. 6.15. Reconstructed images for various channel BERs and rates are shown in Fig. 6.16.

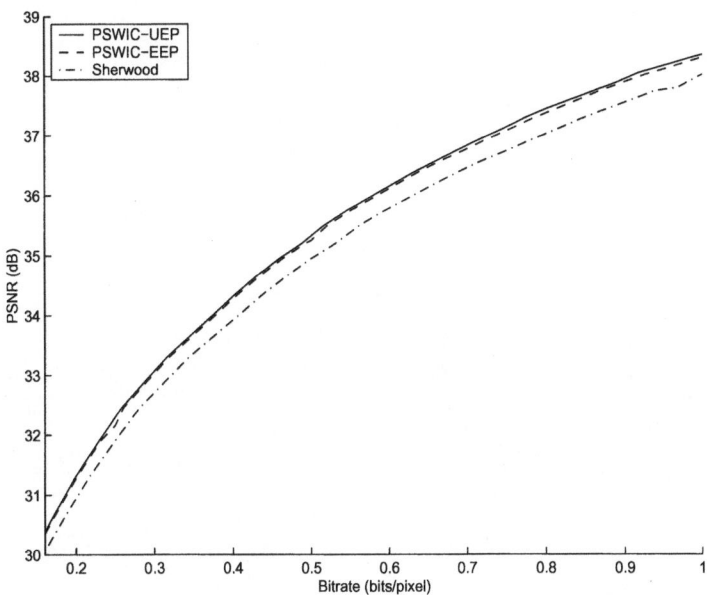

Figure 6.15. Progressive transmission of images over a noisy channel with BER=0.01.

	ERWIC		Sherwood
	EEP	UEP	EEP
0.25 bpp	32.14	**32.19**	31.91
0.5 bpp	35.23	**35.29**	34.96
1.00 bpp	38.28	**38.36**	38.03

Table 6.1. Comparison in terms of PSNR (in dB) of the proposed ERWIC coding scheme for the transmission of images over BSC with BER 0.01.

Another important feature of the ERWIC coder is the fact that even in the case of channel mismatch (i.e. if the channel is noisier than originally estimated) its performance degrades gracefully with the increase of the bit error probability. Conventional robust coders stop at the occurrence of the first uncorrectable

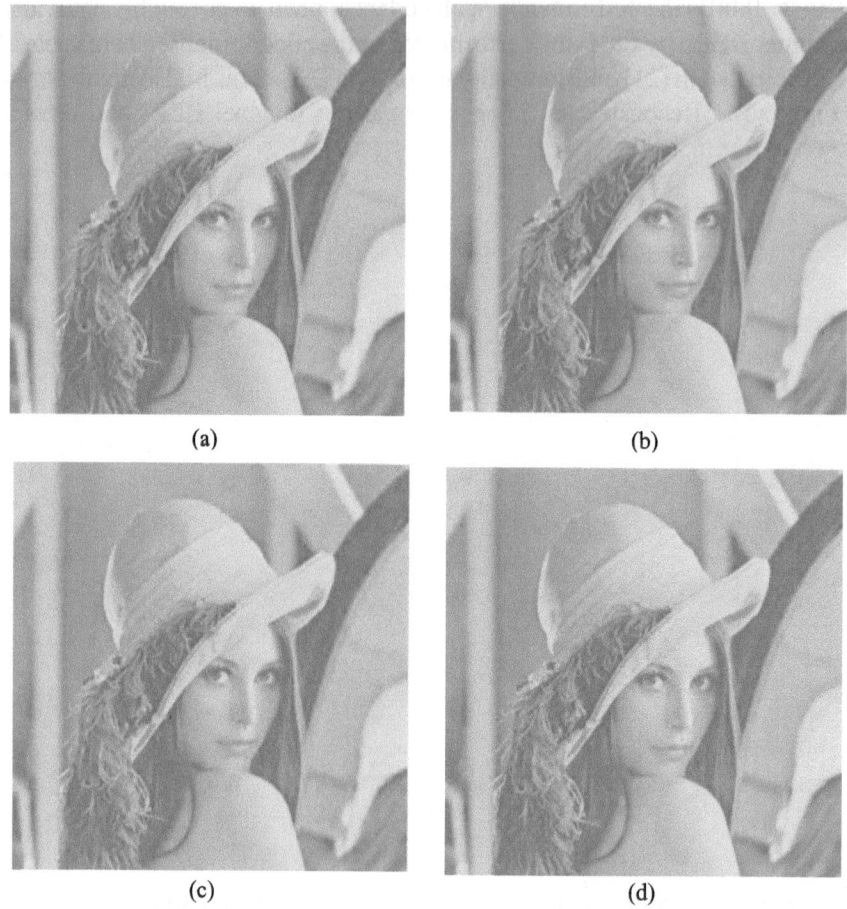

(a)   (b)

(c)   (d)

Figure 6.16. Reconstructed "Lena" when transmitted over noisy channels using the ERWIC algorithm (a) 0.25 bpp, BER=0.001 (33.10 dB). (b) 0.5 bpp, BER=0.001 (36.26 dB). (c) 0.25 bpp, BER=0.01 (32.19 dB). (d) 0.5 bpp, BER=0.01 (35.29 dB).

	ERWIC	Sherwood	Man
0.25 bpp	33.10	**33.16**	31.98
0.5 bpp	**36.26**	36.25	35.08
1.00 bpp	**39.38**	39.34	N/A

Table 6.2. Comparison in terms of PSNR (in dB) of the proposed coding schemes for the transmission of images over BSC with BER 0.001. Equal Error Protection was used with the proposed ERWIC scheme.

error thus resulting in significant abrupt performance decrease due to channel mismatch. Specifically, for the ERWIC coder, the reduction in its performance due to mismatch is much smaller. For example, the average reconstruction quality at 1 bit/pixel with the ERWIC coder is approximately 1.0 dB above the quality achieved by the coder in [24] which employs Embedded Block Coding with Optimized Truncation (EBCOT) source coding (better than Set Partitioning In hierarchical Trees, SPIHT) and the source coder used in the present paper and more powerful (and more complicated as well) turbo codes [25].

Since our source coder performs approximately as well as the SPIHT [8] coder, our superior results can be primarily attributed to the organization of the bitstream in such a way that enables error localization and decoding beyond the point of an uncorrectable error. This feature alone makes our EEP-based coder perform better than state-of-the-art coders based on unequal error protection. Additionally, the careful allocation of protection among layers makes the UEP variants of the proposed scheme even more efficient.

Taking into consideration the experimental results, we reach the conclusion that the bitstreams composed of independent layers appear to be generally more reliable for applications of image transmission over noisy channels. This is due to the ability of such streams to discard corrupted layers and decode the stream beyond the first uncorrectable error.

6.2 Application to Video Transmission

For application to video transmission, a three-level dyadic wavelet decomposition using the 9/7 biorthogonal filter was performed for each frame. The standard "football" test sequence (144 frames, 30 frames/second) was compressed at a variety of rates, ranging from 1 to 6 Mbps. The results for source coding are shown in Fig. 6.17.

Two error-resilient coders were implemented, one applying Equal Error Protection (EEP) to all layers and another which applies different levels of protection to the base layer and the enhancement layers. The concatenated RCPC/CRC codes described in the previous section were used with both schemes. The RCPC codes were generated using a memory-6 mother code (see Fig. 6.A.1).

Rate (Mbps)	Noiseless unprotected	$\epsilon = 0.001$ EEP	$\epsilon = 0.01$ EEP	$\epsilon = 0.01$ UEP
1.00	27.15	26.45	25.85	25.83
2.00	30.23	29.12	28.18	28.24
4.00	35.36	33.46	31.85	32.01
6.00	40.03	37.37	34.92	35.16

Table 6.3. Average reconstruction quality PSNR (in dB) on 144 frames of the "Football" sequence.

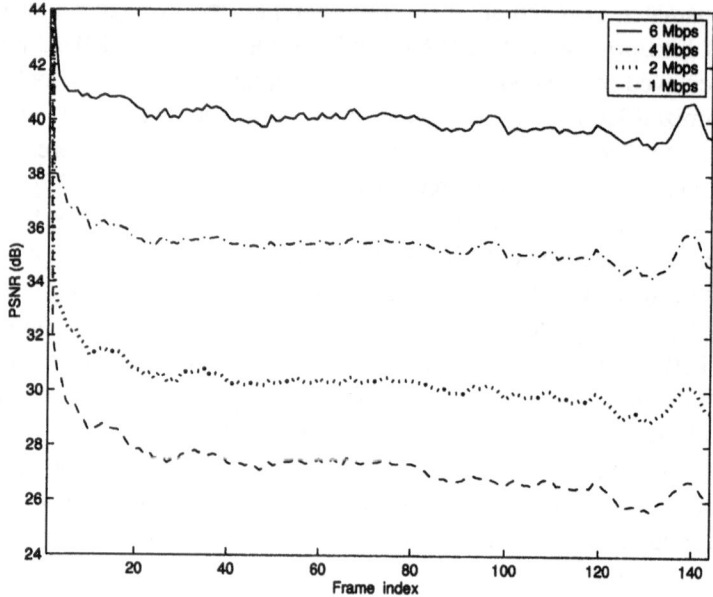

Figure 6.17. Reconstruction quality (source coding only) for 144 frames of the "Football" sequence.

For Bit Error Rate $\epsilon = 0.01$, a rate $\frac{8}{12}$ puncturing matrix was used for the protection of the entire stream in the EEP case whereas in the case of Unequal Error Protection (UEP), the significance and refinement enhancement layers were protected using puncturing matrices with rates determined using the optimization algorithm of Section 5. For $\epsilon = 0.001$, only EEP using very moderate protection was employed given by a rate $\frac{8}{9}$ puncturing matrix. All puncturing matrices are based on the puncturing tables of [4].

Results for the cases mentioned above are reported in Table 1. As shown, the schemes applying UEP performed better than those applying EEP. This gain is more pronounced at higher rates, since the lower rates (base layer) receive equal amounts of protection both in the EEP and the UEP cases. Additionally, due to the underlying highly scalable source coder, the reconstruction quality degrades gracefully when the Bit Error Rate increases.

Reconstructed frames with no protection, base layer protection and full protection are compared in Fig. 6.18. As can be seen, even when Forward Error Correction is not applied at all for some portions of the bitstream, the underlying error-resilient coding scheme is able to decode the frames at a satisfactory level of quality. If full protection is applied, the coder delivers excellent reconstruction quality (Fig. 6.18e).

(a)

(b)

(c)

(d)

(e)

Figure 6.18. "Football" sequence: (a) Original frame. (b) Decompressed frame (source coding/decoding only). (c) Corrupted frame in both the base and the enhancement layer (only motion vectors protected). (d) Corrupted frame in the enhancement layer (only base layer protected). (e) Fully protected frame.

7. Conclusions

In this chapter, error-resilient techniques have been proposed for the efficient transmission of still images and video over unreliable channels. The proposed techniques are applied with multiresolution decomposition of images or video frames. The resulting coders produce scalable bitstreams which can deliver very good quality over a variety of different bandwidths. The error-resilient structure of the stream endows the proposed systems with the capability to localize and discard the corrupted portion of the transmitted information and thus, attain superior reconstruction quality. When combined with forward error correction, the resulting streams are shown to yield very good performance in terms of error resilience and reconstruction quality.

Appendix: Rate Compatible Punctured Convolutional Codes

One of the most commonly used methods in digital communication systems in order to add redundancy for error correction and detection [16] are the Rate Compatible Punctured Convolutional codes [4] ($RCPC$) which are obtained by puncturing a conventional convolutional code.

$RCPC$ codes are described by a memory M mother code of rate $R_c = 1/N$, where N is the number of output bit sequences. Like block codes, an $RCPC$ code is defined by a generator matrix using the formula

$$\mathbf{g} = \{g_{ij}\}, \text{ where } i \in [0,1], j \in [0,(M-1)], \text{ and } g_{ij} \in [0,1] \qquad (6.A.1)$$

where $g_{ij} = 1$ implies connection of register i with output j. The puncturing matrix $\mathbf{a}$ with period P is described as

$$\mathbf{a}(l) = \{a_{ij}\}, \text{ where } i \in [0,(n-1)], j \in [0,(P-1)], \text{ and } a_{ij} \in [0,1] \qquad (6.A.2)$$

where $a_{i,j} = 1$ implies that the corresponding output bit is punctured, otherwise it is retained. The rates of the $RCPC$ code family are determined by the puncturing period P, the output bit streams N and are described as follows

$$R = \frac{P}{P+l}, \ l = 1, ..., (N-1)P \qquad (6.A.3)$$

The derived code rates are between $1/N$ and $P/(P+1)$, where the denominator equals the number of non-zero elements of the puncturing matrix.

The generator matrix of the convolutional encoder used in this chapter is shown below. The implementation of the convolutional encoder is depicted in Fig. 6.A.1.

$$\mathbf{g} = \{g_{ik}\} = \begin{bmatrix} 1 & 1 & 0 & 1 & 1 & 0 & 1 \\ 1 & 0 & 1 & 0 & 0 & 1 & 1 \\ 1 & 0 & 1 & 1 & 1 & 1 & 1 \\ 1 & 1 & 0 & 0 & 1 & 1 & 1 \end{bmatrix}$$

The output of the encoder is punctured (i.e. certain code bits are not transmitted) using the puncturing matrices. The puncturing matrices change the code rate and hence the correction power of the code according to source and channel needs.

Appendix: List-Viterbi Decoding of Convolutional Codes

In the present work, the decoding of convolutional codes is performed using the List-Viterbi Algorithm (LVA) [17], which is a generalized version of the Viterbi Algorithm (VA). It is used in communications systems in conjunction with error detection codes such as Cyclic Redundancy Codes or Reed-Solomon codes and yields superior performance in comparison to that of the simple Viterbi algorithm. The LVA produces a rank-ordered list after a trellis search of the globally best candidate bit sequences. The two different versions of List-Viterbi algorithm are the serial and the parallel. Here, we describe the fast serial List-Viterbi algorithm.

The serial LVA finds the N most probable candidates, one at a time, beginning with the most likely path. This path, considered as the best, is extracted by the simple Viterbi algorithm. The algorithm computes the kth best candidate only if the previously found $k-1$ paths contain errors which are detected using the CRC check. Thus, if the best path appears to have an error, then the algorithm finds the second best path. The paths candidate to be the globally second best, called the locally second best paths, are defined as the ones which, after leaving from the best path

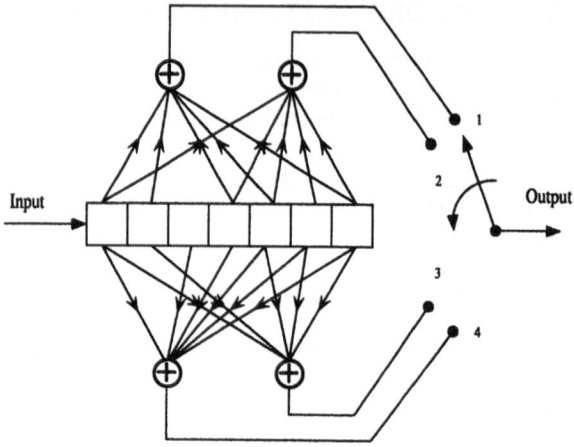

Figure 6.A.1.   Memory-6 convolutional encoder.

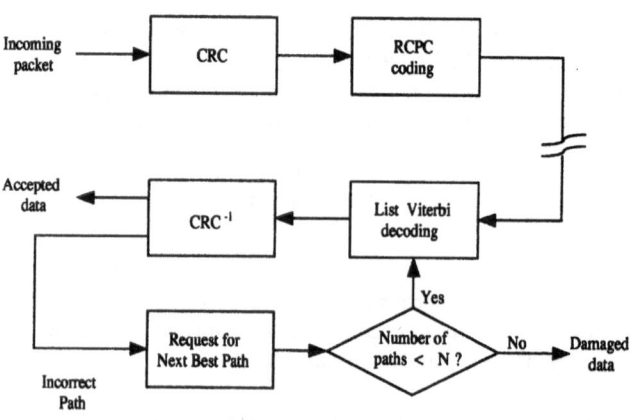

Figure 6.B.1.   Block diagram of List-Viterbi Algorithm.

at some instance, merge with it again in a later instance and never diverge again. These paths are ordered by increasing Hamming distance or Euclidean distance. The first path in this list is assumed as the globally second best path. If the second best path has an error, the algorithm finds the third best path. The algorithm first computes the locally second best paths of the second best path and then compares the best of the locally best paths with the second in the list. The path with the smaller distance is the globally third best path. All other best paths can be found in a similar manner.

Acknowledgments

The authors would like to thank K. E. Zachariadis for his assistance with some of the results for video transmission.

References

[1] Y. Wang, S. Wenger, J. Wen, and A. K. Katsaggelos, "Error resilient video coding techniques," *IEEE Signal Processing Magazine*, vol. 17, pp. 61–82, July 2000.

[2] G. Wornell and V. Poor, *Wireless Communications: A Signal Processing Perspective*. Prentice-Hall, 1998.

[3] C. E. Shannon, "Coding theorems for a discrete source with a fidelity criterion," *IRE National Convention Record, Part 4*, pp. 142–163, 1959.

[4] J. Hagenauer, "Rate-compatible punctured convolutional codes (RCPC codes) and their applications," *IEEE Trans. Communications*, vol. 36, pp. 389–400, April 1989.

[5] I. H. Witten, R. M. Neal, and J. G. Cleary, "Arithmetic coding for data compression," *Commun. ACM*, vol. 30, pp. 520–540, June 1987.

[6] A. Skodras, C. Christopoulos, and T. Ebrahimi, "The JPEG 2000 still image compression standard," *IEEE Signal Processing Magazine*, vol. 18, pp. 36–58, Sept. 2001.

[7] J. M. Shapiro, "Embedded image coding using zerotrees of wavelet coefficients," *IEEE Trans. on Signal Processing*, vol. 41, pp. 3445–3462, Dec. 1993.

[8] A. Said and W. A. Pearlman, "A new fast and efficient image codec based on set partitioning in hierarchical trees," *IEEE Trans. Circuits and Systems for Video Technology*, vol. 6, pp. 243–250, June 1996.

[9] M. Antonini, M. Barlaud, P. Mathieu, and I. Daubechies, "Image coding using wavelet transform," *IEEE Trans. Image Processing*, vol. 1, pp. 205–210, April 1992.

[10] D. Taubman, "High performance scalable image compression with EBCOT," *IEEE Trans. Image Processing*, vol. 9, pp. 1158–1170, July 2000.

[11] S. Cho and W. A. Pearlman, "A full-featured, error-resilent scalable wavelet video codec based on the set partintioning in hierarhical trees (SPIHT) algorithm," *IEEE Trans. Circuits and Systems for Video Technology*, vol. 12, pp. 157–171, March 2002.

[12] G. Cheung and A. Zakhor, "Bit allocation for joint source/channel coding of scalable video," *IEEE Trans. Image Processing*, vol. 9, pp. 340–356, March 2000.

[13] H. Watanabe and S. Singhal, "Windowed motion compensation," in *Proc. SPIE Visual Communications and Image Processing*, vol. 1605, (Boston, USA), pp. 582–589, Nov. 1991.

[14] N. V. Boulgouris, D. Tzovaras, and M. G. Strintzis, "Lossless image compression based on optimal prediction, adaptive lifting and conditional arithmetic coding," *IEEE Transactions on Image Processing*, vol. 10, pp. 1–14, Jan. 2001.

[15] J. F. Arnold, M. R. Frater, and Y. Wang, "Efficient drift-free signal-to-noise ratio scalability," *IEEE Trans. Circuits and Systems for Video Technology*, vol. 10, pp. 70–82, Feb. 2000.

[16] S. Lin and D. J. Costello, *Error Control Coding: Fundamentals and Applications*. Prentice-Hall, 1982.

[17] N. Seshadri and C.-E. Sundberg, "List Viterbi decoding algorithm with applications," *IEEE Journal on Selected Areas in Communications*, vol. 42, pp. 313–323, Feb./March/April 1994.

[18] G. Sherwood and K. Zeger, "Error protection for progressive image transmission over memoryless and fading channels," *IEEE Trans. Communications*, vol. 46, pp. 1555–1559, Dec. 1998.

[19] P. Moo and X. Wu, "Resynchronization properties of arithmetic coding," in *Proc. IEEE Int. Conference on Image Processing*, (Kobe, Japan), Oct. 1999.

[20] G. Castagnoli, J. Ganz, and P. Graber, "Optimum cyclic redundancy-check codes with 16-bit redundancy," *IEEE Trans. Communications*, vol. 38, pp. 111–114, Jan. 1990.

[21] G. Sherwood and K. Zeger, "Progressive image coding on noisy channels," *IEEE Signal Processing Letters*, vol. 4, pp. 189–191, July 1997.

[22] H. Man, F. Kossentini, and M. J. Smith, "A family of efficient and channel error resilient wavelet/subband image coders," *IEEE Trans. Circuits and Systems for Video Technology*, vol. 9, pp. 95–108, Feb. 1999.

[23] V. Chande and N. Farvardin, "Progressive transmission of images over memoryless noisy channels," *IEEE Journal on Selected Areas in Communications*, vol. 18, pp. 850–860, June 2000.

[24] B. Banister, B. Belzer, and T. R. Fisher, "Robust image transmission using JPEG2000 and turbo codes," *IEEE Signal Processing Letters*, vol. 9, pp. 117–119, April 2002.

[25] C. Berrou and A. Glavieux, "Near optimum error correcting coding and decoding: turbo codes," *IEEE Trans. Communications*, vol. 44, pp. 1261–1271, Oct. 1996.

Chapter 7

WAVELET CENTRIC VIDEO CODING AND CODING ARTIFACT CONCEALMENT

Boštjan Marušič, Primož Skočir, and Jurij F. Tasič

University of Ljubljana, Faculty of Electrical Engineering,
Tržaška 25, 1000 Ljubljana, Slovenia
{bostjan.marusic,primoz.skocir,jurij.tasic}@fe.uni-lj.si

Abstract   This chapter introduces an approach to video coding, which is inspired by the multidimensional extensibility of the wavelet transform. A full video coding system
including a post-processing stage for decoded video enhancement is presented
and discussed. The video coding approach builds upon a three-dimensional motion compensated wavelet transform and context based coding of bitplanes. As
such it is a digression from standard hybrid coding. Video post-processing follows a similar three-dimensional philosophy and is based on a three-dimensional
extension of the SUSAN (Smallest Univalue Segment Assimilating Nucleus)
image-filtering algorithm. Empirical evaluations reveal that the proposed video
coding/enhancement approach provides a viable alternative to hybrid video coding.

Although most of the underlying concepts are at least reviewed through the
chapter, the knowledge of basic wavelet theory and basics of (arithmetic) entropy
coding is a prerequisite.

Keywords: Video coding, wavelet transform, context arithmetic coding, SUSAN filter, video
post-processing.

1.   Introduction

Video coding plays a crucial role in modern visual communications, due to
the fact that video is the most bandwidth-consuming multimedia signal transmitted through contemporary communication infrastructure. In order to alleviate
the burden of transmitting raw video sequences, a coder and a decoder performing data compression are inserted into both ends of the communication path.
The coder reduces the input bit-rate by various orders of magnitude, enabling

J.F. Tasič et al. (eds.), Intelligent Integrated Media Communication Techniques, 221-261.

real-time transmission, while paying the price of introducing coding artifacts into the decoded video signal.

Video signals are sequences of temporally adjacent images also referred to as video frames. The frame rate or in other words the frequency of constitutive images (temporal sampling frequency) is usually above 25 fps (frames per second) to avoid flickering of perceived image sequences, although lower rates are regularly used in low bit-rate applications, such as teleconferencing and low-bandwidth internet video broadcast [1, 2, 3]. Due to temporal adjacency the consecutive frames are generally not completely dissimilar, although on a larger scale rough changes between two neighboring frames may occur due to fast camera movement, object occlusions and scene changes. From the statistical point of view, the similarity of temporally adjacent frames results in inter-frame correlation or equivalently correlation between pixels at the same spatial location in adjacent frames. The inter-frame (between consecutive frames) correlation is in general proportional to the frame rate as a higher sampling frequency implies a smaller temporal distance between frames. A similar observation in terms of correlation can also be deduced for the intra-frame case (correlation between pixels that bellong to the same frame). The value of a pixel at a particular spatial location tends to be highly correlated with the values of pixels in the vicinity - an observation that can be easily interpreted using a simple image model. If on a very coarse scale an image can be modeled as a collage of uniform textures glued together by edges, the correlation of pixels corresponding to a uniform image area can be easily deduced. The last observation is not true for edge pixels as by definition edges are areas of abrupt changes in pixel value. Although real-world video frames match the described simple image model only coarsely, pixels from these frames tend to exhibit strong correlation with their temporal and spatial neighborhoods.

Video compression is achieved by discarding repeated (redundant) information as well as irrelevant information which is not perceptible to the user (details masked by highly active textures, for example) or is at least not vitally significant for the overall perception of the decoded video. The removal of redundant information is reversible and leads to lossless video coding. On the other hand the removal of irrelevant information is not reversible (lossy algorithms) resulting in artifacts after decoding. The presence of correlation between adjacent pixels strongly indicates the presence of redundant information in video signals. Video coding algorithms highly rely on statistical dependencies present in the video signal in order to achieve compression. Actually, strong empirical evidence suggests that signal decorrelation leads to a higher degree of compression. Nevertheless, the reader should take into account that despite all there exists no formal proof to support such rationale [4, 5]. Following the before mentioned observation, video coding algorithms apply a (decorrelating) linear transform to video frames in order to reduce the correlation of adjacent pixels within a frame (intra-frame decorrelation) and achieve a higher degree of compression. Only after this step, the irrelevant information is discarded by quantizing the transform coefficients. Eventually the actual decrease in representation budget is achieved via entropy coding. This three-step approach, which forms the basis of most state of the art image coding engines is known as transform coding. The paradigm is clearly distinguishable in JPEG and JPEG2000 [6, 7] still image coding standards.

Video coding adds an additional level of complexity compared to image coding as correlations along the temporal axis are also taken into account in order to achieve higher compression. As differences between adjacent frames of the same scene are due to moving objects, it is reasonable to assume that the currently observed video frame can be at least coarsely approximated by transposing pixels from the previous video frame to new locations in the present one according to apparent movement. The spatial pixel displacement from one frame to the next one is specified by a so-called motion vector, whereas the computation of best fitting motion vectors while the process of finding motion vectors, is referred to as motion estimation. The process of transposing the previous frame pixels in order to obtain an approximation of the current frame is called motion compensation or equivalently motion prediction. In practical applications however, one cannot assign a motion vector to each pixel, as this would actually expand the presentation of a video frame instead of reducing it. The practical approach is to partition the video frame into rectangular areas of pixels called macroblocks, and to assign a single motion vector to the whole block - an approach, which is at the basis of all modern video coding standards.

In order to exploit inter- and intra-frame correlations, the first frame is coded independently as a picture, while the frames that follow are coded based on motion estimation. More precisely, one has to code motion vectors and the difference between the motion-predicted frame and the original current frame. Note that motion compensation (MC) does not at all yield accurate predictions as motion in natural images does not usually occur due to rectangular object transposition - objects have their own complex shapes with motion resulting from rotation, camera pan, object occlusion, etc. Nevertheless the prediction error image is well decorrelated also after block based motion compensation and compresses well using intra-coding techniques. Due to the fusion of two distinctive approaches, namely transform coding and motion compensated prediction for inter and intra-frame correlation removal, the described video coding technique is termed hybrid coding. Hybrid coding has met wide acceptance in international standards as well as in non-standard coders, mainly due to its coding efficiency and intrinsic low delay (only one frame is coded at a time).

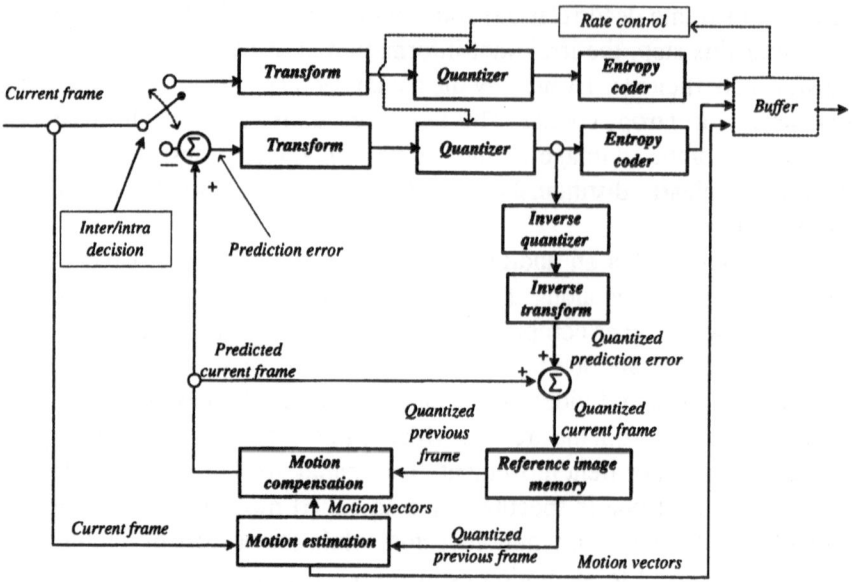

Figure 7.1. A block diagram of a hybrid coder.

1.1   A Note on Video Coding Standards

This short note is included here to facilitate later on the insight in the differences between the proposed approach and standard based coders. Video Coding Standards fall broadly into one of the following two categories: video broadcast and interactive videoconferencing. In the first category there are ISO standards such as MPEG1, which is targeted at CD-ROM video storage at bit-rates around 1Mbps-1.5 Mbps and MPEG2 designed for digital TV as well as HDTV quality video broadcast at rates above 4Mbps. MPEG1 and MPEG2 employ discrete cosine transform (DCT) of 8×8 pixel blocks and half-pixel accurate block motion estimation. Bi-directional motion prediction (future frames can also be used as a reference) is used. Motion vectors and quantized DCT coefficients are coded by variable length coding.

Into the second category fall the ITU H.261, H.263 and H.263+ (version 2) standards. Their application area is low bit-rate videoconferencing. H.261 is targeted to ISDN visual communications at multiples of 64 kbps. The H.263 and H.263+ were initially targeted to rates below 64 kbps, although it soon became apparent that H.263 provides significant improvement over H.261 at any bit rate. In terms of bit rates H.263 achieves 50% savings at lower bit-rates although the savings are not as dramatic at rates over 64 kbps. Compared to MPEG1 it achieves around 30% savings at rates around 1Mbps. H.263 represents today's state of the art for standardized video coding allowing coding a video of any resolution at any bit rate. In essence H.261 and H.263 coders are DCT based hybrid coders with a similar basic structure as the MPEG coders.

1.2 The Role of Video Enhancement

A high degree of compression inevitably leads to decoded image degradation due to irreversible quantization of transform coefficients in the process of transform coding. Especially at very low bit-rates, where coder design has to deal with subtle compromises between coding efficiency and decoded image quality and where a great deal of information has to be consequently discarded to meet the predefined bit-rate, the artifacts in decoded video frames become perceptible to the observer.

In the effort of achieving higher compression-ratios while maintaining the decoded video quality, the video coder designs can follow two diverse paths. On one hand challenging the whole coder concept by introducing new approaches or at least optimizing existing coder building blocks can result in greater coding efficiency, while on the other hand the perceptibility of objectionable artifacts can be reduced by processing decoded frames directly. It is the role of video enhancement to process each decoded frame with the goal of diminishing the coding artifacts so that the decoded image is less objectionable to the human observer and that it degrades gracefully at high compression-ratios. Note that enhancing the video's perceptual (or subjective) quality does not always result in an increase in objective quality such as PSNR, which is the most commonly used quality measurement criterion in video coding literature, as enhancement usually means filtering out some of the information from the decoded frame.

It is well known that the human observer is less tolerant to correlated artifacts than uncorrelated ones (such as Gaussian noise). Unfortunately the artifacts resulting from the quantization of transform coefficients are typically correlated. For example, the quantization of block DCT coefficients results in blocking artifacts (an abrupt change in image intensity at the borders of the DCT block), as well as in another type of less perceptible artifact called ringing effect or mosquito noise. As we will see later on, the wavelet transform produces only ringing effects, which are perceptible as oscillations in texture intensity in the locality of image edges. The ringing effect due to wavelets is much less objectionable than the blocking artifacts of DCT thus allowing wavelet video coders to produce decoded video that degrades gracefully at lower bit-rates.

Video enhancement algorithms incorporate a multitude of intelligent processing approaches in order to achieve their goal. The algorithm design is challenged by having to take into consideration human image perception properties as well as complex models of artifact formation and detection in order to remove or at least partially mask the coding distortions [8, 9, 10, 11].

2.   Context and Overview

2.1   Coding

Modern video coding algorithms have to exhibit excellent coding performance and at the same time exhibit several properties that allow their application in different application domains. For the transmission of video through heterogeneous networks the ability to produce output at any bit rate - also called rate scalability - is of great importance. Error resilience is another vital prerequisite. In the attempt to achieve higher compression ratios while maintaining the quality of decoded video optimization of existing coder building blocks is a valid path to follow, as demonstrated by existing video coding standards that all build upon the hybrid coding paradigm. However, challenging the coder design with new building blocks can be potentially more rewarding. This chapter is primarily devoted to a presentation and discussion of a coding approach that is based on the wavelet transform [12, 7] and is intrinsically different from the hybrid-coding paradigm. This section on the other hand overviews literature in order to describe the approaches we build upon.

Although being widely accepted, hybrid coding suffers from some disadvantages. Noteworthy, the temporal prediction loop in hybrid coders results in frame interdependence, which makes optimal bit allocation a troublesome task [13]. As an alternative to the hybrid model one can think of a non-discriminative decorrelation approach, which treats spatial and temporal correlation equivalently. Instead of the application of a two-dimensional transform followed by motion compensated prediction one can generalize the decorrelating transform to extend its reach also along the temporal axis. A three-dimensional transform thus manipulates a sub-sequence of video frames that can be seen as a pure three-dimensional signal, transforming it into decorrelated transform domain coefficients. As there is no prediction loop, and consequently no frame interdependence, the available bit budget can be trivially [7, 4, 14] allocated to groups of transform coefficients based on their relevance. In addition to simplified bit allocation the pure transform approach offers conceptual simplicity at the cost of a higher coding delay, as an entire group of frames has to be coded simultaneously.

In the process of defining our video-coding algorithm, besides changing the coding paradigm from hybrid coding to three-dimensional transform coding, we also chose to deviate the choice of the decorrelating transform away from the established DCT. Extensive research efforts have been focused on the application of the wavelet transform to visual information coding during the last decade. Consequently there exist sufficient empirical and theoretical evidence

[12, 15, 16, 17, 18, 19, 20, 21, 7] to argue that such transforms present an appropriate tool for image and video signal compression. The quantization of wavelet transform coefficients leads to artifacts that are less objectionable, while at the same time the multiresolution insight into the transformed data enabled by the wavelet transform offers new possibilities for algorithm design. The extension of the wavelet transform to the two- and three-dimensional case is straightforward and hence appropriate for the 3-D transform-coding approach. Although applications of simple (without motion compensation) three-dimensional transform extensions to video coding have not been extensively reported there are some papers that are worth noting [12, 15]. A three-dimensional transform without motion compensation has however been reported to achieve inferior coding efficiency compared to hybrid coding [12, 17, 18, 22, 7]. This is a direct consequence of the specific behavior of the correlation along the temporal axis, which is consistently reduced in areas of intensive motion resulting in large high-temporal frequency transform coefficients. In order to reduce the variance of the temporal high frequency coefficients and consequently increasing the coding efficiency it seems reasonable to perform part of the wavelet transform computation (filtering) not along the temporal axis but along the motion trajectories, i.e. along the motion vectors. In this case the three-dimensional transform is inherently combined with the motion estimation process (but not with motion compensated prediction) leading to an increased coding gain. Literature reports several successful implementations of such approach: the work done by J. R. Ohm [17], the one by Choi and Woods [18], the work done by Hsiang and Woods [23, 24] and the work reported by Pearlman's group in [22, 25, 26].

The idea of Choi and Woods [18] forms an integral part of our proposed algorithm. Choi and Woods initially achieved good coding results within the 3-D transform framework by relying on a complex bit allocation and macroblock optimization technique that leads to variable macroblock size motion estimation [18]. No particular attention has been paid to the entropy coding process. The computational complexity of their approach might therefore be objectionable for real-time video applications due to complex (macroblock size) optimizations.

On the other hand one might conceive a video coder based on a similar motion compensated transform, but using fixed macroblock motion estimation and at the same time employing more sophisticated entropy coding approaches that would compensate for the coding deficiency of a fixed-macroblock-size transform, hopefully resulting in a coding efficiency comparable to the one achieved by Choi and Woods [18] at fraction of their computational complexity. An approach in such direction was reported in a paper by W. A. Pearlman and others [22] in which a simplified three-dimensional transform from [18] was coupled with a three-dimensional extension of the well known SPIHT

[16] algorithm, which is essentially a complex quantization/entropy coding approach that relies on non-linear parent-child dependencies in the hierarchy of the wavelet transform coefficients. The obtained results were comparable or superior to the ones by MPEG 2 [27]. Another paper from the field of three-dimensional coding that is worth mentioning is the one by Taubman and Zakhor [21] in which a complex and extremely effective context based arithmetic coding of three-dimensional wavelet coefficients is presented. The method uses a three-dimensional transform without block motion estimation, but on the other hand acknowledges the specific correlation behavior along the temporal axis by sending one global motion vector per frame in order to align the frames before transform computation. Obviously this leads to optimal performance only when motion is a consequence of camera pan. On the other hand the context-based arithmetic coding approach can compress the video signal at an arbitrary bit rate, meeting the rate exactly and thus making it ideal for heterogonous and/or best effort transmission environments.

2.2   Enhancement

There are two strategies commonly adopted to reduce compression artifacts. For a given decoding algorithm and a bit rate constraint, the quality of recon-structed images can be enhanced by pre-filtering techniques [8] that remove unnoticeable details in source images so that less information has to be coded. Post-processing [9] on the other end is adopted to remove compression artifacts without increasing the bit rate or modifying the coding procedure. Most com-monly 2-D spatial processing is used but in the case of video the extension to the temporal axis usually increases the post-processing efficiency [10].

Video processing filters can in general be classified on the basis of the di-mensionality of the filter region of support (kernel dimensionality)[10]. More specifically, there are 1-D temporal, 2-D spatial and 3-D spatio-temporal filters. These can be further classified into non-motion compensated and motion com-pensated approaches. Motion compensation is frequently used in video process-ing algorithms, as it is a feasible way for removing temporal non-stationarities from video sequence, from which the filtering algorithm can gain [10, 11]. Temporal and spatio-temporal filters take advantages of the correlations in the temporal or both spatial and temporal directions. These filters are in general extensions of 2-D filters that consequently inherit most of their shortcomings, i.e. sensitivity to both temporal and spatial non-stationarities. In order to over-come this, there are two possible ways to go. The first consists of introducing different forms of temporal and spatial adaptivity into the filtering algorithm, so that the filter can adapt itself to the local statistics encountered in the sig-nal. Motion compensation on the other hand performs a similar role along the temporal axis by increasing the temporal stationarity of the video sequence.

In addition one can introduce complex image models on which to base the adaptivity of the algorithm [8, 9, 10, 11]. From the abundance of reported approaches, let us mention filtering based on local image statistic [28] and HMRF (Huber Markov Random Fields) image models [29] that were reported to be very effective in removing ringing artifacts [10, 29]. On the other hand, standardized approaches (video codecs based on DCT) usually use adaptive smoothing with FIR filters, mainly because of the low computational complexity.

Further in this chapter a post-processing algorithm, which is based on a Gaussian image model is thoroughly discussed. It has a built-in degree of adaptivity, at both spatial and temporal domain levels. Since it effectively removes wavelet-ringing artifacts it is included as an optional building block in the herein described wavelet video coding chain.

3.   Approach

3.1   Overall System Concept

The approach described in this chapter combines the 3-D transform by Choi and Woods [18] using fixed macroblocks with the context based arithmetic coding of bitplanes proposed by Taubman, which is sometimes also referred to as Layered Zero Coding (LZC) [30]. The temporal transform consists of three stages of a motion compensated temporal transform. The obtained temporal subbands are then decomposed with a 2-D wavelet transform and subsequently encoded using context arithmetic coding of coefficient bitplanes. Following the motion compensated three-dimensonal transform this chapter includes the description of the context-based algorithm. After giving a slightly different interpretation from the one in [30], some implementation details are discussed. Attention is paid to a prescanning technique for all-zero subband coding and to the application of adaptive arithmetic coding models (histograms) that necessitate no prior knowledge of the source statistics and at the same time deliver at least comparable performance to the one of the original algorithm. The system concept is shown in Figure 7.2, while the detailed explanation of various building blocks follows.

In order to achieve perceptually more pleasant decoded video, post-processing is included in the decoder. However, due to complexity constraints only a simple, yet effective, approach based on simple distortion models can be chosen. No pre-processing of any kind is included into the system in order to keep the computational burden low.

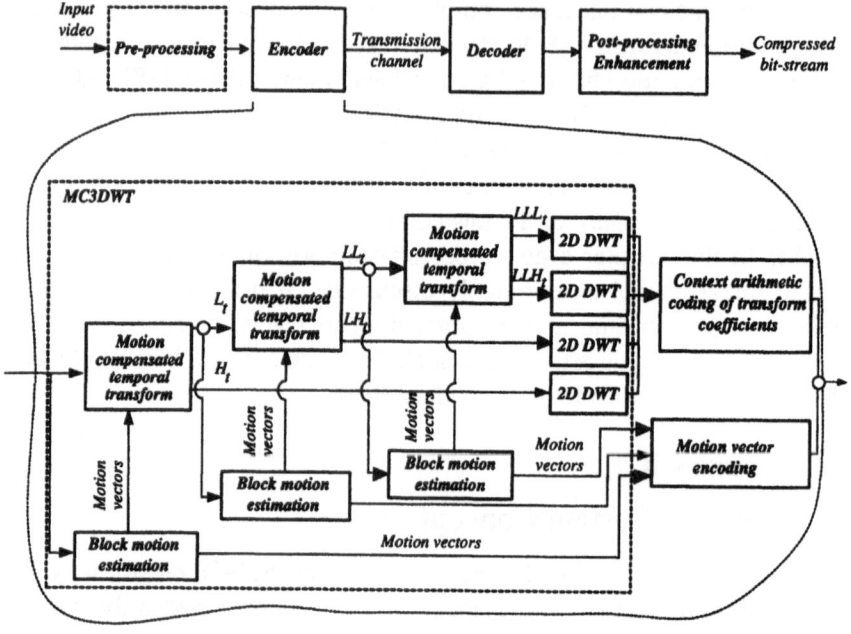

Figure 7.2.   Block diagram of the proposed video coding/enhancement system.

3.2   Motion Compensated Three Dimensional Wavelet Transform - MC3DWT

The wavelet transform decomposes the input signal into transform coefficients that are organized in groups called subbands [31, 21, 7]. A separable three-dimensional wavelet transform of a subsequence of a video signal $S[i, j, t]$ (where i and j are spatial indexes and t is the temporal index) can be computed by consecutively filtering (followed by decimation) the signal along each of the three coordinate axes independently [7]. In our case the temporal filtering is not performed along the temporal axis but along the motion trajectories that are given by the motion vectors $MV(i, j) = [d_i(i, j), d_j(i, j)]^T$, where $d_i(i, j)$ and $d_j(i, j)$ are the horizontal and vertical components of the motion vector at spatial location (i, j), respectively. In this case the adjacent pixels that take place in the computation are more likely to be correlated resulting in a low variance temporal high frequency subband (Figure 7.6). As transform computation along the temporal axis differs from the computation along the spatial axes, we will describe the transform computation in two phases called temporal and spatial transform.

3.2.1   Motion compensated temporal transform - MCTT.   Let us define the MCTT in the simplest case where it operates on a pair of consecutive frames yielding two subbands: the temporal low frequency subband $L_t[i, j]$ and the temporal high frequency subband $H_t[i, j]$. The definition is given by Equation (7.1). A transform operating on more than two consecutive frames has been shown by Ohm [17] to have higher computational requirements without significantly increasing the coding gain; therefore a two-frame temporal transform is used. A two-tap filter pair with coefficients $[\frac{1}{\sqrt{2}}, \frac{1}{\sqrt{2}}]$ and $[\frac{1}{\sqrt{2}}, -\frac{1}{\sqrt{2}}]$, i.e. the Haar transform [7] is used in the transform computation. Let us interpret the basic two frame temporal transform. By examining Equation (7.1) closely one can see that the transform computes a scaled average and scaled difference of the input frames. The idea is to remove as much information as possible from the difference image, so that it can be compressed using few resources (bits). In order to achieve this, the difference is not computed by subtracting the values of the pixels at the same spatial location, but by subtracting pixels that lie on the same motion trajectory. In this way the difference is likely to result close to zero (See Figure 7.3). Also note that the number of input pixels is equal to the number of output transform coefficients - in other words: the transform is non-expansive.

$$L_t [i - d_i(i, j), j - d_j(i, j)] =$$
$$\frac{1}{\sqrt{2}} \cdot (S [i, j, t+1] + S [i - d_i(i, j), j - d_j(i, j), t]) \qquad (7.1)$$

$$H_t [i, j] = \frac{1}{\sqrt{2}} \cdot (S [i, j, t+1] - S [i - d_i(i, j), j - d_j(i, j), t])$$

Block motion estimation (BME) is used to estimate the motion vector for each rectangular block of pixels. This indispensable simplification however leads to problems in terms of transform computation, as the calculation given by Equations (7.1) is not always feasible for all spatial indexes i and j. This is due to the fact that a one-to-one correspondence between pixels in consecutive frames cannot always be established. It is quite likely that macroblocks at block matching overlap result in multiple connections between pixels (Figure 7.4). Moreover it is also possible that some of the pixels from a previous frame $S[i, j, t]$ are never found to match pixels in the current frame $S[i, j, t+1]$ (Figure 7.4). Both these cases are termed unconnected as opposed to the case when a one-to-one mapping exists and the transform can be computed following Equation (7.1).

In order to guarantee perfect reconstruction, also the values of the unconnected pixels have to be preserved. In the case of a two-tap transform the values of the unconnected pixels can be inserted into the transform subbands as follows. For an unconnected pixel in the previous frame, the pixel value is

Figure 7.3. The effect of the MCTT on two frames of the "Foreman" sequence a), b). Images c) = L_t and d) = H_t represent the result of the transform without motion compensation, while f) = L_t and g) = H_t represent the result of the transform with motion compensation. Image e) is a map of connected/unconnected pixels (dark gray = connected, black = unconnected from current frame, light gray = unconnected from previous frame).

scaled and inserted in the low frequency band (7.2) as such pixels are likely to be uncovered ones. On the other hand for an unconnected pixel in the current frame $S[i, j, t + 1]$ a displaced frame difference is taken (7.3) as this is likely to lower the value of the resulting coefficient. In both cases (7.2) (7.3) scaling is required so that the statistics of the unconnected pixels match the statistics of the transform coefficient obtained by the Haar transform from the connected case.

$$L_t [i, j] = \frac{2}{\sqrt{2}} \cdot S [i, j, t] \tag{7.2}$$

$$H_t [i, j] = \frac{1}{\sqrt{2}} \cdot (S [i, j, t + 1] - S [i - d_i(i, j), j - d_j(i, j), t]) \tag{7.3}$$

The inverse of the MCTT can be trivially obtained by manipulation of equations (7.1), (7.2) and (7.3). Also, note that in order to guarantee perfect reconstruction, the motion vectors used have to be restricted in order to point only to pixels inside the previous frame otherwise information is lost.

The coding gain can be considerably increased if one employs half-pixel accurate motion estimation in the MCTT computation [17] (gain of ca. 2 dB in PSNR for progressive sequences). In this case, motion vectors are obtained by performing the block matching on interpolated reference images (by a factor of two). Formulas equivalent to (7.1), (7.2) and (7.3) for the half-pixel accuracy case are given by (7.4), (7.5) and (7.6) respectively. Note that symbol $\tilde{S}$ represents interpolated frames, while $\overline{d_i(i,j)}$ and $\overline{d_j(i,j)}$ represent the nearest integer value (flooring) of the half pixel accurate motion vector.

$$L_t\left[i - \overline{d_i(i,j)}, j - \overline{d_j(i,j)}\right] =$$

$$\frac{1}{\sqrt{2}} \cdot \left(\begin{array}{c} \tilde{S}\left[i - \overline{d_i(i,j)} + d_i(i,j), j - \overline{d_j(i,j)} + d_j(i,j), t+1\right] \\ + S\left[i - \overline{d_i(i,j)}, j - \overline{d_j(i,j)}, t\right] \end{array} \right) \qquad (7.4)$$

$$H_t[i,j] = \frac{1}{\sqrt{2}} \cdot (S[i,j,t+1] - \tilde{S}[i - d_i(i,j), j - d_j(i,j), t])$$

$$L_t[i,j] = \frac{2}{\sqrt{2}} \cdot (S[i,j,t]) \qquad (7.5)$$

$$H_t[i,j] = \frac{1}{\sqrt{2}} \cdot (S[i,j,t+1] - \tilde{S}[i - d_i(i,j), j - d_j(i,j), t]) \qquad (7.6)$$

3.2.2 Hierarchy of temporal transform computation. To decompose a group of frames using the motion compensated temporal transform described above, one has to apply the transform to consecutive pairs of frames in the group. If the group consists of $l = 2N$ frames this yields N temporal low frequency subbands L_t and N H_t subbands. Due to the nature of transform computation, the L_t subbands inherit the resemblance with the original frames, while the H_t subbands are almost zero-valued. Therefore, one might apply the two-pair transform again on the L_t pairs yielding LL_t and LH_t subbands as shown in Figure 7.6. In general, if $l = m \cdot 2^k$ the temporal MC compensated transform can be applied k times. Note that the number of transform iterations defines the fundamental coding delay (if k iterations are applied, at least 2^k) frames have to be coded together). As reported in [17] more temporal transform stages result in higher coding gain. However, care has to be taken when applying several temporal transform stages consecutively. First, one has to account for the reconstruction error due to the imperfect inverse transform when half-pixel accuracy motion vectors are used. Choi and Woods report that two transform stages seem a good compromise between increase in coding gain and added reconstruction error [18]. On the other hand Ohm [17] uses four stages yielding 40 - 45 dB PSNR when no quantization is applied. Taubman [21] found that three transform stages represent an overall optimal choice for his transform framework.

Second, the computational complexity of the investigated video coder is governed by the motion estimation procedure. Thus the number of motion vectors has to be kept equal to - or smaller than - the number of motion vectors

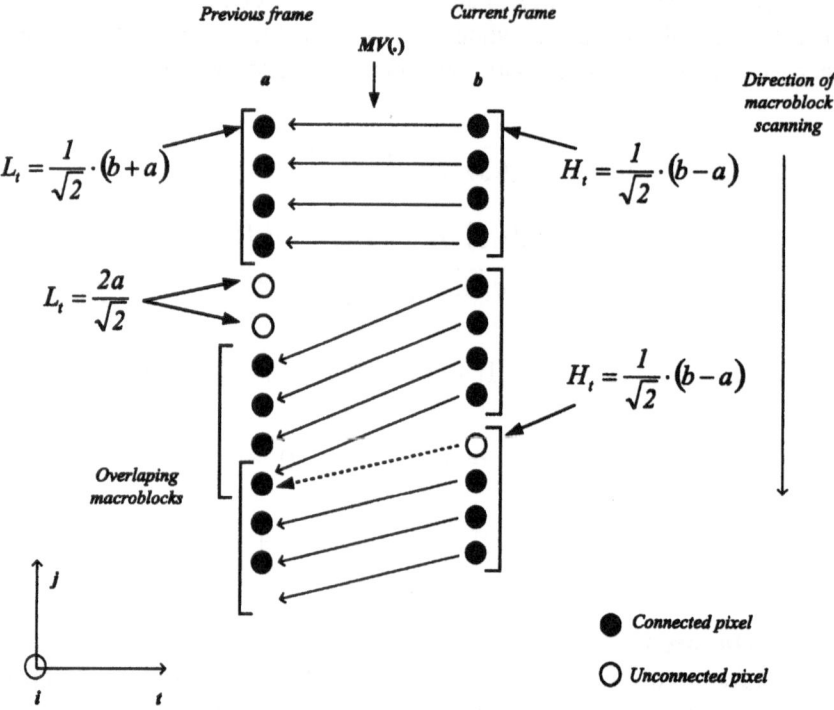

Figure 7.4. Generation of connected and unconnected pixels, and temporal motion compensated transform.

in hybrid coders, in order to allow for a fair comparison with hybrid coders. Note that in the case of two temporal transform stages, the number of motion vectors is equal to the number of vectors a hybrid coder has to send, in order to code a group of frames of the same size. However, two transform stages in most cases do not provide adequate coding efficiency. For the abovementioned reasons, the temporal transform computation is organized as follows below and not as shown in Figure 7.6. The transform is applied to a group of pictures (GOP) of length 8 enabling three stages of the temporal transform. The procedure is shown in Figure 7.5. Note that the number of motion vectors for three stages of the MCTT is equal to one in hybrid coders. However, empirical tests reveal that the entropy of motion vectors from three stages of MCTT is generally larger than the entropy of motion vectors in hybrid coders. This can be attributed to the fact that motion vectors from stages 2 and 3 of the MCTT carry information about global motion, while motion vectors in hybrid coders reflect motion from frame to frame.

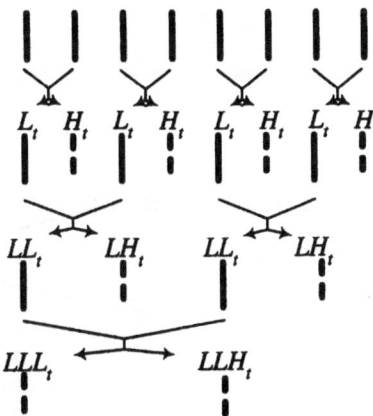

Figure 7.5. Hierarchy of MCTT computation. Three iterations of MCTT are applied to eight input frames. The dotted frames represent the MCTT output.

3.2.3   MC3DWT.   After applying two stages of the temporal motion compensated transform, the obtained subbands are decorrelated in the temporal direction. The application of a spatial two-dimensional wavelet transform to the temporal transform subbands removes the remaining spatial correlation. Usually four to six 2-D transform iterations are applied to sufficiently decorrelate the temporal subbands. Note that the maximum number of transform iterations is in general restrained by the input signal size. The filter pair used for the 2-D wavelet transform computation is the Villasenor's 9/7 biorthogonal filter pair proposed in [32]. This filter choice results in elevated compression performance and yields less perceptible artifacts. The application of the motion compensated temporal transform followed by a 2D spatial wavelet transform can be seen as a three dimensional transform of a three dimensional input - we chose to label this transform as MC3DWT. The result of MC3DWT of four consecutive frames from the "Salesman" sequence is shown in Figure 7.6. Note that the pixels in temporal subbands of types LH_t and H_t are fairly uncorrelated to their spatial neighborhood, so that in principle further spatial decorrelation would not be required on these. However, for reasons explained later, a 2D wavelet transform is applied to all the temporal subbands as discussed above.

3.3   Context Based Bitplane Coding

The application of the MC3DWT transforms a subsequence of input video frames into data that can be more easily compressed. Note however, that with the transform output being real valued, the actual number of bits for the precise representation of the coefficients has actually increased. Also note that up to this point no irreversible processing has been applied - the MC3DWT guaran-

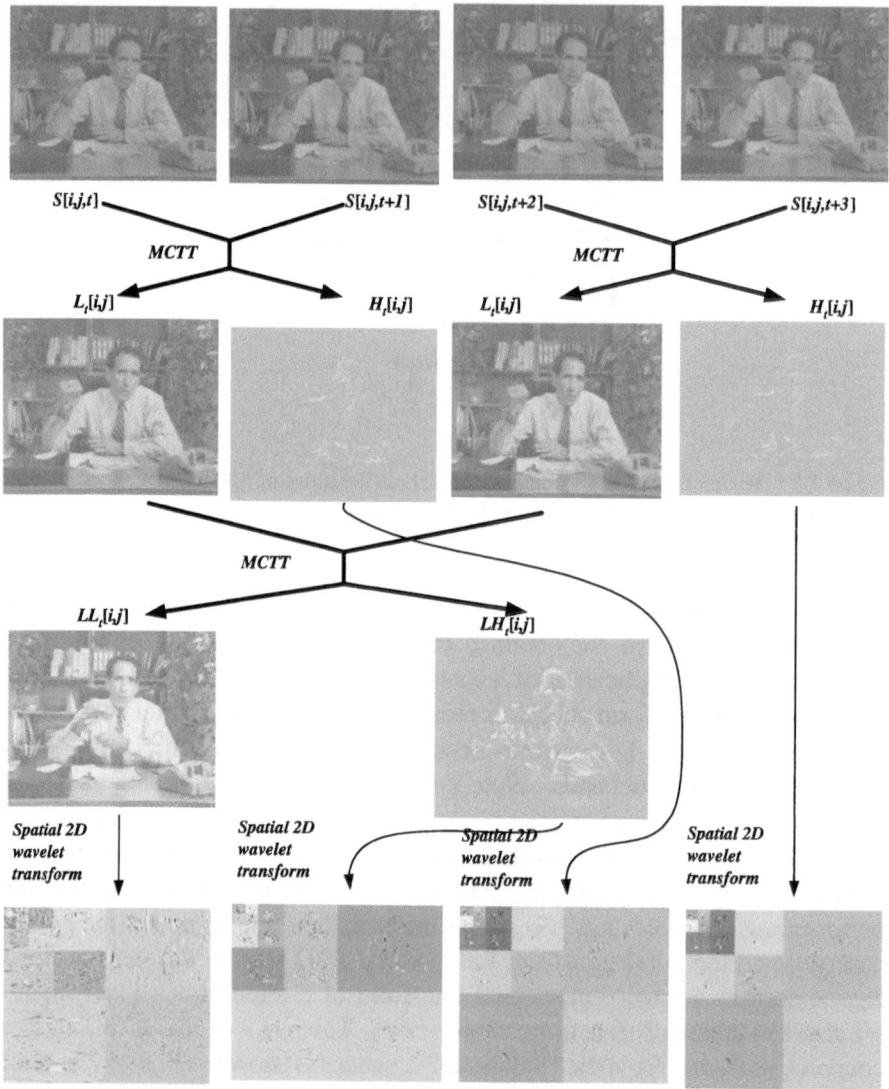

Figure 7.6. Computation of the MC3DWT of four frames from the "Salesman" sequence. Two stages of the temporal transform with motion compensation are followed by 4 stages of two 2-D wavelet transforms on the temporal subbands.

tees perfect reconstruction (if integer pixel accurate MVs are used). In order to achieve compression the coefficients have to be quantized and entropy encoded. The process of quantization, that is the mechanism by which irrelevant information is discarded, is irreversible, hence distortions are introduced at this stage.

Besides achieving the highest quality at a predefined bit-rate, it is also desired that the algorithm is scalable at least in terms of bit rate. In other words, one would like to use the same coding algorithm at arbitrary bit-rates always resulting in optimal coding efficiency.

For the purpose of entropy coding, context arithmetic coding is used. This is essentially a multi-model arithmetic coding [30, 19, 33] approach in which for each input symbol one tries to select the model (histogram) that best approximates the predicted probability of the symbol being coded, taking into account some prior knowledge of the symbol source. The transform coefficients are not quantized in a standard manner. Instead, a successive approximation quantizer producing a bit per coefficient at each quantization iteration is used. The bits for all coefficients that belong to a single successive approximation quantizer iteration are grouped together to form a bitplane. This bitplane organization permits to achieve a particular quality by sending only a subset of all available bitplanes. Note that the coding of a particular bitplane can be stopped at any arbitrary location in the bitplane, in order to meet the predefined bit rate exactly. As a consequence part of the coefficients may be coded at 1 bit coarser accuracy so as to meet the bit rate exactly. Besides the exact rate allocation, there is another advantage in coding the transform coefficients bitplanewise. In this case symbols entering the input of the entropy coder can take only two different values resulting in fast model adaptation, which is important if the number of models is large. Although the input video frames are transformed with a decorrelating transform, there are still some inherent dependencies between the transform coefficients that can be used to predict symbol values. The context arithmetic coding used for the compression of MC3DWT coefficients relies heavily on these dependencies.

3.3.1   Context arithmetic coding.

Arithmetic coding [33] is an entropy coding approach that maps a large number of input symbols into a long binary sequence that uniquely defines the input symbols. The output binary sequence is the compressed representation of the input. By applying arithmetic coding the entropy of the symbol source is achieved [33].

The arithmetic coding approach clearly separates the modeling of the probability density function of the input source from the coding process. The objective of the source modeling is to estimate to the highest possible accuracy the probability of the symbol currently being coded. The arithmetic coding process entropy encodes a symbol sequence $x_i = \{x_1, x_2, \ldots, x_n\}$ with the probability of appeareance of the sequence at the coder input $p(x_1, x_2, \ldots, x_n)$. The arithmetic coder encodes the input sequence with the minimal possible number of bits, which equals $-\log_2(p(x_1, x_2, \ldots, x_n))$, and corresponds to the information content of the sequence. In other words, the responsibility of the source modeling is to estimate the probability $p(x_1, x_2, \ldots, x_n)$. By observing that the symbols x_i are not necessarily independent the probability of the symbol sequence can be written as in (7.7). By taking into account the symbol dependencies and writing the sequence probability with a product of conditional probabilities the number of bits after entropy coding of the sequence can be considerably lowered.

$$p(x_1, x_2, \ldots, x_n) = \prod_{i=1}^{n} p(x_i | x_1, \ldots x_{i-1}) \qquad (7.7)$$

The estimation of the conditional probabilities conditioned on all previous symbols is typically not feasible in practical cases due to the large number of possible combinations. The sequence of symbols that preceded the symbol currently being coded, on which the current symbol probability is conditioned, is called context. In order to decrease the number of conditional probabilities to a practically manageable extent, the number of conditioning contexts has to be reduced. This can be done by approximating the conditional probability by conditioning it only to N previous symbols as shown in (7.8). Note that the number of conditional probabilities is in this case proportional to B^N, where B represents the number of distinct values each input symbol can take. From this perspective, the application of successive approximation quantization again reveals to be advantageous as it maps a real-valued coefficient source into a set of bitplanes or in other words in a source with $B = 2$.

$$p(x_i | x_1, \ldots x_{i-1}) \approx p(x_i | x_{i-N}, \ldots x_{i-1}) \qquad (7.8)$$

The above idea of conditional probability estimation summarizes the essence of context arithmetic coding. The entropy coder uses multiple probability models. The coding model that best approximates the conditional probability of a symbol is selected. The selection is based on the context, i.e. on values of previously coded symbols. In other words, the symbol value is predicted from previously coded symbols and the model that best approximates its probability is used in the coding process. Using the same approach, the decoder can select the right model from previously decoded symbols and consequently there is no need to communicate this information in to the decoder. Note that the models are not necessarily fixed. Adaptive probability models can be used that learn on the fly the source statistics without prior knowledge. In other words, with every new input symbol, the probability models are updated so that they reflect the statistics of all encountered symbols. As an advantage, adaptive models adapt quickly to non-stationarities in the signal leading to higher coding efficiency. Figure 7.7 provides a graphic representation of the context arithmetic coding principle.

There is however an inherent drawback of adaptive models in conjunction with context arithmetic coding. The accuracy of the model is proportional to

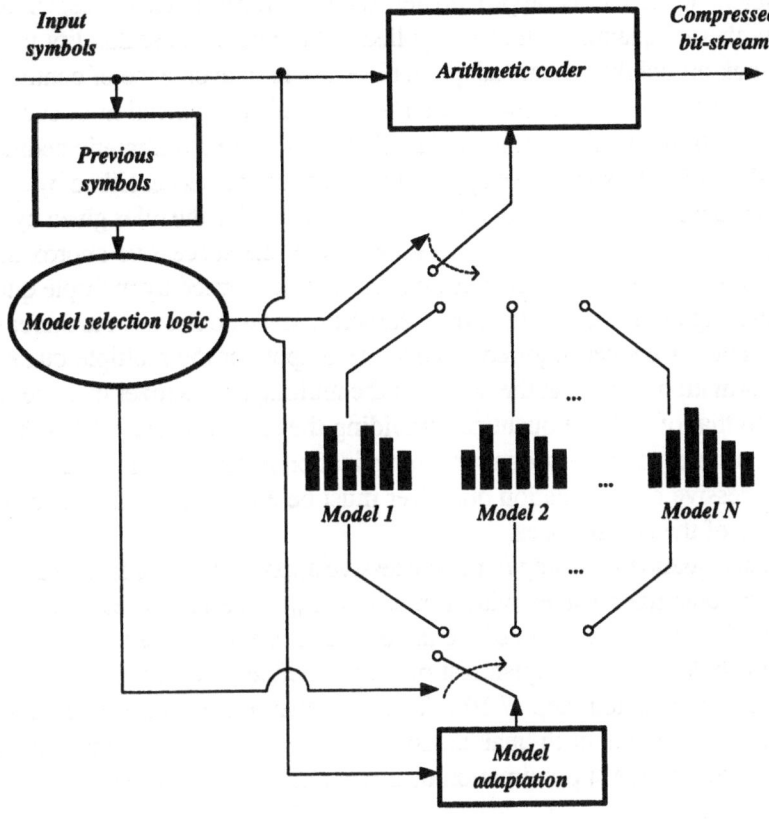

Figure 7.7. Context arithmetic coding.

the number of symbols coded with that model. While this does not present a problem in quickly adapting single model arithmetic coding, it can lead to a phenomenon known as context dilution [19] in multi-model arithmetic coding. The phenomenon of context dilution (no model predicts accurately the probability) is a consequence of poor probability estimation due to slow model adaptation. In order to take advantage of context arithmetic coding and avoid inaccurate models one has to use a reasonably small number of models.

3.3.2 Successive approximation quantization. Successive Approximation Quantization forms and integral part of the context-based entropy coding approach as proposed in this chapter. A more in-depth discussion of context arithmetic coding of wavelet transform coefficients follows in the next subsections.

Successive approximation quantization decomposes each real-valued transform coefficient into a sequence of bits (bitplanes) from which the original coefficient value can be approximated with arbitrary precision depending on the number of quantization steps applied. The obtainable scalability property and quick probability model adaptation, resulting in avoidance of context dilution, are the previously mentioned advantages of the application of successive approximation quantization in conjunction with context arithmetic coding.

At this point the successive approximation quantizer as described by Shapiro in [20] is applied. This kind of quantizer conforms with the rules given by Taubman [21] for nonexpansive quantization. Note that successive approximation quantization can be interpreted as quantization performed by multiple quantizers featuring different quantization accuracies. If we do not want to increase the representation budget required to code the outputs of the multiple quantizers, the information content at the output of the multilayer quantizer must be less or equal to that of a single quantizer providing the same accuracy. This theoretically translates to the restriction that quantization intervals of finer quantizers of the successive approximation quantizer must be embedded in the quantization intervals of the coarser ones.

At each quantization step n, the successive approximation quantization compares the quantized values with a threshold T_n. The first threshold T_0 is set to half of the absolute value of the largest transform coefficient $w[i, j, t]$ (7.9). The threshold in the n-th quantization step is defined recursively as half of the previous quantization step (7.10). Thus the quantization threshold decreases by a factor of two at each quantization step, clearly indicating the connection between successive approximation quantization and bitplane encoding.

$$T_0 = \frac{\max_{i,j,t} |w[i, j, t]|}{2} \tag{7.9}$$

$$T_{n+1} = \frac{T_n}{2} \tag{7.10}$$

The implementation of the successive approximation quantizer necessitates two more values $u_h[i, j, t]$ in $u_l[i, j, t]$ for each transform coefficient. These values define the lower and higher bound of an uncertainty interval within which the quantized value $w_q[i, j, t]$ lies. The quantized value is always defined as the center of the uncertainty interval. The values are initialized as given by equations (7.11).

$$
\begin{aligned}
u_h[i, j, t] &= 2T_0 \\
u_l[i, j, t] &= 0 \\
w_q[i, j, t] &= \frac{u_l[i,j,t] + u_h[i,j,t]}{2}
\end{aligned} \tag{7.11}
$$

The output bit $b_n[i, j, t]$ at every transform coefficient location in the n-th quantization step indicates whether the coefficient value is above or bellow half of the uncertainty interval (7.12).

$$b_n\left[i,j,t\right] = \begin{cases} 1, & w[i,j,t] \geq \frac{u_l[i,j,t]+u_h[i,j,t]}{2} = w_q[i,j,t]; \\ 0, & w[i,j,t] < \frac{u_l[i,j,t]+u_h[i,j,t]}{2} = w_q[i,j,t] \end{cases} \qquad (7.12)$$

As soon as the quantizer outputs the bit, the uncertainty interval bounds are updated to reflect the increased accuracy provided by that bit (7.13).

$$\begin{array}{ll} \text{if} \quad b_n\left[i,j,t\right]=0 \quad \text{then} \quad u_l\left[i,j,t\right]=u_l\left[i,j,t\right]+T_n \\ \text{if} \quad b_n\left[i,j,t\right]=1 \quad \text{then} \quad u_h\left[i,j,t\right]=u_h\left[i,j,t\right]-T_n \end{array} \qquad (7.13)$$

The process of successive approximation quantization is depicted in Figure 7.8. It should be clear that the described quantization approach works only on unsigned values, so for the coding purpose, the signs have to be coded separately.

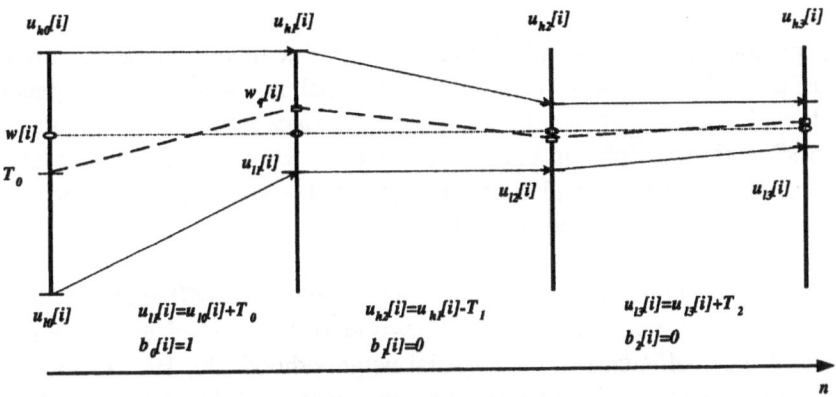

Figure 7.8. Successive approximation quantization of transform coefficients. The boundaries of the uncertainty interval are modified according to the output bit in order for the center of the interval to approximate the input value with increased accuracy after each quantization step. Note that a monodimensional case is depicted in the figure.

All the bits $b_n[i,j,t]$ for $0 \leq i \leq N-1, 0 \leq j \leq M-1$ and $0 \leq t \leq U-1$ together form a three-dimensional bitplane. The symbols N and M stand for the spatial resolution along abscissa and ordinate axes respectively, while U is the number of video frames that are jointly coded (8 for the described MC3DWT).

3.3.3   Wavelet transform coefficient dependencies and coding model selection.
Context arithmetic coding heavily relies on accurate conditional entropy estimate. In case of adaptive arithmetic coding, the source statistics are based on a small number of already coded symbols. In order to achieve an accurate estimation of the conditional entropy, it is instructive to take into

account the actual dependencies between the coded symbols. Note that it is not always the optimal solution to condition the probability of a symbol based on N successive previously coded symbols. In general, the source correlation pattern might dictate a more sophisticated form of conditioning context. In other words, it is advisable to use a conditioning context that consists of previously coded symbols that are highly correlated with the current symbol. For this reason we generalize the definition of the conditional probability of a symbol x_i by introducing a context function $\kappa(i)$, which is in general an arbitrary function of previously coded symbols. The function that best takes into account the interdependencies of the successive approximation quantization bitplanes will be discussed shortly and referred to as context selection. In this more general setting, the equation of sequence probability approximation has the form (7.14).

$$p(x_1, x_2, \ldots, x_n) \approx \prod_{i=1}^{n} p\left(x_i | \kappa(i)\right) = \prod_{i=1}^{n} p\left(x_i | \kappa(x_1, \ldots x_{i-1})\right) \quad (7.14)$$

In practice $\kappa(i)$ can be interpreted as the index of the probability model featuring the best probability estimate of symbol x_i. Although very general and flexible, this general setting is problematic since it allows for a multitude of possible choices defining the context function $\kappa(i)$. Therefore, one has to resort to heuristics based on intuitive insight into the source statistics and to try to empirically find a suitable solution.

3.3.4   Dependencies between wavelet transform coefficients.   The magnitude of wavelet transform coefficients tends to be correlated to the magnitude of the coefficients in their spatial neighborhood [30, 19, 21, 34]. Accordingly, a coefficient is likely to be above a threshold if a certain number of its neighboring coefficients are. This is especially true for coefficients in the spatial high frequency subbands. If a particular coefficient represents an image edge, its neighbors probably also do. If a coefficient is not part of an edge, it is probably near zero valued, and so are its neighbors.

Wavelet transform coefficients exhibit also a particular form of inter-subband dependency that has been widely acknowledged in the wavelet image coding literature [16, 20], leading to the application of the zero-tree data structure. Although the net effect of taking into account the inter-subband dependencies in the context arithmetic framework is modest (0.1 dB - 0.4 dB in image coding tests) these interdependencies are reflected in the definition of our proposed context selection, as there is no additional computational burden associated with it.

There is a third type of dependency that can be exploited in the model selection mechanism, namely inter-bitplane dependency. In our setting, we do not code the coefficients directly, but we instead proceed bitplane-wise. Due to the way the bitplanes are obtained, it is reasonable to expect dependencies between bits at the same spatio-temporal locations in different bitplanes [19, 21]. These are also termed structural dependencies.

3.3.5 The binary map and context selection.

In order to assess the structural dependencies, i.e. the inter-bitplane dependencies, the context function will be computed based on a value of a particular binary map χ_n, which aggregates the values of previously coded bitplanes. The local and parent-child dependencies are on the other hand reflected in the choice of locations from map χ_n that define the context function.

Note that the binary map χ_n is dynamic and changes its value according to all the bitplanes coded previously and according to the already coded bits in the current bitplane.

The binary map is first initialized to zero and is updated during the first successive approximation step $n = 0$ to reflect the values of already coded bits from the first bitplane as follows from (7.15).

$$\chi_0[i,j,t] = \begin{cases} b_0[i,j,t], & \text{if } b_0[i,j,t] \text{ already coded;} \\ 0, & \text{if } b_0[i,j,t] \text{ not coded yet.} \end{cases} \quad (7.15)$$

When coding bitplanes from higher successive approximation passes ($n \neq 0$), the binary map merges values from all previously coded bitplanes and the values of the already coded bits of the current bitplane in order to reflect the structural dependencies present in the signal. The map χ_n is obtained as a "logical or" of all previously coded bitplanes b_n. This process is described by equation (7.16).

$$\chi_n[i,j,t] = \begin{cases} b_n[i,j,t] \vee \chi_{n-1}[i,j,t] = \\ b_n[i,j,t] \vee b_{n-1}[i,j,t] & \text{if } b_n[i,j,t] \text{ already coded;} \\ \vee b_0[i,j,t], \\ \\ \chi_{n-1}[i,j,t] = \\ b_{n-1}[i,j,t] \vee b_{n-2}[i,j,t] & \text{if } b_n[i,j,t] \text{ not coded yet.} \\ \vee b_0[i,j,t], \end{cases}$$

$$(7.16)$$

The value of the binary map changes with each coded bit and in this way enables the computation of the context function based on all previously coded bits (see Figure 7.9).

The context function $\kappa(i,j,t)$ takes into account the local correlations ($F_n[i,j,t]$) as well as the parent child-correlations ($F_p[i,j,t]$), while structural correlations are take into consideration by basing the context function on the binary map, is given by equation (7.17). The context function and the consequential context selection mechanism in (7.17) are a result of empirical observations. Note however, that it does not necessarily represent a globally optimal solution. The problem with context selection comes from the fact that no formal procedure for optimal context function definition exists.

$$\kappa(i,j,t) \;=\; \left\{ \begin{array}{ll} (F_n[i,j,t] + F_p[i,j,t]) & \Longleftarrow \quad \chi_n[i-1,j,t] = 0 \\ 256 & \Longleftarrow \quad \chi_n[i-1,j,t] \neq 0 \end{array} \right\}$$

$$\begin{aligned} F_n[i,j,t] \;=\;& \chi_n[i-1,j,t] + 2 \cdot \chi_n[i,j-1,t] \\ +\;& 4 \cdot \chi_n[i+1,j,t] + 8 \cdot \chi_n[i,j+1,t] \\ +\;& 16 \cdot (\chi_n[i-1,j-1,t] \vee \chi_n[i+1,j-1,t]) \\ +\;& 32 \cdot (\chi_n[i-1,j+1,t] \vee \chi_n[i+1,j+1,t]) \end{aligned}$$

$$\begin{aligned} F_p[i,j,t] \;=\;& 64 \cdot \chi_n\left[\tfrac{i}{2},\tfrac{j}{2},t\right] \\ +\;& 128 \cdot \left(\begin{array}{l} \chi_n[\tfrac{i}{2},\tfrac{j}{2},t] \vee \chi_n[\tfrac{i}{2}+1,\tfrac{j}{2},t] \\ \vee \chi_n[\tfrac{i}{2}-1,\tfrac{j}{2},t] \vee \chi_n[\tfrac{i}{2},\tfrac{j}{2}+1,t] \\ \vee \chi_n[\tfrac{i}{2},\tfrac{j}{2}-1,t] \end{array} \right) \end{aligned}$$

$$(7.17)$$

Note that the definition in (7.17) takes into account eight neighbors of the current coefficient along with his parent coefficient and its four immediate neighbors. The number of probability models would be too high if each of the mentioned coefficients would trigger the selection of a different model. For this reason, values are grouped together with a logical-or operator, consequently decreasing the number of different outputs of the context function, and thus avoiding model inaccuracy. Note that the structural (inter bitplane) correlations are intrinsically taken into account in the definition of the binary map (7.16).

There is one more thing that is worth noting. If a coded bit in any of the previous bitplanes had a value of one, the value of the binary map would be set to one at that location and would never change in latter quantization passes. As a consequence, all future bits at this particular location are coded using a single probability model with index 256. This is reasonable as once a coefficient has been quantized to a non-zero value the probability of the two possible values a bit in a bitplane can take are approximately equal (Figure 7.9). Another interpretation of the proposed context computation is that we use a more complex model selection mechanism for the bits that are expected to be zero-valued and use a single model for the rest of the bits. One can think of the zero valued bits as the bits that carry information about the location of a transform coefficient. After a location of a coefficient has been encoded one has to code the value bits, which are in this case coded using a single model with index 256. Context arithmetic coding of bitplanes can be thus seen as a means of effective coefficient location coding (Figure 7.9).

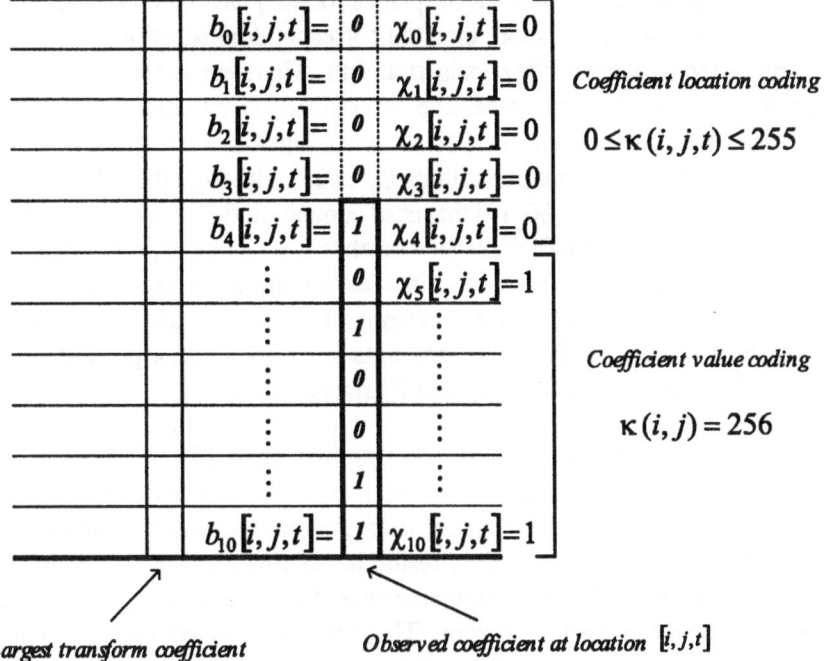

$$Coefficient\ location\ coding$$
$$0 \le \kappa(i,j,t) \le 255$$

$$Coefficient\ value\ coding$$
$$\kappa(i,j) = 256$$

Largest transform coefficient *Observed coefficient at location* $[i,j,t]$

Figure 7.9. Coefficient location coding vs. value coding and model selection. The displayed binary map values are shown before the corresponding bit is coded.

There is another important aspect of the proposed context selection mechanism. The value of the context function and consequently the value of the currently coded bit are predicted not only from a causal neighborhood of that coefficient (from the already coded bits in the current bitplane) but also from non-causal locations. Since the binary map χ_n is used in context function evaluation it is possible to take into account the coefficient values also at non-causal locations. If a bit in a current bitplane has not been coded yet this simply means that the prediction is based on a coarser resolution value of a corresponding transform coefficient which still conveys a lot of information about the neighborhood of a currently coded bit to greatly enhance model selection.

3.3.6   Sign coding.   As discussed above the successive approximation quantizer operates on unsigned values, therefore the coefficient signs are not embedded into the bitplanes at the quantizer output and have to be coded separately. It seems irrational to transmit the signs of all coefficients, as not all of them will be reconstructed to a non-zero value. Whether a small transform coefficient is coded or not depends on the predefined bit budget, and consequently on the number of successive approximation quantization passes. If a coefficient value is never coded, i.e. if all the bitplanes for all quantization passes are equal to zero at a particular location, then the corresponding transform coefficient is reconstructed as zero and the sign is irrelevant. Therefore the sign value is transmitted only after the first successive approximation bit equals one. The sign information is embedded into the bit stream along with the bitplane information using a single model adaptive arithmetic coder, although there have been reported implementations [30, 19] that used context modeling for sign coding. As signs of neighboring coefficients are correlated, it can be expected that context modeling of sign symbols leads to increased coding performance. However, these speculations have not been confirmed by our experimental results. This can be attributed to the fact that signs consume only a minor percentage of the whole bit budget at typical video coding rates.

3.3.7   Subband prescanning and the indication bit.   A fundamental property of the wavelet transform is energy compaction. After the application of the transform, most of the signal energy is packed into a small number of transform coefficients that mainly reside in the low frequency subband. As a result, during the early successive approximation passes most of the high frequency subbands are equal to zero. Therefore it seems reasonable to prescan parts of the currently coded bitplane that correspond to a transform subband and issue an indication bit signaling the significance of that subband. If the bitplane contains at least one non-zero bit in a particular subband, then an indication bit of value one is output to the bitstream signaling that what follows in the bitstream is the encoded part of the current bitplane corresponding to that particular subband. On the contrary, when the subband is completely zero during the current quantization pass, then a zero indication bit is output indicating that what follows in the bitstream is the indication bit for the next scanned subband. This prescanning approach considerably increases coding efficiency, but at the same time results in asymmetry between coding and decoding, that is felt as being not critical. Notice also that the importance of symmetry/asymmetry between encoding and decoding situation is application dependent.

3.4   Post-Processing

Coders inevitably introduce artifacts into decoded video sequences. This is especially true at low bit rates. In order to improve the objective quality measure (PSNR) as well as the visual perception of the decoded video, post-processing is introduced into the video coding chain. Although post-processing functions can be introduced directly in the coding chain (for example in the prediction loop of hybrid coders) [35], we opt for the approach in which the enhancement post-processing stage is inserted at the end of the codec chain as in (Figure 7.2). In this case, the post-processing can be an optional building block that does not affect the codec design as such.

This chapter introduces two related post-processing approaches. The first one explicitly separates temporal filtering from filtering along the spatial axes of the video frames (Figure 7.10). Of course such a choice is motivated by advantages that a two stage filtering scheme exhibits in comparison with a pure 3-D filtering approach. Coarsely speaking, the second stage might in principle improve the results of the first filtering stage. The second post-filtering approach presented is a filter with a pure 3-D kernel.

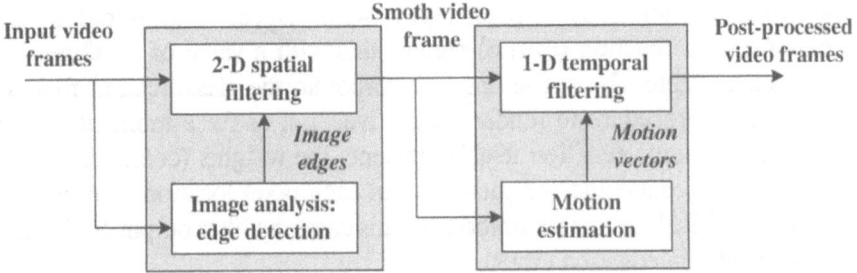

Figure 7.10.   Block diagram of the two-stage post-processing building block.

3.4.1   Spatial filtering with the SUSAN algorithm.

Removing coding artifacts while at the same time preserving image structure is a crucial prerequisite of filters employed in the post-processing stage. While linear filtering is inadequate due to its inability to preserve image detail [36], one might opt for adaptive filtering and algorithms based on complex image models [10, 36]. Although yielding good results, such complex approaches are unable to implement in real-time, which makes them an inappropriate choice for video enhancement. Nonetheless, complex image-model based approaches might well be suited for image enhancement where computational demands are considerably lower.

In order to achieve the desired post-processing effectiveness and minimally affect the original image structure while keeping the computational demands low, we chose to rely on the following assumption. Consequently, a classification procedure is required in order to group picture elements corresponding to a certain similarity criterion.

A modification of linear filtering that intrinsically incorporates pixel classification in order to preserve image detail is the SUSAN (Smallest Univalue Segment Assimilating Nucleus) filter [37]. Being a modification of linear filtering SUSAN exhibits low computational requirements, while being effective due to its adaptive nature. Let us first introduce the rationale behind SUSAN filtering and its mode of operation. Then the application of the SUSAN filter and its modifications to perform video post-processing will be discussed.

Pixel classification can be based on the comparison of pixel intensity values with a predefined threshold. Let us define the classification function $g(m, n)$ as shown in (7.18).

$$g(m, n) = \begin{cases} 1, & |S[i - m, j - n] - S[i, j]| \leq T; \\ 0, & |S[i - m, j - n] - S[i, j]| > T. \end{cases} \qquad (7.18)$$

The classification function compares the intensity of the pixel $S[i, j]$ with the intensity of a displaced pixel (displacement $[m, n]$). If the absolute pixel difference is less than a predefined threshold T, than the pixel $S[i - m, j - n]$ is said to be homogenous to pixel $S[i, j]$ with respect to threshold T, and the classification function $g(m, n)$ signals this with a value of 1. Otherwise $g(m, n)$ equals zero. Now one has to incorporate the classification function into a filtering procedure to render it adaptive. Let us for a moment assume that the classification function itself represents the weights (coefficients) of a linear filter. Of course the weights adapt at each pixel location to reflect the homogenity of the local neighborhood. In this case, the filter output S_S is given by the convolution equation (7.19).

$$S_S[i, j] = \sum_{m,n} g(m, n) \cdot S[i - m, j - n] \qquad (7.19)$$

The output of the SUSAN filter with weights defined by equation (7.18) is heavily influenced by the choice of the threshold T. Note that this directly translates into filter sensitivity to image details and artifacts. On the other hand a simple averaging filter with low attenuation at higher frequencies in not effective in reducing artifacts with substantial high frequency contents. For this reason the filter weights should correspond to a transfer function with greater attenuation at higher frequencies, which translates into smoother impulse response in the spatial domain. Accordingly, the filter weights are chosen to correspond to an exponential function as shown in (7.20). Besides providing a greater artifact reduction, this definition also lowers the sensitivity of the filter output to the value of threshold T.

$$g(m, n) = e^{-\left(\frac{S[i-m,j-n]-S[i,j]}{T}\right)^2} \qquad (7.20)$$

The adaptivity of the algorithm to the local neighborhood can be further improved by classifying the pixels not only according to their intensity differences, but also taking into account the pixel distances $r^2 = m^2 + n^2$. Doing so, Equation (7.20) can be reformulated into (7.21). The effect of the pixel differences and the pixel distances on the filter output is defined by parameters T and σ respectively.

$$g(m, n) = e^{-\frac{m^2+n^2}{2\sigma^2}} \cdot e^{-\left(\frac{S[i-m,j-n]-S[i,j]}{T}\right)^2} \qquad (7.21)$$

In order for the filter not to amplify or attenuate the signal, filter attenuation must be set to the value of one. This is done by normalizing filter weights according to.

$$g(m,n) = \frac{g(m,n)}{\sum\limits_{m,n} g(m,n)} \tag{7.22}$$

Another conjecture that can be proven empirically comes from the fact that not taking into account the value of the pixel at the position of the output pixel can greatly enhance the filter output sensitivity to impulse noise that may be present in the signal. Combining together equations (7.19), (7.20), (7.21) and (7.22) transforms equation (7.19) into (7.23). Note that in some circumstances the value of the denominator in (7.23) can reach zero. In this case the output cannot be computed and a median filter [36] is used instead.

$$s_S[i,j] = \frac{\sum\limits_{(m,n)\neq(0,0)} S[i-m,j-n] \cdot e^{-\frac{m^2+n^2}{2\sigma^2} - \frac{(S[i-m,j-n]-S[i,j])^2}{T^2}}}{\sum\limits_{(m,n)\neq(0,0)} e^{-\frac{m^2+n^2}{2\sigma^2} - \frac{(S[i-m,j-n]-S[i,j])^2}{T^2}}} \tag{7.23}$$

The described filer structure is flexible enough to be adjusted either to a 1-D or a 3-D case. For the first post-processing approach (separate spatial and temporal stage) the filter from equation (7.23) is used for the purpose of spatial filtering. The temporal filtering stage is described below.

3.4.2   Temporal filtering.   The filtering along the temporal axis of the video signal additionally removes coding artifacts in homogenous areas where it is most perceptible. In order to avoid any contra-productive temporal filtering, motion estimation is incorporated into the filtering process. Doing this leads to following a rationale similar to the one adopted to perform the MCTT along motion trajectories. The filter has to take into account that the location of video objects might have changed and that many other pixel transformations might have occurred while transiting from one frame to the other. It is therefore reasonable to track the movement of homogenous regions from frame to frame, and to perform the filtering along motion curves defined by motion vectors $MV(i,j,t) = [d_i(i,j,t), d_j(i,j,t)]^T = [d_i^{(t)}, d_j^{(t)}]^T$ where t is the temporal index of the observed frames and T denotes matrix transposition. Note that unlike in the case of MCTT computation, the temporal filtering can be performed along more than two video frames. A 1-D SUSAN filter is applied along the temporal axis. If we assume that the spatially filtered frames S_F are at the input, then the output of the temporal filter can be expressed by equation (7.24).

$$s_T[i,j,t] = \frac{\sum_l S_F\left[i-d_i^{(t+l)}, j-d_j^{(t+l)}, t-l\right] \cdot e^{-\frac{r^2}{2\sigma^2} - \varepsilon[i,j,t]}}{\sum_l e^{-\frac{r^2}{2\sigma^2} - \varepsilon[i,j,t]}}; \text{ where}$$

$$\varepsilon[i,j,t] = \frac{\left(S_F\left[i-d_i^{(t+l)}, j-d_j^{(t+l)}, t-l\right] - S_F[i,j,t]\right)^2}{T^2}$$

(7.24)

The temporal filtering equation (7.24) is similar in structure to the spatial filtering equation (7.23) with some notable differences. Besides the direction of filtering along the motion trajectories as defined by motion vectors, also the notion of pixel distance has to be redefined. In this case the value r is defined by the motion vector $[d_i, d_j]^T$ and the distance between frames l, i.e r^2 = $d_i^2 + d_j^2 + l^2$. Furthermore, the value of the current pixel is also taken into consideration while computing the filter output. This was not true in the spatial filtering case where the current pixel was omitted from the computation in order to increase filter stability in presence of impulse noise. By letting the current input pixel influence the output pixel, the smoothing degree gets increased. Although impulse noise is rarely encountered in wavelet image coding, empirical evidence has suggested the described setting in which spatial filtering is made less susceptible to impulse noise. It is also important to note that motion vectors required by the post-processing stage are already available after decoding the bit stream containing MC3DWT coefficients that include the block motion vectors required to compute the (inverse) transform. The computational complexity of the post-processing stage is thus substantially lowered. Another point concerning the implementation of the described temporal filtering is worth noting. As in the computation of MCTT, it is not always possible to establish a one-to-one relationship between pixels in consecutive frames. Consequently it is even harder to connect pixels on the same motion trajectory along multiple frames. Therefore in case of unconnected pixels, the temporal filtering is not applied, leaving these pixels intact. This is not the only drawback of the application of block based motion estimation. If the accuracy of the motion estimation algorithm is not adequate, this can lead to temporal filtering of pixels that are not correlated, leading to distortion due to post-processing.

3.4.3   3-D filtering.

As mentioned before, the two-stage filter can be combined into a pure 3-D filter. Although a very similar approach is taken, the 3-D filter is in most cases not as effective as the two-stage filter. This comes from the fact that the second stage of the two-stage filter already operates on substantially enhanced values, whereas the 3-D filter operates on original input samples all the time. The 3-D filter weights are obtained by extending the definition of the 2-D weights from equation (7.21), so that it encompasses the temporal dimension as well (7.25). Note that motion vectors are also taken into account. The distance r is in this case defined as $r^2 = (m + d_i^{(t-l)})^2 + (m + d_j^{(t-l)})^2 + l^2$

$$g(m,n,l) = \frac{e^{-\frac{r^2}{2\sigma^2}} - \frac{\left(S\left[i-m-d_i^{(t-l)}, j-n-d_j^{(t-l)}, t-l\right] - S[i,j,t]\right)^2}{T^2}}{\sum_{(m,n,l)} e^{-\frac{r^2}{2\sigma^2}} - \frac{\left(S\left[i-m-d_i^{(t-l)}, j-n-d_j^{(t-l)}, t-l\right] - S[i,j,t]\right)^2}{T^2}} \qquad (7.25)$$

The filter output S_F is in this case a 3-D convolution given by equation (7.26).

$$s_F[i,j,t] = \sum_{(m,n,l)} g(m,n,l) \cdot S\left[i - m - d_i^{(t-l)}, j - n - d_j^{(t-l)}, t - l\right]) \qquad (7.26)$$

4. Experiment

In order to evaluate the proposed video coding/enhancement framework the described algorithms have been thoroughly scrutinized. More precisely, the results are reported for four monochrome sequences in SIF format (352 × 240, 30 fps) and CIF (352 × 288, 25 fps): Coastguard (SIF), Football (SIF), Table tennis (CIF) and Foreman (CIF). The sequences are coded at five different bit rates in the range from 256 kbit/s to 1.0 Mbps. The quality of the decoded video has been evaluated objectively (PSNR) as well as from a perceptual point of view. Although PSNR does not always correlate well with perceptual image quality [7, 4], it has been found to match subjective evaluations in the tested cases. For this reason, since they are sufficiently illustrative, only PSNR values are reported.

The obtained PSNR values (MC3DWT) have been compared to the ones achieved by MPEG 2 [38, 27], 3D SPIHT [25, 26] and a simple "hybrid" coder. The hybrid wavelet based coder that uses context arithmetic coding of bitplanes (HYBRID). The performance assessment of the proposed coding algorithm from the comparison with MPEG 2 seems problematic, as MPEG 2 is not optimized for monochrome sequences. In addition, the rate control in MPEG 2 cannot produce the prescribed bit-rate exactly, although the rate has always been shown to be close enough to allow for fair comparison. For this reason a simple hybrid coder with block motion compensation was also used in the comparison. The "hybrid" coder is based on a two-dimensional wavelet transform along with context adaptive arithmetic coding. The group of frames contains 16 frames. The first frame is intra coded, the remaining 15 frames are coded using prediction (P frames only). The MPEG 2 coder is operating in the main profile, the GOP is 12 frames with the usual IBPBP...

frame pattern. Both hybrid coders (MPEG and "hybrid") and the proposed coder use half-pixel accurate motion estimation. While the range of motion vectors is limited to -15 to +15 in the proposed and the "hybrid" coders, the search range for MVs is left to its default value for MPEG. Bit allocation in the "hybrid" coder allocates twice the number of bits to the intra-coded frame than the inter-coded frames. Note that the "hybrid" coder achieves the desired bit-rate exactly. The 3D SPIHT algorithm was tested with the publicly available software implementation from the authors of 3D-SPIHT. The average PSNR results of the tests are grouped in Tables 7.1 to 7.4.

Test sequence	Algorithm	PSNR (dB) 256 kbit/s	PSNR (dB) 384 kbit/s	PSNR (dB) 512 kbit/s
Coastguard averaged over 288 frames	Proposed algorithm	26, 23	27, 71	28, 66
	MPEG 2	26, 14	27, 70	28, 83
	3D SPIHT	25, 50	26, 54	27, 35
	Hybrid	25, 88	27, 28	28, 38
Δ(proposed-MPEG 2)		0, 61	0, 09	0, 01
Δ(proposed-3D SPIHT)		1, 24	0, 73	1, 17

Test sequence	Algorithm	PSNR (dB) 768 kbit/s	PSNR (dB) 1 Mbit/s	PSNR (dB) 1.5 Mbit/s
Coastguard averaged over 288 frames	Proposed algorithm	29, 90	31, 49	33, 04
	MPEG 2	30, 39	31, 49	33, 35
	3D SPIHT	28, 67	29, 64	30, 73
	Hybrid	29, 62	31, 05	32, 63
Δ(proposed-MPEG 2)		$-0, 17$	$-0, 49$	0, 00
Δ(proposed-3D SPIHT)		1, 31	1, 23	1, 85

Table 7.1.   Average PSNR results for the Costguard sequence.

In order to assess the decoded image quality subjectively, Figure 7.11 and Figure 7.12 provide frame number 10 from the Football sequence decoded with MPEG 2 and the proposed algorithm at 256 kbit/s and 768 kbit/s. It can be seen that annoying blocking artifacts result from MPEG 2 coding while the proposed algorithm does not suffer from this type of degradation, although BME is inbuilt into the algorithm. On the other hand one may observe the typical image degradation that is a result of quantization of wavelet coefficients and is visually perceived as edge ringing and image smoothing. Obviously this type of degradation is much less noticeable than the blocking effects.

To evaluate the post-processing stage, the proposed 3-D extensions of the SUSAN filter have been tested on four decoded sequences: Coastguard, Foreman, Hall monitor and Container. All sequences were in SIF (352×240, 30 fps) format. The proposed filters (2D+T, 3-D SUSAN) have been compared

Test sequence	Algorithm	PSNR (dB) 256 kbit/s	PSNR (dB) 384 kbit/s	PSNR (dB) 512 kbit/s
Football averaged over 64 frames	Proposed algorithm	26, 99	18, 15	28, 98
	MPEG 2	27, 71	28, 89	29, 75
	3D SPIHT	27, 57	29, 05	30, 59
	Hybrid	27, 61	28, 88	29, 83
Δ(proposed-MPEG 2)		0, 14	−0, 72	−0, 74
Δ(proposed-3D SPIHT)		0, 11	−0, 58	−0, 90

Test sequence	Algorithm	PSNR (dB) 768 kbit/s	PSNR (dB) 1 Mbit/s	PSNR (dB) 1.5 Mbit/s
Football averaged over 64 frames	Proposed algorithm	30, 60	31, 49	32, 90
	MPEG 2	31, 35	32, 23	35, 39
	3D SPIHT	32, 36	33, 53	32, 39
	Hybrid	31, 32	32, 39	33, 65
Δ(proposed-MPEG 2)		−0, 77	−0, 75	−0, 74
Δ(proposed-3D SPIHT)		−1, 61	−1, 76	−2, 04

Table 7.2. Average PSNR results for the Football sequence.

Test sequence	Algorithm	PSNR (dB) 256 kbit/s	PSNR (dB) 384 kbit/s	PSNR (dB) 512 kbit/s
Table tennis	Proposed algorithm	29, 11	31, 14	32, 67
	MPEG 2	27, 96	30, 7	32, 31
	3D SPIHT	30, 12	31, 97	33, 42
	Hybrid	28, 39	30, 52	31, 93
Δ(proposed-MPEG 2)		1, 52	1, 15	0, 44
Δ(proposed-3D SPIHT)		−0, 64	−1, 01	−0, 83

Test sequence	Algorithm	PSNR (dB) 768 kbit/s	PSNR (dB) 1 Mbit/s	PSNR (dB) 1.5 Mbit/s
Table tennis	Proposed algorithm	34, 54	35, 84	38, 65
	MPEG 2	34, 56	36, 8	38, 38
	3D SPIHT	35, 84	37, 56	39, 6
	Hybrid	33, 86	35, 16	37, 98
Δ(proposed-MPEG 2)		−0, 77	−0, 75	−0, 74
Δ(proposed-3D SPIHT)		−1, 61	−1, 76	−2, 04

Table 7.3. Average PSNR results for the Table tennis sequence.

to the standard deringing filter [39]. The aim of post-processing is to improve the perceptual quality of decoded images. This does not always translate in

Test sequence	Algorithm	PSNR (dB) 256 kbit/s	PSNR (dB) 384 kbit/s	PSNR (dB) 512 kbit/s
Foreman averaged over 288 frames	Proposed algorithm	31, 44	33, 11	34, 21
	MPEG 2	29, 99	32, 81	34, 29
	3D SPIHT	31, 04	32, 78	33, 9
	Hybrid	31, 05	32, 78	33, 89
Δ(proposed-MPEG 2)		1, 50	1, 45	0, 30
Δ(proposed-3D SPIHT)		0, 44	0, 40	0, 33

Test sequence	Algorithm	PSNR (dB) 768 kbit/s	PSNR (dB) 1 Mbit/s	PSNR (dB) 1.5 Mbit/s
Foreman averaged over 288 frames	Proposed algorithm	35, 99	36, 90	39, 01
	MPEG 2	36, 05	37, 16	38, 94
	3D SPIHT	34, 99	36, 73	38, 52
	Hybrid	31, 32	32, 39	33, 65
Δ(proposed-MPEG 2)		−0, 08	−0, 06	−0, 26
Δ(proposed-3D SPIHT)		0, 31	0, 25	0, 07

Table 7.4.   Average PSNR results for the Foreman sequence.

an increase in PSNR. However in this case, the application of the proposed filters results also in an increase in PSNR. The increase in PSNR for the tested sequences at various bit rates can be found in Table 7.5, Table 7.6, Table 7.7 and Table 7.8.

Foreman	256 kbit/s	384 kbit/s	512 kbit/s	768 kbit/s	1 Mbit/s	1,5 Mbit/s
Deringing	0, 09	0, 1	0, 12	0, 13	0, 15	0, 18
2D+T	0, 16	0, 22	0, 25	0, 24	0, 09	−0, 2
3-D	0, 34	0, 48	0, 59	0, 75	0, 79	0, 82

Table 7.5.   Average increase in PSNR (dB) after post-processing of the Foreman sequence.

Container	256 kbit/s	384 kbit/s	512 kbit/s	768 kbit/s	1 Mbit/s	1,5 Mbit/s
Deringing	0, 12	0, 12	0, 05	−0, 21	−0, 56	−1, 2
2D+T	0, 39	0, 27	0, 18	−0, 77	−1, 69	−3, 43
3-D	0, 33	0, 32	0, 31	−0, 44	−1, 18	−2, 65

Table 7.6.   Average increase in PSNR (dB) after post-processing of the Container sequence.

As far as the post-processing methods are concerned the two-stage (2D+T) filter is a clear winner. However, the reader should note the fact that post-

a)

b)

Figure 7.11. Decoded frame number 10 from the Football sequence at 256 kbit/s using MPEG 2 a) and the proposed algorithm b).

processing is only applicable at lower bit-rates where image degradation is considerable. The results in Tables 7.5 to 7.8 clearly show, that post-processing becomes counterproductive at bit-rates above 768 kbps in our case. It can be

a)

b)

Figure 7.12.   Decoded frame number 10 from the Football sequence at 768 kbit/s using MPEG 2 a) and the proposed algorithm b).

conjectured that post-processing is not applicable at higher bit-rates as it tends to remove image detail, while there is no need for artifact removal.

Hall monitor	256 kbit/s	384 kbit/s	512 kbit/s	768 kbit/s	1 Mbit/s	1,5 Mbit/s
Deringing	0,19	0,1	−0,01	−0,43	−0,55	−1,05
2D+T	0,48	0,33	−0,01	−1,13	−1,53	−2,88
3-D	0,44	0,36	0,15	−0,71	−1,01	−2,15

Table 7.7. Average increase in PSNR (dB) after post-processing of the Hall monitor sequence.

Coastguard	256 kbit/s	384 kbit/s	512 kbit/s	768 kbit/s	1 Mbit/s	1,5 Mbit/s
Deringing	0,06	0,06	0,04	0,03	−0,07	−0,15
2D+T	0,21	0,29	0,1	0,05	−0,52	−1,14
3-D	0,14	0,16	−0,06	−0,18	−0,77	−1,38

Table 7.8. Average increase in PSNR (dB) after post-processing of the Coastguard sequence.

In order to assess the post-processed image quality subjectively, Figure 7.13 provides details of the frame number 10 belonging to the "Hall monitor" and "Foreman" sequences obtained by the proposed algorithm at 256 kbit/s. It can be seen that the annoying blocking artifacts that are common to MPEG (DCT based) coding are not generated. However similar blocking artifacts might arise at much lower bit-rates as block motion estimation is built in the proposed algorithm. On the other hand one may observe the typical image degradation that is a result of quantization of wavelet coefficients and is visually perceived as edge ringing and image smoothing. Obviously this type of degradation is much less noticeable than the blocking effects. With the application of the presented post-processing filters the visibility of these artifacts can be additionally reduced.

5. Conclusion

In this chapter general approaches to video coding have been discussed and an approach based on MC3DWT and context based arithmetic coding of bitplanes was presented and extensively discussed. Experimental results revealed the effectiveness of the proposed 3-D transform based approach when compared to hybrid-paradigm based coders and to the more advanced 3D SPIHT coder.

Post-processing methods were also discussed and two low complexity artifact removal algorithms were presented. The proposed coding algorithm in conjunction with the described enhancement stages provide a viable alternative to hybrid video coding at higher bit-rates. However due to the application of the 3-D transform, the algorithm can not achieve very low bit-rates, therefore its application domain is in moderate and high bit-rate video coding. Moreover, due to the intrinsic delay introduced by the application the 3-D transform (due

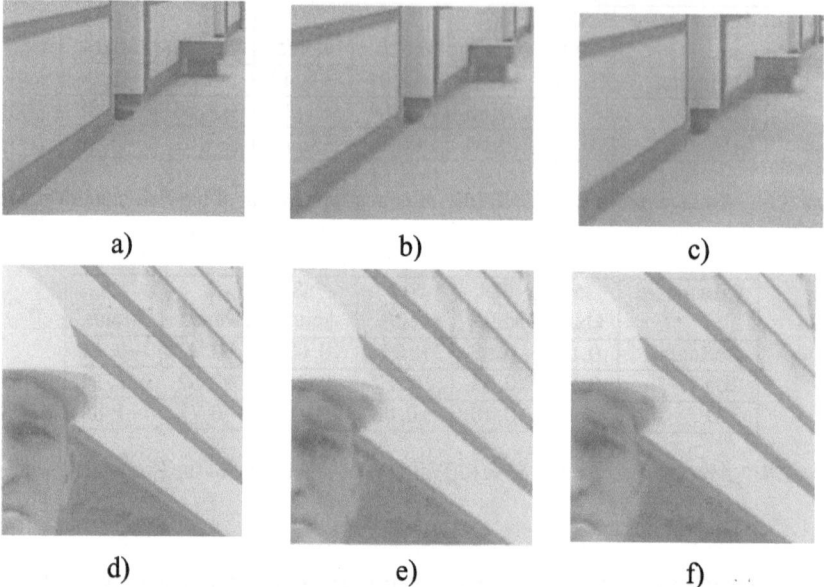

a)   b)   c)

d)   e)   f)

Figure 7.13.   Coding and enhancement results. The images show the effect of video coding and enhancement on a detail from the "Hall monitor" and "Foreman sequences" coded at 256 kbps. Images a) and d) represent the original input frame, while b) and e) represent the decoded video frame using the MC3DWT based codec. Perceptual quality enhancement due to post-processing by the two-stage (2D+T) filtering approach is visible from c) and f).

to a group of frames coded together), the proposed algorithm is suitable for applications where the introduced delay can be tolerated.

From the approach described there is an important lesson to be learned. Although improvements of standard video coders are usually achieved by building on top of the hybrid-coding paradigm this is not the only way to achieve improved video coding performance. Also the introduction of an alternative paradigm 3-D transform based paradigm as an alternative to hybrid coding can provide substantial increase in performance, as demonstrated by our approach.

Furthermore, a video post-processing stage that forms an integral part of our proposed approach, significantly improves the quality of decoded video. The later is true in terms of perceived quality (subjective quality) as well as in terms of peak signal-to-noise ratio (objective quality measure).

More specifically two low complexity artifact removal algorithms were presented in this chapter. The proposed coding algorithm in conjunction with the described enhancement stages provides a viable alternative to hybrid video coding at higher bit-rates. However due to the application of the 3-D transform, the algorithm can achieve very low bit-rates, therefore its application domain is in moderate and high bit-rate video coding.

References

[1] G. Cote, B. Erol, M. Gallant, and F. Kossentini, "H.263+: Video coding at low bit rates," *IEEE Trans. on Circuits and Systems for Video Technology*, vol. 8, November 1998.

[2] A. M. Tekalp, *Digital Video Processing*. Prentice-Hall, 1995.

[3] A. Bovik, *Handbook of Image and Video Processing*. Academic Press, 2000.

[4] R. M. G. A. Gersho, *Vector Quantization and Signal Compression*. Boston: Kluwer Academic Publishers.

[5] R. J. Clarke, *Transform Coding of Images*. Academic Press, 1985.

[6] B. Furht, "A survey of multimedia compression techniques and standards. Part I: JPEG standard," *Real-Time Imaging Journal*, vol. 1, pp. 49–67, April 1995.

[7] J. K. M. Vetterli, *Wavelets and Subband Coding*. Englewood Cliffs, New Jersey: Prentice Hall, 1995.

[8] N. Vasconcelos and F. Dufaux, "Pre and post-filtering for low bit-rate video coding," in *ICIP'97*, (Santa Barbara, California), 1997.

[9] M. Y. Shen and C. C. J. Kuo, "Review of postprocessing techniques for compression artifact removal," *Journal of Visual Commun. and Image Representation*, vol. 9, no. 1, 1998.

[10] J. C. Brailean, R. P. Kleihorst, S. Efstratiadis, A. K. Katsaggelos, and R. L. Lagendijk, "Noise reduction filters for dynamic image sequences: A review," *Proceedings of the IEEE*, vol. 83, September 1995.

[11] R. Lagendijk, P. M. B. van Roosmalen, and J. Biemond, *The Image and Video Processing Handbook*, ch. Video Enhancement and Restoration. 1999.

[12] S. Mallat, *A Wavelet Tour of Signal Processing*. Academic Press, 1998.

[13] K. Ramchandran, A. Ortega, and M. Vetterli, "Bit allocation for dependent quantization with applications to multiresolution and MPEG video coders," *IEEE Transactions on Image Processing*, Aug. 1994.

[14] Y. Shoham and A. Gersho, "Efficient bit allocation for an arbitrary set of quantizers," *IEEE Trans on ASSP*, vol. 36, no. 9, 1988.

[15] C. I. Podilchuk, N. S. Jayant, and N. Farvardin, "Three-dimensional sub-band coding of video," *IEEE Trans. on Image Proc.*, vol. 4, February 1995.

[16] A. Said and W. A. Pearlman, "A new and efficient image codec based on set partitioning in hierarchical trees," *IEEE Trans. on Systems and Circuits for Video Technology*, vol. 6, June 1996.

[17] J. R. Ohm, "Three-dimensional subband coding with motion compensation," *IEEE Trans. on Image Proc.*, vol. 3, September 1994.

[18] S. Choi and J. W. Woods, "Motion-compensated 3-D subband coding of video," *IEEE Trans. on Image Proc.*, vol. 8, pp. 155–167, Feb. 1999.

[19] X. Wu, "High-order context modeling and embedded conditional entropy coding of wavelet coefficients for image compression," in *Proceedings of Thirty-First Asilomar Conference on Signals, Systems & Computers*, vol. 2, pp. 1378–1382, 1997.

[20] J. M. Shapiro, "Embedded image coding using zerotrees of wavelet coefficients," *IEEE Trans. on Signal Proc.*, vol. 41, December 1993.

[21] D. Taubman and A. Zakhor, "Multi-rate 3-D subband coding of video," *IEEE Transactions on Image Proc*, vol. 3, pp. 572–588, Sept. 1994.

[22] W. A. Pearlman, B. Kim, and Z. Xiong, *Wavelet Image and Video Compression*, ch. Embedded Video Subband Coding with 3D SPIHT. Boston: Kluwer Academic Publishers, 1998.

[23] S. Hsiang and J. W. Woods, "Invertible three-dimensional analysis/synthesis system for video coding with half pixel accurate motion compensation," in *SPIE VCIP*, 1999.

[24] S. Hsiang and J. W. Woods, "Embedded video coding using invertible motion compensated 3D subband/wavelet filter bank," *Image Communication*, vol. 16, no. 8.

[25] B. J. Kim and W. A. Pearlman, "An embedded wavelet video coder using set partitioning in hierarhical trees," in *Proc. Data Compression Conference*, 1997.

[26] B. J. Kim, Z. Xiong, and W. A. Pearlman, "Low bit-rate scalable video coding with 3D SPIHT," *IEEE Trans. CSVT*, vol. 10, 2000.

[27] J. L. Mitchell, W. B. Pennebaker, C. E. Fogg, and D. J. LeGall, *MPEG video compression standard*. Chapman & Hall, 1996.

[28] J. S. Lee, "Digital image enhancement and noise filtering by use of local statistics," *IEEE Transactions on Pattern Analysis and Machine Intelligence*, vol. PAMI-2, March 1980.

[29] T. P. O'Rourke and R. L. Stevenson, "Improved image decompression for reduced transform coding artifacts," *IEEE Transactions on Circuits and Systems for Video Technology*, vol. 5, December 1995.

[30] D. Taubman, "High performance scalable image compression with EBCOT," *IEEE Trans. on Image Proc.*, March 1999.

[31] G. Karlsson and M. Vetterli, "Three dimensional subband coding of video," in *Proc. ICASSP*, April.

[32] J. D. Villasenor, B. Beltzer, and J. Liao, "Wavelet filter evaluation for image compression," *IEEE Transactions on Image Processing*, vol. 4, August 1995.

[33] I. H. Witten, R. Neal, and J. Cleary, "Arithmetic coding for data compression," *Comm. ACM*, vol. 30, pp. 520–540, June 1987.

[34] X. Wu, "Lossless compression of continuous-tone images via context selection, quantization, and modeling," *IEEE Trans. on Image Processing*, vol. 6, pp. 656–664, May 1997.

[35] Y. L. Lee, H. W. Park, D. S. Park, and Y. S. Kim, "Loop filtering for reducing blocking effects and ringing effects," New Proposal Q15-E-22, KAIST Samsung Electronics Co., Whistler, Canada, July 1998.

[36] I. Pitas and A. N. Venetsanopoulos, *Nonlinear Digital Filters*. Kluwer Academic Press, 1990.

[37] S. M. Smith and J. M. Brady, "SUSAN-A new approach to low level image processing," Technical Report TR95SMS1c, Oxford University, 1995.

[38] "MPEG 2 video codec (test model TM 5)."
http://www.mpeg.org/MPEG/MSSG/.

[39] "Information technology-coding of audio-visual objects: Visual," Committee Draft 14496-2, ISO/IEC, Tokyo, March 1998.

Chapter 8

BIT-RATE CONTROL FOR THE JPEG ALGORITHM

Javier Bracamonte, Michael Ansorge, and Fausto Pellandini
Institute of Microtechnology, University of Neuchâtel,
Rue A.-L. Breguet 2, 2000 Neuchâtel, Switzerland.
{javier.bracamonte, michael.ansorge, fausto.pellandini}@unine.ch

Abstract   JPEG is an international industry standard algorithm for compressing continuous-tone still images. JPEG which stands for Joint Photographic Expert Group is currently one of the most popular encoding formats used for storing and transmitting images. JPEG is a variable coding bit-rate method and in this chapter two algorithms for controlling the produced compression ratio are reported. The bit-rate control techniques feature an excellent accuracy, a low computational complexity, and a full compliance with the JPEG standard bitstream. Furthermore, the application of these algorithms does not affect the quality of the decompressed images, i.e., to reach the target compression ratio there are no additional losses other than those inherent to JPEG. The most performant of the two bit-rate control techniques has been successfully implemented in a consumer electronics industrial prototype.

Keywords: Image compression, JPEG, bit-rate control.

1.   Introduction

One of the characteristics of the JPEG algorithm is that of producing a compression bit rate that varies depending on the level of information content of the image that is being coded. As a result, two images with the same spatial and gray-level resolutions—i.e., same width, height, and number of bits per pixel—but with different degrees of image detail, will result into two JPEG-

J.F. Tasič et al. (eds.), Intelligent Integrated Media Communication Techniques, 263-301.
© 2003 *Kluwer Academic Publishers.*

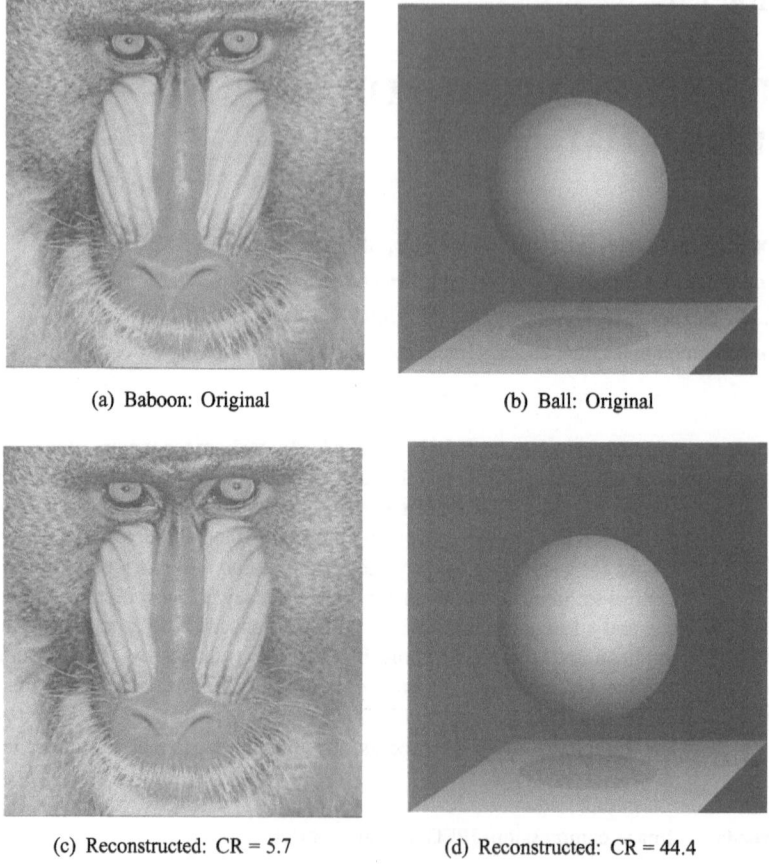

(a) Baboon: Original   (b) Ball: Original

(c) Reconstructed: CR = 5.7   (d) Reconstructed: CR = 44.4

Figure 8.1.   JPEG compression: different CRs for different images.

compressed bitstreams that, in general, do not have the same number of bits. These two images are thus coded with different bit rates.[1]

Figures 8.1(a) and (b) show two images of 512x512 pixels that have largely-different degrees of information content. The compression with JPEG of the image in Figure 8.1(a), which is an image that contains a significant amount of detail, results in a bitstream whose number of bits corresponds to a compression bit rate of 1.40 bit/pixel (bpp). The reconstructed image for this bit rate is shown in Figure 8.1(c). The compression bit rate for the image in Figure 8.1(b), which is an image that contains substantially less detail than the image in Figure 8.1(a), is 0.18 bpp; its reconstructed version is shown in Figure 8.1(d). In terms of

---

[1]In the image compression literature, the compression (bit) rate is used, alike the compression ratio (CR), to characterize the performance of a compression algorithm. When the input images are of 256 gray levels—i.e., 8 bits per pixel—the compression bit rate (CBR) is related to the compression ratio (CR) by: CBR=8/CR.

compression ratio (CR), JPEG is coding the former image with a compression ratio of 5.7, while it is coding the latter, using the very same set of JPEG parameters, with a CR of 44.4: a large difference! It is worth observing that in Figures 8.1(c) and (d), the higher CR for the image in Figure 8.1(b) is not obtained to the detriment of the quality of the decompressed image. Actually, in terms of PSNR (Peak Signal to Noise Ratio), the quality of the reconstructed image in Figure 8.1(d) is even 15.22 dB *better* than that of the reconstructed image in Figure 8.1(c). This is reminded since it is always possible to increase the compression ratio in exchange of a loss in the quality of the reconstructed images; nevertheless, in this example this is not the case nor the point that it is wanted to be emphasized.

This image-depended, variable compression bit-rate characteristic of JPEG represents a main drawback for some applications. In digital still cameras for example, it is important to be able to predict the size of the compressed pictures in order to control the allocation of the solid-state memory of the camera. Being able to control the resulting compression bit rates is also required in networking applications that involve the transmission of image/video signals. In these networks, the time required to transmit a compressed image is generally fixed or not allowed to exceed a certain value; precluding in consequence, the use of non-controllable bit-rate schemes.

In this chapter, two bit-rate control methods for the JPEG algorithm are presented. These techniques permit to produce, with an excellent accuracy, a user- or system-requested compression ratio (or its equivalent compression bit rate) of JPEG-coded images. The presented methods [1] have several advantages with respect to other bit-rate schemes that have been previously reported [2, 3, 4, 5]. Some of these advantages are:

- A full compliance with the JPEG algorithm. This feature gives the system all the advantages inherent to *standard* algorithms; as, for example, a widespread use of the compressed images.

- A very low computational complexity. This is an important characteristic in order to produce high-speed, cost-effective hardware implementations, and also to keep power consumption as low as possible.

  The bit-rate control system proposed in [2], for example, requires too many iterations to converge. While the systems in [3, 4, 5] must execute many computations for each 8x8 block of pixels in order to check if the allocated bits per block have not been exceeded.

- By their own principle, the application of the bit-rate control mechanisms described in this chapter, does not cause any further degradation to the compressed images. The quality of the decompressed images is

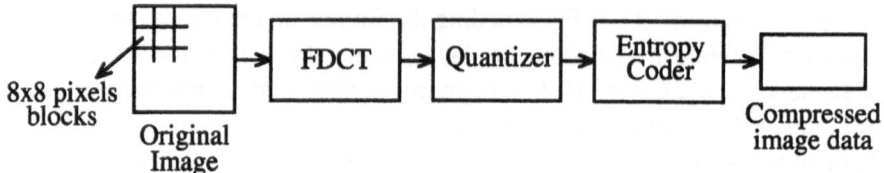

Figure 8.2.   Baseline JPEG algorithm.

exactly the same that would be given by JPEG (with the corresponding parameters) in the absence of the bit-rate control system.

The lack of this feature is a major drawback of the systems reported in [3, 4, 5], which must suppress high frequency information in order to make the bit rate converge to the target value.

- Finally, the methods proposed in this chapter, provide an excellent accuracy for setting the compression ratio, and the largest application operational range.

The remainder of this chapter is organized as follows. Section 2 illustrates JPEG's variable compression rate nature and Section 3 shows how it is possible to modulate the output compression ratio given by JPEG. In Section 4 a study of the characteristic of the compression ratio vs. a modulating parameter is studied. The general scheme of the bit-rate control method is given in Section 5 and a particular implementation of this scheme is described in Section 6. An optimized implementation is reported in Section 7, and its results are given in Section 8. A detailed computational complexity analysis is made in Section 9. Section 10 briefly describes JPEG2000 and its position with respect to JPEG, particularly in hardware implementations. Finally, the chapter is closed with a summary in Section 11.

2.   JPEG: A Variable Compression Bit-Rate Algorithm

2.1   Baseline JPEG Algorithm

The baseline sequential JPEG algorithm is depicted in Figure 8.2. The input image is subdivided into blocks of 8x8 pixels. Sequentially, each of these blocks is processed as follows:

- A Forward Discrete Cosine Transform (FDCT) operation transforms the 8x8 pixels from the original image into 8x8 coefficients in the frequency domain. The goal of this transformation is to decorrelate the original data and redistribute the signal energy among only a small set of transform coefficients within the low frequency zone. Before applying the forward

DCT, the value of the 64 pixels of the original block, is level-shifted by -128 (i.e., 128 is subtracted from each pixel value).

- Based on a psychovisual analysis [6, 7], a normalization array can be defined. Its purpose is to quantize those DCT coefficients that are visually significant with relatively short quantization steps, while using large quantization steps for those coefficients which are less important.

- Though the quantization of the DCT coefficients is the main mechanism of data compression, additional data compression can be obtained by entropy coding the output of the quantizer. The entropy coding block in Figure 8.2 comprises three lossless operations.

  - *DC DPCM:* The value of the DC coefficient[2] is proportional to the average value (brightness) of the original block of pixels. Since the average brightness of two consecutive 8x8-pixel blocks is generally very similar, each DC coefficient is DPCM[3] encoded using the DC coefficient of the previous 8x8-pixel block as the prediction value.

  - *AC Run-length:* Many AC coefficients become zero after the quantization operation. Lining the quantized 8x8 elements of the 2-D DCT matrix up into a 64-element 1-D vector, by following a zigzag pattern [8], produces long sequences (or runs) of zero-valued coefficients. These long runs of zeros can be more efficiently represented by using run-length coding [9].

  - *Huffman Coding:* Finally, the symbols produced at the output of both the DPCM and run-length operations are Huffman encoded [10]. Both DC and AC Huffman tables are defined respectively for the DC (DPCM-coded) and the AC (run-length coded) coefficients. While both Huffman and arithmetic coding are defined by the general JPEG standard for this last entropy coding stage, the baseline system requires Huffman coding only. This has two advantages: first, Huffman coding demands a less complex hardware implementation than arithmetic coding; and second, the arithmetic QM-coder suitable for JPEG is covered by several patents; its commercial implementation, in consequence, requires the payment of license fees. It is reported in [11] that the use of arithmetic in place of Huffman coding improves JPEG's compression ratio by 5% to 10%.

---

[2] The top-left element $X(0,0)$, from the array X that represents the 2-D DCT coefficients, is referred to as the DC coefficient, the remaining 63 are referred to as the AC coefficients.
[3] Differential Pulse Code Modulation.

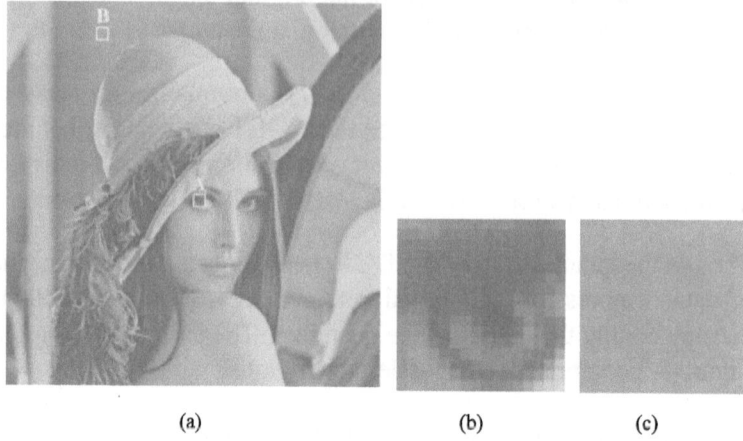

Figure 8.3. (a) Two different image-activity blocks. (b) Block A. (c) Block B.

2.2   The Effect

This section presents an example to illustrate the image-dependent, variable-rate mechanism inherent to JPEG.

In Figure 8.3 two 8x8-pixel blocks, A and B, are specified within the test image Lena (only for the purpose of highlighting the difference in the content of image detail, the two blocks in Figure 8.3 are indeed 16x16-pixels blocks, the center region of which are the real 8x8-pixel blocks). Each block will be coded separately, following the JPEG scheme of Figure 8.2. The outcome of each of the operations will be shown and at the end of the algorithm, the two resulting bitstreams will be compared. Block A has been selected as to contain more image detail than block B and should (due to the JPEG mechanism we are discussing) require more bits to be coded.

The values of the 64 pixels of block A are given by the matrix A in Equation 8.1.

$$A = \begin{bmatrix} 77 & 69 & 69 & 70 & 71 & 67 & 67 & 67 \\ 74 & 70 & 72 & 67 & 64 & 66 & 66 & 65 \\ 67 & 72 & 68 & 68 & 64 & 66 & 68 & 66 \\ 65 & 72 & 71 & 71 & 62 & 65 & 69 & 90 \\ 68 & 76 & 79 & 65 & 61 & 56 & 58 & 99 \\ 78 & 93 & 95 & 69 & 59 & 56 & 56 & 71 \\ 73 & 108 & 106 & 92 & 68 & 60 & 60 & 83 \\ 70 & 105 & 110 & 105 & 90 & 84 & 83 & 102 \end{bmatrix} \qquad (8.1)$$

The result after the 2-D DCT has been applied to the array A (whose elements are previously offset by -128) is given in Equation (8.2). It can be seen that most of the energy is concentrated in only a few low frequency—i.e., the top left corner region—coefficients.

$$
\mathbf{X} = \begin{bmatrix}
-431 & 25 & 11 & -48 & -3 & -16 & 4 & -6 \\
-52 & -16 & 6 & 40 & 8 & 15 & 0 & 5 \\
35 & 9 & -20 & 4 & -5 & 4 & -6 & 0 \\
-18 & 19 & 11 & -2 & 2 & -3 & 6 & 6 \\
15 & -26 & 2 & -3 & 11 & -3 & 2 & -1 \\
-1 & 6 & -3 & -2 & -1 & 1 & -5 & -4 \\
-1 & 1 & -4 & 9 & -4 & 2 & -1 & 3 \\
-2 & -1 & 0 & -4 & 4 & -4 & -1 & -1
\end{bmatrix} \tag{8.2}
$$

The quantization of the 2D-DCT coefficients is carried out by dividing one-to-one, each element of the matrix $\mathbf{X}$ by the corresponding element of a normalization array, which in this example is given by the matrix $\mathbf{N}$ in Equation (8.3).

$$
\mathbf{N} = \begin{bmatrix}
16 & 11 & 10 & 16 & 24 & 40 & 51 & 61 \\
12 & 12 & 14 & 19 & 26 & 58 & 60 & 55 \\
14 & 13 & 16 & 24 & 40 & 57 & 69 & 56 \\
14 & 17 & 22 & 29 & 51 & 87 & 80 & 62 \\
18 & 22 & 37 & 56 & 68 & 109 & 103 & 77 \\
24 & 35 & 55 & 64 & 81 & 104 & 113 & 92 \\
49 & 64 & 78 & 87 & 103 & 121 & 120 & 101 \\
72 & 92 & 95 & 98 & 112 & 100 & 103 & 99
\end{bmatrix} \tag{8.3}
$$

The quotients of these 64 divisions are rounded to their nearest integers, giving as a result the matrix $\mathbf{Q}$ below,

$$
\mathbf{Q} = \begin{bmatrix}
-27 & 2 & 1 & -3 & 0 & 0 & 0 & 0 \\
-4 & -1 & 0 & 2 & 0 & 0 & 0 & 0 \\
2 & 1 & -1 & 0 & 0 & 0 & 0 & 0 \\
-1 & 1 & 0 & 0 & 0 & 0 & 0 & 0 \\
1 & -1 & 0 & 0 & 0 & 0 & 0 & 0 \\
0 & 0 & 0 & 0 & 0 & 0 & 0 & 0 \\
0 & 0 & 0 & 0 & 0 & 0 & 0 & 0 \\
0 & 0 & 0 & 0 & 0 & 0 & 0 & 0
\end{bmatrix}. \tag{8.4}
$$

Reordering the elements of the matrix $\mathbf{Q}$ by following a zigzag pattern (serpentine path from the top-left to the bottom-right coefficient) yields the 64-element vector $\mathbf{z}$,

$$
\mathbf{z} = [-27 \quad 2 \quad -4 \quad 2 \quad -1 \quad 1 \quad -3 \quad 0 \quad 1 \quad -1 \quad 1 \quad 1 \\
-1 \quad 2 \quad 0 \quad 0 \quad 0 \quad 0 \quad 0 \quad -1 \quad \underbrace{0 \dots 0}_{44}].
$$

The value of the DC coefficient of the previous transformed and quantized 8x8-pixel block was found to be -25. The DPCM operation over the DC coefficients, simply substitutes the first element of the vector $\mathbf{z}$ by -2 [= -27 - (-25)], becoming

$$d = \begin{bmatrix} -2 & 2 & -4 & 2 & -1 & 1 & -3 & 0 & 1 & -1 & 1 & 1 \\ & & & -1 & 2 & 0 & 0 & 0 & 0 & 0 & -1 & \underbrace{0 \ldots 0}_{44} \end{bmatrix}. \quad (8.5)$$

A zero run-length coding of the AC coefficients in vector **d** results in a sequence of pairs of values (*run,coeff*) given by the expression in Equation (8.6). The first value, *run*, indicates the number of zeros that precedes the second value, *coeff*, in the sequence represented by elements of vector **d**. The DC coefficient is not concerned by this operation, a *run* = 0 has been associated to its value in Equation (8.6) just to allow a generalization of the explanations that follow below.

$$r = \begin{bmatrix} (0,-2) & (0,2) & (0,-4) & (0,2) & (0,-1) & (0,1) \\ (0,-3) & (1,1) & (0,-1) & (0,1) & (0,1) & (0,-1) \\ (0,2) & (5,-1) & \text{EOB} \end{bmatrix}. \quad (8.6)$$

The symbol EOB stands for End-Of-Block. Since the compressed data of all the 8x8-pixel blocks which make up the complete input image are concatenated into a single bitstream, an EOB symbol allows the decoder to detect the boundaries of each 8x8-pixel block. In this case EOB is also implicitly indicating that the coefficients following the fourth -1 in Equation (8.5) are all zeros.

Though a statistical analysis could generate an (optimal) Huffman code for each possible combination of values of (*run,coeff*), this would produce a Huffman table with a too large number of entries. By representing *coeff* in a different format, the number of Huffman codes can be reduced. Thus, the value of *coeff* is broken down into two components: category (*cat*) and complement (*cmp*). Table 8.1 shows some of the different ranges for the values of *coeff* defined by JPEG, both for the (DPCM-ed) DC and AC coefficients. Each range is identified with a category number *cat*. For the AC coefficients, *coeff* can never be zero, therefore the appearance of the Non-Available (N/A) symbol in Table 8.1.

The value of *cat* only specifies a range of values. In order to completely specify *coeff* a complementary number *cmp* is required. The binary representation of *cmp* is given by the least significant *cat* bits of *coeff*, that is, the wordlength of *cmp* is given by the value of *cat*. Furthermore when *coeff* < 0, the value of *cmp* is decremented by one. This last operation is needed to represent without ambiguity both positive and negative number, in the same range of magnitude, with a same number of bits.[4]

Following these manipulations on *coeff*, the pairs (*run,coeff*) in Equation (8.6) can be rewritten as (*run/cat,cmp*), in which by expressing *cmp* as a binary num-

---

[4]In two's complement representation, with n bits, it is possible to represent integers in the asymmetric-to-zero range $[-2^{n-1}, 2^{n-1} - 1]$.

Table 8.1. Baseline JPEG coefficient categories [12].

Range	DC Category	AC Category
0	0	N/A
-1,1	1	1
-3,-2,2,3	2	2
-7,...,-4,4,...,7	3	3
-15,...,-8,8,...,15	4	4
-31,...,-16,16,...,31	5	5
-63,...,-32,32,...,63	6	6
⋮	⋮	⋮
-4095,...,-2048,2048,...,4095	C	C

ber yields,

$$\mathbf{c} = \begin{matrix} [(0/2,01) & (0/2,10) & (0/3,011) & (0/2,10) & (0/1,0) \\ (0/1,1) & (0/2,00) & (1/1,1) & (0/1,0) & (0/1,1) \\ (0/1,1) & (0/1,0) & (0/2,10) & (5/1,0) & \text{EOB}]. \end{matrix} \qquad (8.7)$$

The last part of the JPEG's entropy coding operation consists in looking a binary code up in a Huffman table for each of the *run/cat* combinations in Equation (8.7). For this example the JPEG Huffman tables reported in Annex K of the JPEG standard document [12] has been used; samples of these tables are reproduced on Table 8.2. This operation generates the following bitstream:

01101/0110/100011/0110/000/001/0100/11001/000/001/001/000/

0110/11110100/1010

where for readability of the example a slash has been inserted between the binary codes associated with each of the elements of the vector **c**. The last code, namely "1010", is the binary code that corresponds to the EOB symbol.

The final result indicates that the original 8x8-pixel block requires 62 bits to be coded. The representation of each pixel of matrix **A** in Equation (8.1) requires 8 bits, which makes a total of 512 bits to represent the complete block. JPEG is thus coding this block with a compression ratio[5] of 512/62 = 8.26.

Block *B* will be coded by following the same steps as for coding block *A*. At this point the reader might find it useful to compare the intermediate results of the encoding of both blocks. The values of the 64 pixels of block *B* are given

---

[5]The compression ratio is defined as the quotient of the number of bits of the uncompressed original image divided by the number of bits of the compressed image.

Table 8.2.   Samples of a Huffman table for the DC and AC coefficients [12].

Category (DC coefficients)	Code Length (cat & cmp)	Code Word
0	2	00
1	4	010
2	5	011
3	6	100
⋮	⋮	⋮
11	20	111111110

Run/Category (AC coefficients)	Code Length (cat & cmp)	Code Word
0/0	4	1010 (=EOB)
0/1	3	00
0/2	4	01
0/3	6	100
...	...	...
0/A	26	1111111110000011
1/1	5	1100
1/2	7	11011
...	...	...
1/A	26	1111111110001000
2/1	6	11100
2/2	10	11111001
...	...	...
2/A	26	1111111110001110
3/1	7	111010
3/2	11	111110111
...	...	...
3/A	26	1111111110010101
4/1	7	111011
4/2	12	1111111000
...	...	...
4/A	26	1111111110011101
5/1	8	1111010
5/2	13	11111110111
⋮	⋮	⋮
F/A	26	1111111111111110

by the following matrix:

$$B = \begin{bmatrix} 134 & 133 & 139 & 141 & 134 & 134 & 135 & 137 \\ 133 & 134 & 136 & 136 & 136 & 136 & 135 & 133 \\ 138 & 135 & 137 & 140 & 137 & 139 & 138 & 135 \\ 132 & 138 & 137 & 141 & 139 & 137 & 137 & 136 \\ 135 & 138 & 140 & 142 & 136 & 139 & 138 & 137 \\ 136 & 135 & 134 & 135 & 140 & 138 & 139 & 140 \\ 136 & 136 & 138 & 138 & 137 & 136 & 135 & 141 \\ 137 & 136 & 140 & 140 & 138 & 139 & 138 & 140 \end{bmatrix}. \quad (8.8)$$

The result of the application of a 2D-DCT on **B** (after subtracting 128 from each of its elements, as required by JPEG) is the matrix **X**,

$$X = \begin{bmatrix} 72 & -4 & -6 & -4 & 1 & 2 & 1 & -2 \\ -6 & 3 & -3 & 0 & -1 & 3 & -1 & -1 \\ -2 & 0 & 2 & -2 & 2 & 1 & 3 & 1 \\ -2 & -2 & 2 & -2 & 1 & -1 & 0 & 1 \\ 3 & 1 & -3 & -5 & -1 & 2 & 0 & -2 \\ 0 & 0 & 0 & -2 & 2 & -1 & -1 & 0 \\ 3 & -2 & 2 & 3 & 1 & 2 & 0 & 1 \\ 3 & 3 & 1 & -2 & 0 & 3 & 2 & -2 \end{bmatrix}. \quad (8.9)$$

The smaller amount of image activity of block *B* in comparison with block *A*, is reflected in the transform domain, by the fact that the AC coefficients in Equation (8.9) have smaller magnitudes than the corresponding AC coefficients in Equation (8.2). This difference is even more emphasized after the quantization.

The quantization of the 2D-DCT coefficients in **X** with the normalization array **N** given in Equation (8.3), produces the following matrix,

$$Q = \begin{bmatrix} 5 & 0 & -1 & 0 & 0 & 0 & 0 & 0 \\ -1 & 0 & 0 & 0 & 0 & 0 & 0 & 0 \\ 0 & 0 & 0 & 0 & 0 & 0 & 0 & 0 \\ 0 & 0 & 0 & 0 & 0 & 0 & 0 & 0 \\ 0 & 0 & 0 & 0 & 0 & 0 & 0 & 0 \\ 0 & 0 & 0 & 0 & 0 & 0 & 0 & 0 \\ 0 & 0 & 0 & 0 & 0 & 0 & 0 & 0 \\ 0 & 0 & 0 & 0 & 0 & 0 & 0 & 0 \end{bmatrix}. \quad (8.10)$$

Notice the higher number of zero-valued coefficients in this matrix with respect to the matrix in Equation (8.4). As previously stated, this is a clear indication of the different "image-complexity" of the two original block of pixels.

The zigzag reordering of the elements of matrix $\mathbf{Q}$ produces the following vector,

$$\mathbf{z} = \begin{bmatrix} 5 & 0 & -1 & 0 & 0 & -1 & \underbrace{0\ldots0}_{58} \end{bmatrix}.$$

The value of the DC coefficient of the previous transformed and quantized 8x8-pixel block was found to be 4. The DPCM operation over the DC coefficients gives the vector $\mathbf{d}$ below,

$$\mathbf{d} = \begin{bmatrix} 1 & 0 & -1 & 0 & 0 & -1 & \underbrace{0\ldots0}_{58} \end{bmatrix}. \tag{8.11}$$

A zero run-length coding of the sequence of numbers represented by vector $\mathbf{d}$ results in the following vector,

$$\mathbf{r} = \begin{bmatrix} (0,1) & (1,-1) & (2,-1) & \text{EOB} \end{bmatrix}. \tag{8.12}$$

The decomposition of the elements of vector $\mathbf{r}$ into the representation (*run/cat,cmp*) ((*cat,cmp*) for the DC coefficient) as explained at the end of page 270, yields,

$$\mathbf{c} = \begin{bmatrix} (0,1) & (1/1,0) & (2/1,0) & \text{EOB} \end{bmatrix}. \tag{8.13}$$

By searching in the JPEG Huffman tables the binary code that correspond to each of the elements of vector $\mathbf{c}$ the following bitstream is obtained:

0101/11000/111000/1010

again for illustration purposes, a slash has been inserted between the binary codes associated with the different symbols of the vector $\mathbf{c}$. The resulting compression ratio is 26.95 for the block B, which only requires 19 bits in compressed form: approximately one third of the number of bits needed to code the block A.

This example shows how an image that is abound in details, and thus contains a high proportion of high-activity A-like blocks, will require more bits to be coded than a relatively less *elaborated* image.

2.3   The Cause

It is interesting to point out that the Discrete Cosine Transform operation is not the cause of the variable-rate feature of JPEG (as it is sometimes implied in the literature). There do exist image compression schemes, based on the DCT that yield a non-varying compression ratio. Zonal coding [13] for example, is based on the same compression principle as JPEG, that is, on the application of a block-based linear transform such as the DCT. Yet zonal coding can produce the same compression bit rate, regardless of the content of the input image (all the same with a coding efficiency that is less than JPEG's).

The cause of the variable compression rate in JPEG is due to the contributions of two factors; albeit any of the two factors, on its own, is enough to render varying, the output bit rate of the system. The first one is the use of a run-length coding to represent the non zero-valued, normalized, 2D-DCT coefficients. The second, is the use of variable-length Huffman codes to represent the DPCM and run-length encoded coefficients. In Figure 8.2 these two elements are both part of the block denominated entropy coder.

In general, any image or video compression system that includes an entropy coder (e.g., Huffman or arithmetic coding) is expected to deliver a variable compression bit rate. For example, video compression standards MPEG-1, MPEG-2, H.261, and H.263, all include entropy coders at the output stage, and thus, they all deliver variable compression bit rates. A bit-rate smoothing buffer [14, 15] is commonly used in order to regulate the output rate of these compression systems. The main inconvenience of this buffer is its contributed delay to the latency of the compression system.

3. Modifying the Compression Ratio

Let us suppose that one would have preferred to code the block B in the previous example with a higher compression ratio—i.e., with a lower bit rate. This would have been possible by means of scaling the normalization array by a constant factor, before the one-to-one division operation.

For example, quantizing the 2D-DCT coefficients of the block B, given by $\mathbf{X}$ in Equation (8.9), with the normalization array $\mathbf{N}$ in Equation (8.3) previously multiplied by a factor of two would result in the following matrix,

$$\mathbf{Q}_1 = \begin{bmatrix} 2 & 0 & 0 & 0 & 0 & 0 & 0 & 0 \\ 0 & 0 & 0 & 0 & 0 & 0 & 0 & 0 \\ 0 & 0 & 0 & 0 & 0 & 0 & 0 & 0 \\ 0 & 0 & 0 & 0 & 0 & 0 & 0 & 0 \\ 0 & 0 & 0 & 0 & 0 & 0 & 0 & 0 \\ 0 & 0 & 0 & 0 & 0 & 0 & 0 & 0 \\ 0 & 0 & 0 & 0 & 0 & 0 & 0 & 0 \\ 0 & 0 & 0 & 0 & 0 & 0 & 0 & 0 \end{bmatrix}. \tag{8.14}$$

The effect of having doubled the magnitude of the coefficients of the normalization array $\mathbf{N}$ is clear: more 2D-DCT coefficients have been zeroed out in matrix $\mathbf{Q}_1$ than in matrix $\mathbf{Q}$ in Equation (8.10). The effect at the output of the DPCM (the previous quantized DC coefficient was found to be 2) and run-length coders is reflected in the vector $\mathbf{d}_1$, which displays a smaller number of non-zero elements than the corresponding vector $\mathbf{d}$ in Equation (8.11). This change induces a large reduction in the number of bits required to represent the

Table 8.3.   Least squares error approximation of the CR-SF characteristic.

Image	Size (pixels)	LSE Approximation	Slope m	R^2 Value
Airplane	512 x 512	$y = 4.730x + 8.359$	4.730	0.9920
Baboon	512 x 512	$y = 3.194x + 2.736$	3.194	0.9996
Barbara	512 x 512	$y = 3.710x + 5.591$	3.710	0.9986
Boats512	512 x 512	$y = 4.630x + 6.856$	4.630	0.9971
Lena	512 x 512	$y = 5.604x + 9.924$	5.604	0.9922
Peppers	512 x 512	$y = 5.223x + 9.033$	5.223	0.9881
Bridge	256 x 256	$y = 3.226x + 3.056$	3.226	0.9997
Camera	256 x 256	$y = 4.278x + 6.409$	4.278	0.9931
Lena256	256 x 256	$y = 4.347x + 6.473$	4.347	0.9959

elements of vector $\mathbf{d}_1$,

$$\mathbf{d}_1 = [0 \;\; \underbrace{0 \ldots 0}_{63}]. \tag{8.15}$$

After the category and complement representation, followed by Huffman coding, the output bitstream for the vector $\mathbf{d}_1$ contains 6 bits which corresponds to a compression ratio CR of 85.33, in comparison with a CR of 26.95 for the non-scaled normalization array case. In this instance, the increase of the compression ratio is obtained in exchange for a degradation of the quality of the decompressed block. This because the higher the number of coefficients that are *forced* to zero during the quantization stage, the lower the quality of the decompressed images.

4.   Modeling the CR-SF Characteristic

In Section 3 it was shown how by scaling the normalization array one can modify the compression bit rate of the coded images. The question that remains at this point is how to find the constant factor by which the normalization array should be scaled, so that one obtains a given target compression ratio. This is a major issue, since without any effective strategy the determination of the correct scale factor could take numerous iterations. A solution to this issue is discussed in this and the next sections.

The study began by carrying out intensive JPEG-compression computations on a set of the most popular test images used in image coding research. This set which is listed in Table 8.3 contains images of different spatial resolutions and different scene-complexity content. The procedure was to let the scale factor (SF) change over a large range of values and measure the corresponding obtained compression ratios (CR). The objective being to analyze the characteristic of the curves Compression Ratio–Scale Factor for each of the images on the set.

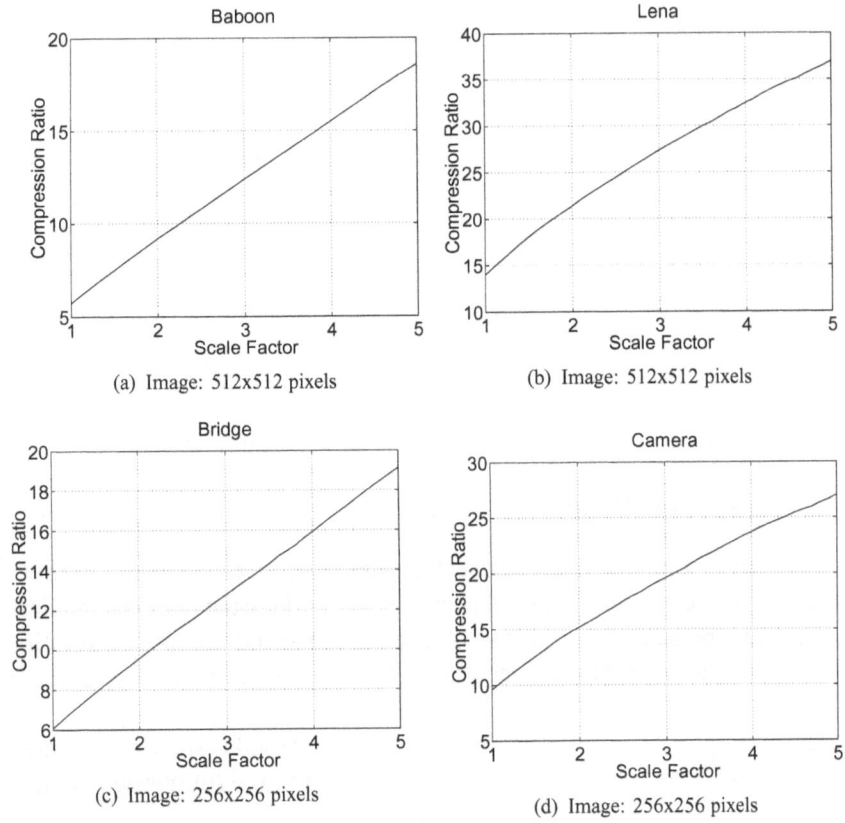

Figure 8.4. Compression ratio – Scale factor characteristic.

The obtained CR-SF characteristic is shown in Figure 8.4 for two different image sizes. It is interesting to note the strong linear behavior of these four curves. Even more interesting and significative for the study under question, was the fact of finding out that the same linear behavior is featured by the CR-SF curves of all the images in the study set.

A linear approximation seems thus a plausible choice to model the CR-SF characteristic. Table 8.3 shows the results of a linear model built with the classical least squares error LSE method. The R-square value metric[6] is also reported, in order to show the excellent correlation between the real curve and the model (for a perfect fit, $R^2 = 1$). Figure 8.5 shows both the real curve and its linear approximation for the two worst cases of the images in the study set.

---

[6]$R^2 = 1 - \text{SSE/SST}$, with SSE$= \sum (x_i - \hat{x}_i)^2$, and SST$= (\sum x_i^2) - (\sum x_i)^2/n$; $n=$ number of data, $x=$real data, $\hat{x}=$model data.

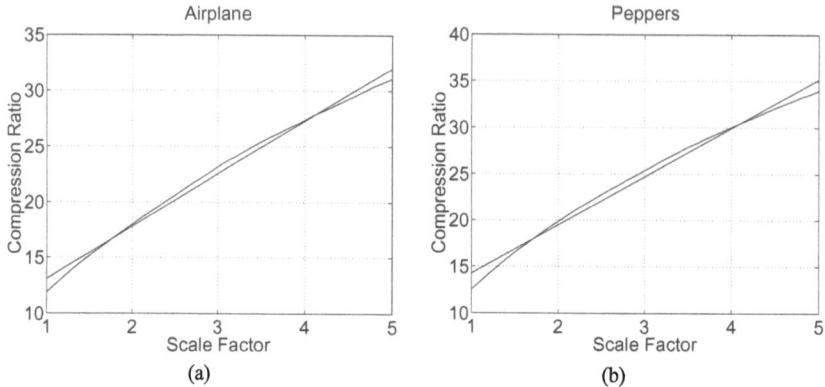

Figure 8.5.   Worst cases LSE approximation of the CR-SF characteristic.

5.   General Bit-Rate Control Scheme

The question raised in the last section was how, given a target compression ratio CR_T, one can predict the scale factor SF_T by which the JPEG normalization matrix must be scaled in order to obtain the requested compression ratio. In this section the general scheme of a method that figures out this problem is presented.

The principle of the proposed bit-rate control method is shown in Figure 8.6. The different blocks of the diagram are described in the following paragraphs.

- Parameter Estimation. To model the CR-SF characteristic with a linear function $y = mx + b$, two parameters are required, namely, the slope m and the intercept b. As can be noticed from Table 8.3 these values are proper for each of the images. It is reasonable to expect that the value of these parameters is related in some way to the information content of the images, and accordingly, one can consider executing some kind of pre-analysis over the image in order to estimate the most suitable values of m and b for a given image.

- Computation of SF_T.   The previously determined linear model $cr = m \cdot sf + b$, is used to compute the scale factor target SF_T. That is, given CR_T, its associated scale factor can be found straightforwardly by two simple operations, $SF_T = (CR_T - b)/m$.

- JPEG. The value SF_T is used to scale the normalization array; after which the JPEG compression properly said is executed on the input image.

Depending on the technique used to estimate the parameters m and b, different bit-rate control implementations are viable for the general scheme in

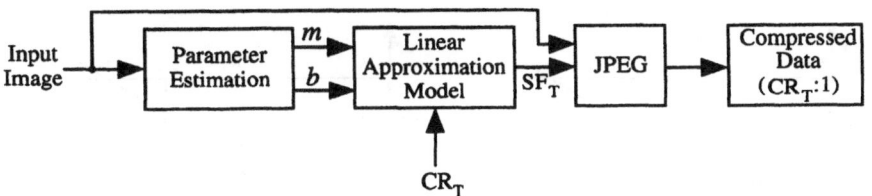

Figure 8.6.   General scheme of the bit-rate control method.

Figure 8.6. Sections 6 and 7 describe two different methods that were developed.

In both methods, the constraint on the accuracy was defined as to obtaining a relative error, between the target compression ratio and the actual compression ratio given by the system, whose absolute value should be less than five percent.

5.1   Range of the Scale Factor

Looking at Figure 8.4 on page 277, one observes that the defined range of operation for the scale factor is the interval $1 \leq sf \leq 5$. This range was chosen for the following reasons. For the lower limit $sf = 1$, the quality of the reconstructed images is fairly good (as witness the pictures in Figures 8.1(c) and (d)), and so very much appropriated for the most common applications. The upper limit $sf = 5$, was selected by experimentation, concluding that when JPEG-compressing images with a scale factor greater than 5, the quality of the reconstructed images starts to be seriously compromised.

It is important to remark that a small change in the scale factor can have a dramatic effect on the quality of the reconstructed images, and thus, in spite of the relatively short scale factor variation range, the associated range of quality of the reconstructed images is extremely large.

6.   Bit-Rate Control by Average-Slope Linear Approximation

In this method the parameter estimation block in Figure 8.6 is replaced by one or two JPEG coding operations. That is, the image is initially compressed with JPEG using a scale factor of 1, for the normalization matrix. Then, the resulting compression ratio is used to (implicitly) estimate the parameter b of the linear model.

By looking over the slope value m for the different linear approximations in Table 8.3, one observes that its range of variation, among the different images, is relatively narrow. This simple observation suggested the construction of an average-slope linear model, in which the value of m is given by the average value of the slopes reported in Table 8.3.

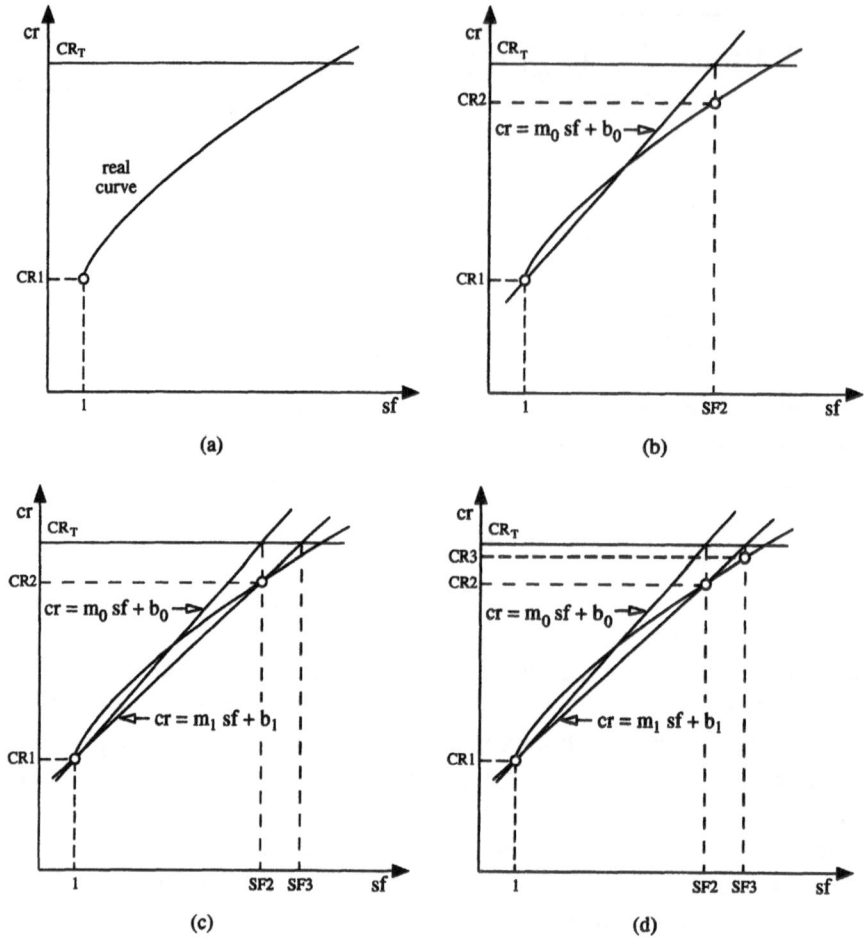

Figure 8.7.   Linear approximation method.

In some instances, the average-slope linear model might not be accurate enough, and thus, a slope-adjustment is necessary, as described in the following section.

6.1   Description of the Algorithm

The mechanism of operation of this method is shown in Figure 8.7. The different steps of the algorithm are:

1 Enter the target compression ratio CR_T.

2 Compress the input image using a scale factor of 1 for the JPEG normalization matrix. Let us call CR1 the resulting compression ratio (Figure 8.7(a)).

- If the absolute value of the relative error between CR_T and CR1 is less than 0.05 then the process stops here. Otherwise it will continue in step 3.

3 With the point (1,CR1) and a given slope m_0, which is determined as the average of the slopes values given in Table 8.3, the value of b_0 is implicitly defined and so is the linear model $cr = m_0 \cdot sf + b_0$. Use this model to find the scale factor that corresponds to a compression ratio $cr = CR_T$. Let us call SF2 this computed scale factor (Figure 8.7(b)).

4 Compress the input image using a normalization array scaled with the value SF2. Let us call CR2 the resulting compression ratio (Figure 8.7(b)).

- If the absolute value of the relative error between CR_T and CR2 is less than 0.05 then the process stops here. Otherwise it will continue in step 5.

5 The fact that CR2 failed to approximate CR_T, at the end of the step number 4, indicates that the average-slope model is not fitting accurately the CR-SF characteristic of the input image. In this case a slope (and an intercept) adjustment are necessary in order to satisfy the maximum relative error constraint of plus/minus five percent.

Both (1,CR1) and (SF2,CR2) are points on the real CR-SF curve. As such, they are appropriate to develop a more accurate approximation that is given by $cr = m_1 \cdot sf + b_1$. In this updated model, find the scale factor that corresponds to a compression ratio cr = CR_T. Let us call SF3 this computed scale factor (Figure 8.7(c)).

6 Finally, using SF3 to scale the JPEG normalization array, compress the input image. Let us call CR3 the resulting compression ratio (Figure 8.7(d)).

6.2 Validation of the Average-Slope Linear Model

Exhaustive testing with pictures from and out of the initial set of images, showed that the value of CR3 is an excellent approximation of CR_T (absolute value of the relative error < 0.05). The results are given in Figures 8.8–8.10 for several images. The graphs in Figures 8.10(c) and (d) belong to a picture that was not part of the initial set of test images (listed in Table 8.3).

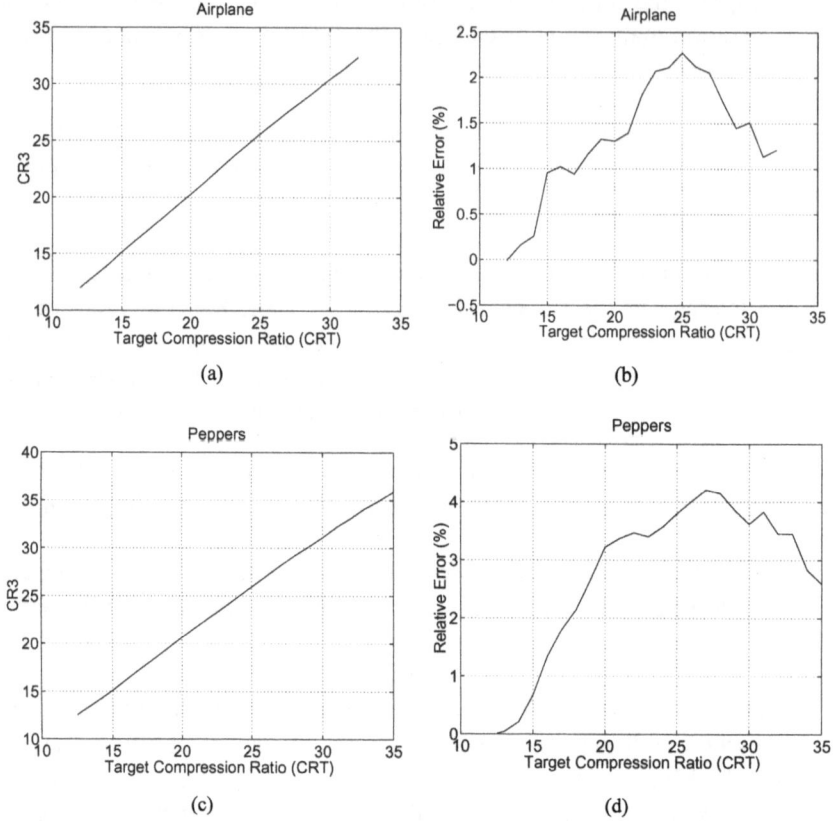

Figure 8.8.   Results for the images Airplane and Peppers.

For each of the graphs, the first value on the x-axis corresponds to the target compression ratio whose associated scale factor is $sf = 1$. The last value corresponds to a scale factor of 5.

6.3   Improving the Average-Slope Linear Model Method

Though the bit-rate control method described in the last section gives very good results, it is nevertheless susceptible of improvement. The following are two important ameliorations that are worthy of consideration.

1 *Reduce the number of iterations.* When the initial average-slope model does not fit accurately the CR-SF curve, an extra JPEG iteration is required in order to compute CR3 with an updated model $cr = m_1 \cdot sf + b_1$ (Figure 8.7(c) and (d)). It would be of great advantage, from a computational complexity point of view, to make the bit-rate control system converge to the value of CR_T—with the same predefined accuracy—within

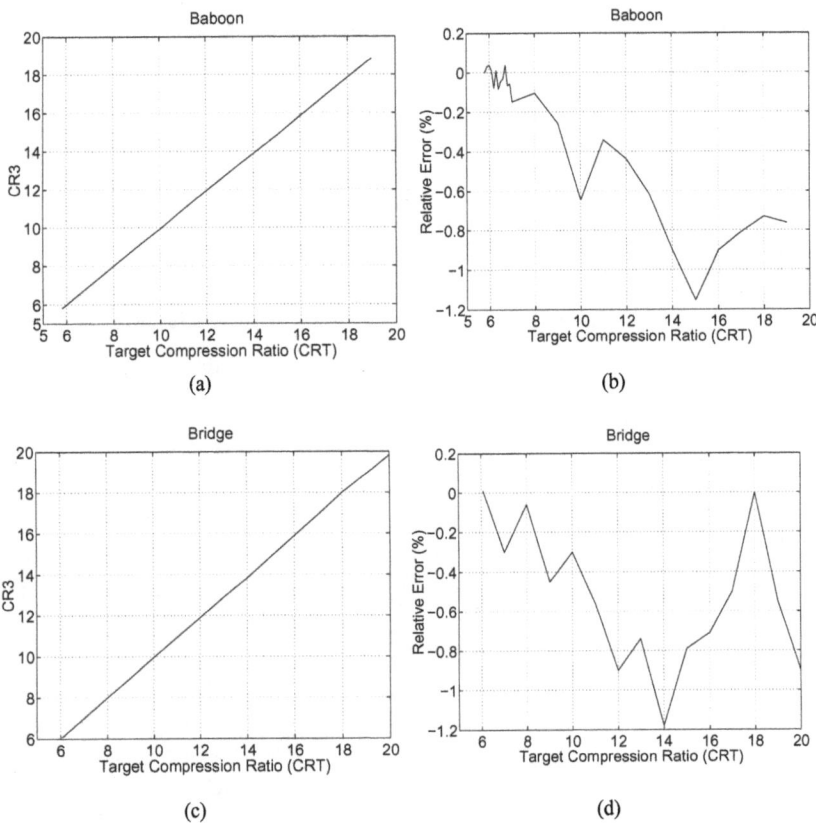

Figure 8.9. Results for the images Baboon and Bridge.

a maximum of two JPEG operations. That is to say that CR_T should be accurately approximated, with no further computations beyond the calculation of CR2. This goal is achieved with the *two-pass method*, whose results are reported in Section 8.1.

2 *Extended range of operation for the scale factor.* Though the range of operation previously defined for the scale factor, is fairly enough for a large number of applications, it would be interesting to have a system that is able to control the compression bit rate of JPEG for the largest "practical" range of applications. This goal is achieved with the *three-pass method*, whose results are reported in Section 8.2. In this method, the range $1 \leq sf \leq 5$ is extended in both directions.

Experimentally, the extended scale factor range, $0.5 \leq sf \leq 15$, was considered for the following reasons.

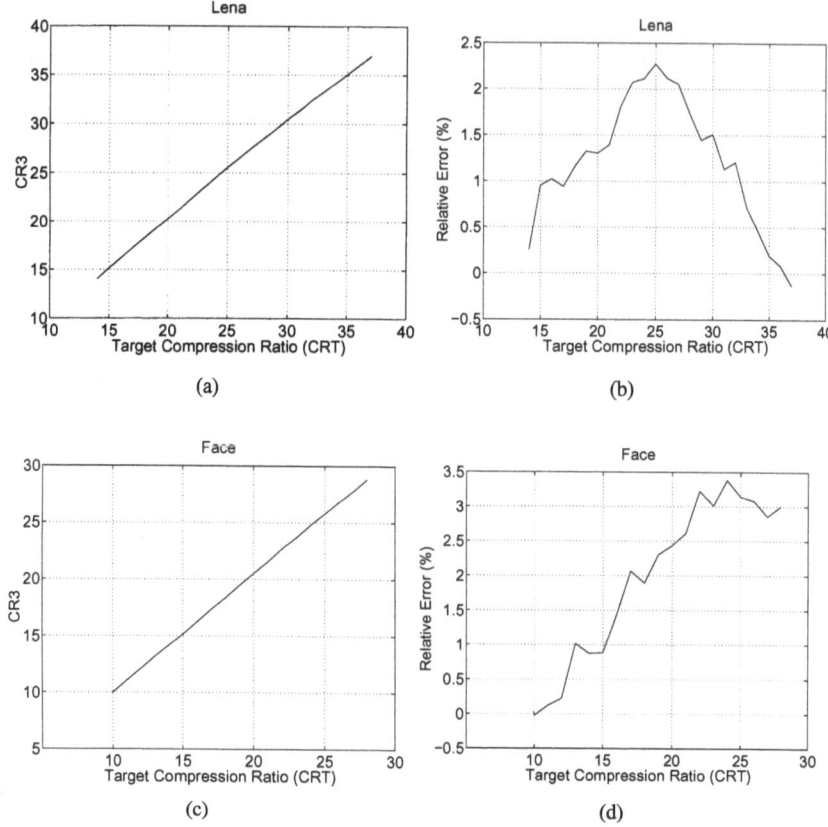

Figure 8.10.   Results for the image Lena and Face.

- *Lower limit.* The quality of the reconstructed images given by default ($sf = 1$) by JPEG is in general satisfactory; still, if necessary, this quality can be upgraded by using a scale factor that is less than the unity. In effect, when the normalization array is scaled by a factor that is less than one, the effect on the resulting compression bit rate would be the opposite than in the example given in Section 3, that is, a decrease of the compression ratio. This loss of compression efficiency is however obtained in favor of an improvement of the quality of the reconstructed images. The reason for choosing $sf = 0.5$ as the lower limit, is because at this value the original and the reconstructed images are normally indistinguishable ([16], Chapter 4). Making thus worthless any attempt to improve the quality of the decoded images below a scale factor of 1/2.

- *Upper limit.* Experiments on a large set of images proved that the quality of the reconstructed images beyond a scale factor of

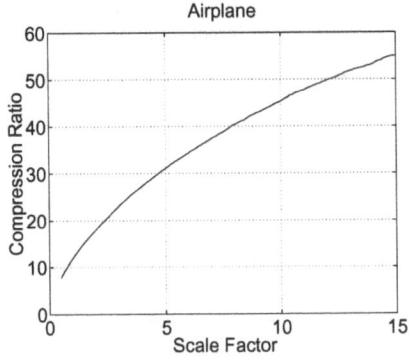

Figure 8.11. Compression ratio – Scale factor characteristic.

15, is strongly objectionable. Decoded images with $sf > 15$, are generally useless for most applications (artwork excluded).

7. Bit-Rate Control by Piecewise-Linear Approximation

7.1 Piecewise-Linear Model

The CR-SF characteristic, on the extended scale factor range, for all the test images was analyzed. It was interesting to note that the behavior among the curves was again very similar; nonetheless, a single average-slope model for the entire new range was no longer valid.

A typical CR-SF characteristic is given in Figure 8.11 for the range $0.5 \le sf \le 15$. The second bit-rate control method consists in approximating this curve not with a single, but with multiple straight lines whose slopes change in function of their region in the scale factor space. Furthermore, the slope associated with each region, is not constant but linearly depends on the value of the compression ratio obtained with a scale factor $sf = 2$.

Six different regions of the CR-SF curves were found to be suitable for separate modeling. The intervals of the scale factor that correspond to these regions are: $[0.5, 1)$, $[1, 1.5)$, $[1.5, 2)$, $(2, 5]$, $(5, 10]$ and $(10, 15]$. The interval $0.5 \le sf \le 2.0$ is the interval in which the first derivate of the CR-SF curve is changing at the highest rate, thus, this region was divided into shorter sub-intervals than the less derivate-changing region $2 < sf \le 15$. The point at $sf = 2$ is, as will be explained below, the starting step of the algorithm, for this reason it is not included in any of the regions above.

Table 8.4.   Slope determination for the different regions.

Region No.	Scale Factor Range	$m : CR1$ Relation
1	$0.5 \leq sf < 1.0$	$m_1 = 0.4939 \cdot CR1 - 0.7064$
2	$1.0 \leq sf < 1.5$	$m_2 = 0.3947 \cdot CR1 - 0.3122$
3	$1.5 \leq sf < 2.0$	$m_3 = 0.2894 \cdot CR1 + 0.6224$
4	$2.0 < sf \leq 5.0$	$m_4 = 0.1565 \cdot CR1 + 1.6517$
5	$5.0 < sf \leq 10.0$	$m_5 = 0 \cdot CR1 + 3.0175$
6	$10.0 < sf \leq 15.0$	$m_6 = -0.1098 \cdot CR1 + 3.8832$

It can be noticed that the modeling of the CR-SF characteristic has become more complex with respect to the *single* straight-line model $y = mx + b$. That is, now the system must be able to predict the slope m_i and the intercept b_i for the different six regions that have been defined.

The problem of the slope estimation for each of the different regions of the piecewise-linear model, was solved by studying the relationship between the value of the compression ratio (at different scale factors) and the value of the slope for each of the regions. It turned out that a strong linear relationship exists between the slope and the compression ratio CR1 obtained with a scale factor $sf = 2$. Hence, the estimation parameters block in Figure 8.6 is replaced by a JPEG operation in which the normalization array is scaled by a factor of 2. The relation $m : CR1$ for the different regions is given in Table 8.4.

The good fitting of the piecewise-linear model to the real curve is shown in Figure 8.12 for the two worst cases of all the images in the original set.

7.2   Description of the Algorithm

The functioning of the piecewise-linear approximation method is shown in Figure 8.13. Though very simple, the complete description of the algorithm is rather repetitive and lengthy due to the multiple conditional branchings. In this section the method is reported for the particular case when the target compression ratio lies in the region $5 < sf \leq 10$. This case involves all the main operations of the procedure and perfectly illustrates the mechanism of the algorithm; which is given in full description in the Appendix, at the end of this chapter.

The different steps of the algorithm are:

1 Enter the target compression ratio CR_T.

2 Compress the input image using a scale factor of 2, for the JPEG normalization array. Let us call CR1 the resulting compression ratio (Figure 8.13(a)).

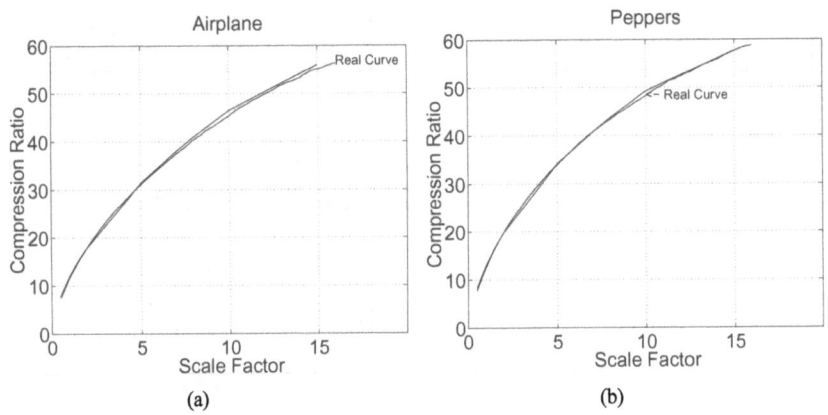

Figure 8.12. Piecewise-linear LSE approximation.

- If the absolute value of the relative error between CR1 and CR_T is less than 0.05 then the process stops here. Otherwise it will continue in step 3.

3 If $(CR_T > CR1)$ then go to step 4 otherwise go to step 11.

4 Insert the value of CR1 in the equation m : CR1 of region No. 4 in Table 8.4, in order to find the slope of the linear model for the region $2 < sf \leq 5$. Having the value m_4 and the point (2,CR1), the intercept b_4 is implicitly defined and so is the linear model $cr = m_4 \cdot sf + b_4$. Use this linear model to find the scale factor that corresponds to a compression ratio $cr = CR_T$. Let us call SF2 this computed scale factor (Figure 8.13(a)).

- If (SF2 > 5) then go to step 5 otherwise go to step 10.

5 Use the linear model $cr = m_4 \cdot sf + b_4$, in order to find the compression ratio CR_V that corresponds to a scale factor $sf = 5$. This operation defines the point (5,CR_V) (Figure 8.13(b)).

Insert the value of CR1 in the equation m : CR1 of region No. 5 in Table 8.4 in order to find the slope m_5 of the linear model for the region $5 < sf \leq 10$. Having the value m_5 and the point (5,CR_V), the intercept b_5 is implicitly defined and so is the linear model $cr = m_5 \cdot sf + b_5$. Use this linear model to find the scale factor that corresponds to a compression ratio $cr = CR_T$. Let us call SF2 this computed scale factor (Figure 8.13(b)). The previous value of SF2 is overwritten, becoming SF2′.

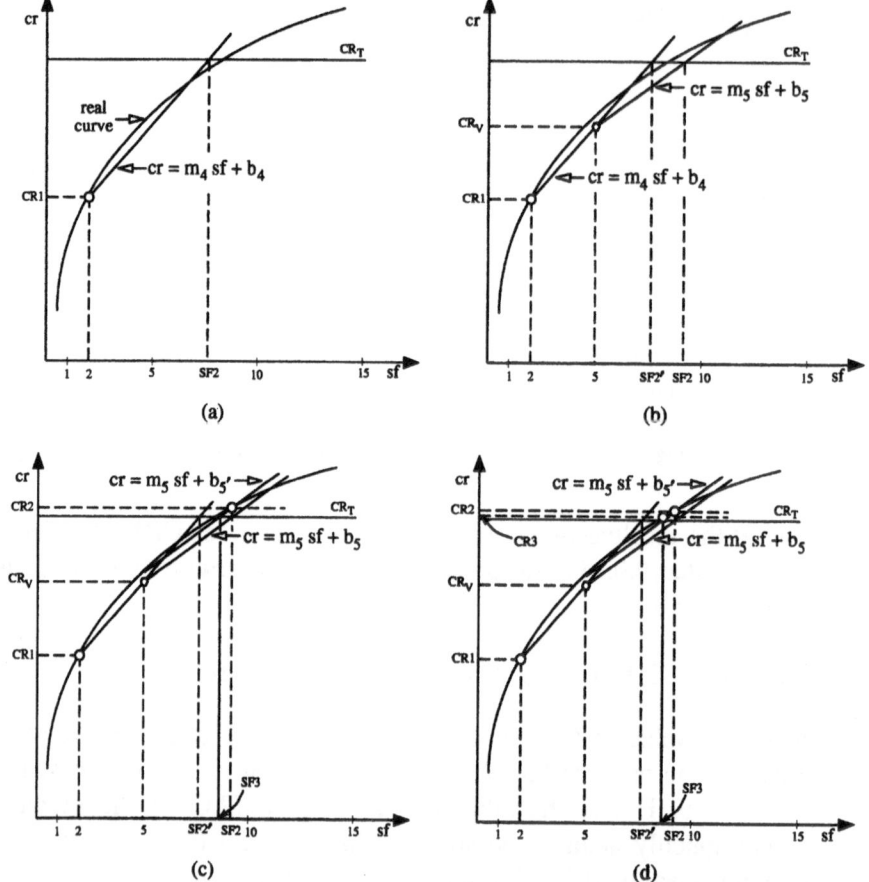

Figure 8.13. Piecewise-linear approximation method.

- If (SF2 > 10) then go to step 9 otherwise go to step 6.

6 Compress the input image using a normalization array scaled by the value of SF2. Let us call CR2 the resulting compression ratio (Figure 8.13(c)).

  - If the absolute value of the relative error between CR2 and CR_T is less than 0.05 then the process stops here. Otherwise it will continue in step 7.

7 In order to appropriately approximate CR_T an intercept adjustment is necessary. The slope m_5 and the point (SF2,CR2) implicitly define a more accurate model $cr = m_5 \cdot sf + b_{5'}$. In this updated model, find the scale factor that corresponds to a compression ratio $cr = CR_T$. Let us call SF3 this computed scale factor (Figure 8.13(c)).

8 Finally, using SF3 to scale the JPEG normalization array, compress the input image. Let us call CR3 the resulting compression ratio (Figure 8.13(d)).

9 This step is explained in the complete description of the algorithm in the Appendix.

10 This step is explained in the complete description of the algorithm in the Appendix.

11 This step is explained in the complete description of the algorithm in the Appendix.

8.   Validation of the Piecewise-Linear Model

The piecewise-linear approximation method can be implemented in two forms. A simple and yet suitable for a large number of applications, two-pass method, ends after the calculation of CR2 in Figure 8.13(c). The more general three-pass method implementation, operates on a larger scale factor range, as described in the previous section, and requires an extra quantization and an extra entropy coding operation with respect to the two-pass method (since the 2-D DCT coefficients are always the same regardless of the number of iterations, the 2-D DCT operation is executed only once).

8.1   Two-Pass Method

Exhaustive testing with pictures from and out of the initial set of images, showed that when the piecewise-linear model algorithm operates in the range $1 \leq sf \leq 5$ the value of CR2 is an excellent approximation of CR_T, within the defined relative error constraint. The results are given in Figures 8.14 to 8.16 for several images. The graphs in Figures 8.16(c) and (d) belong to a picture that was not part of the initial set of test images.

8.2   Three-Pass Method

Exhaustive testing with pictures from and out of the initial set of images, showed that the value of CR3 is an excellent approximation of CR_T (absolute value of the relative error < 0.05), all along the range $0.5 \leq sf \leq 15$. The results are given in Figures 8.17 to 8.19 for several images.

9.   Computational Complexity

Due to its better performance with respect to the average-slope linear model, the following discussion will be limited to the computational complexity of the piecewise-linear model.

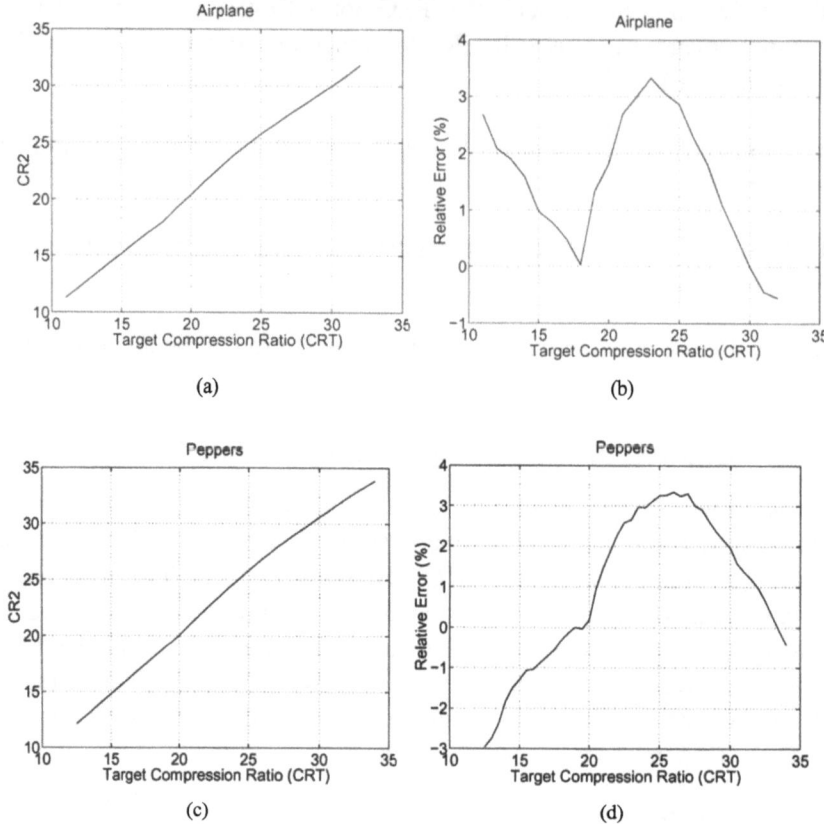

Figure 8.14.   Results for the images Airplane and Peppers (two-pass method).

The computational complexity analysis will be separated into two parts. Part one, particular requirements, concerns the computational resources that are required to find the target scale factor and that are proper to the method proposed in this chapter. Part two, general requirements, concerns the computational resources that are required due to the repetitive computations executed during the iterations.

To simplify, in the following discussion the terms DCT, quantization, and entropy coding, will be used to indicate the respective global operations applied over the entire image (and not over a pixel or block of pixels). For example, for an image of $(M \times N)$ pixels, *one* quantization will stand for the execution of the corresponding $(M \times N)$ rounded divisions (or multiplications).

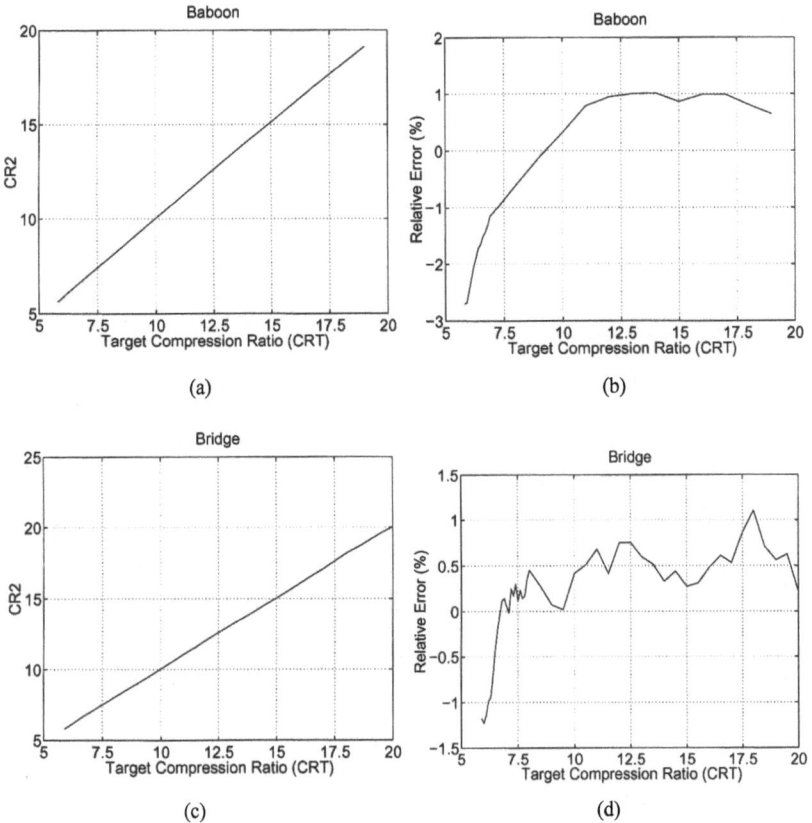

Figure 8.15. Results for the images Baboon and Bridge (two-pass method).

9.1 Two-Pass Method

- *Particular requirements*: To find the slope m_i that best fits the CR-SF characteristic in the regions 2, 3 and 4 defined in Table 8.4, a function $m_i = m_{i0} \cdot CR1 + b_{i0}$ must be evaluated. In order to store the different m_{i0}'s and b_{i0}'s, six units of permanent memory are required. Working memory, is required to temporarily store the values of CRT, CR1, and SF2.

  The number of operations will be computed with respect to the worst-case (from a computational complexity point of view), which happens when the final compression ratio lies in the region $1.0 < sf \leq 1.5$. One comparison indicates that the target region is either region No. 3 or 2. Given the value of CR1, the slope m_3 is calculated with the appropriate equation given in Table 8.4, requiring 1 addition and 1 multiplication.

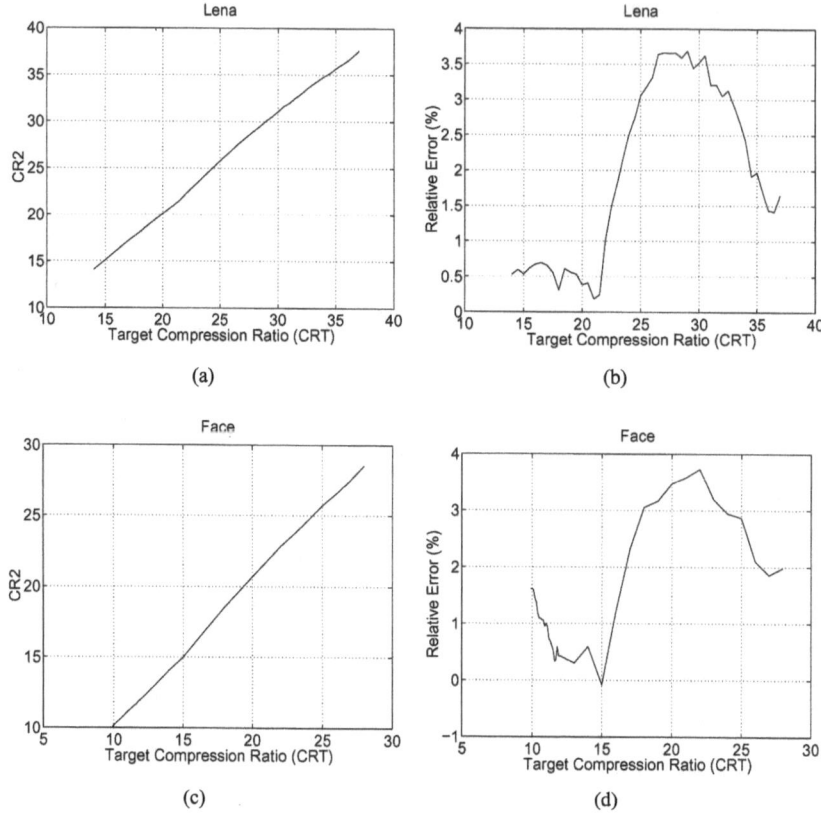

Figure 8.16.   Results for the images Lena and Face (two-pass method).

The first SF2 (later named SF2′) is computed as $SF2 = 2 + (CR_T - CR1)/m_3$. One more comparison indicates that the target region is region No. 2. The computation of CR_{II} (i.e., the CR that corresponds to a scale factor $sf = 1.5$) demands 1 division, 1 addition and 1 subtraction. In order to find the slope m_2, 1 addition and 1 multiplication are required.

Finally, to find the second (and target) SF2, 1 division, 1 addition and 1 subtraction are necessary.

- *General requirements*: One JPEG operation (DCT + Quantization + Entropy coding) is required in order to compute CR1. Once the target scale factor SF2 has been determined, the normalization matrix is scaled, requiring 64 multiplications. Then one re-quantization of the 2-D DCT coefficients with the *new* normalization array is executed. Finally, an additional entropy coding operation is required to produce the compressed bitstream.

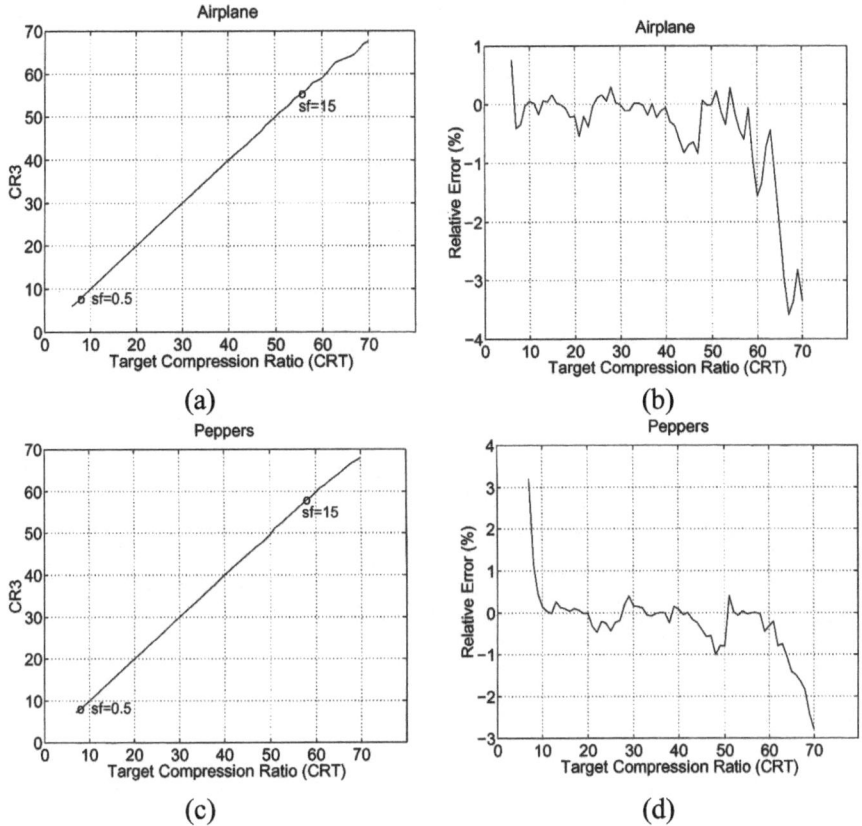

Figure 8.17. Results for the images Airplane and Peppers (three-pass method).

It is important to note that the 2-D DCT operation, which is the most computationally intensive operation of JPEG, could, if memory is not an issue, be executed only once. Accordingly, working memory (RAM or bank of registers) would be required in order to store the 2D-DCT coefficients for the time necessary to execute the second iteration. The size of this image buffer obviously depends on the spatial resolution of the input image.

Table 8.5 summarizes the total number of operations and memory requirements for the piecewise-linear approximation two-pass method. The number of operations given is per image, which for the quantization and entropy coding depends on the space resolution of the input image. The operations due to the default JPEG compression are not shown. If the implementation option described in the previous paragraph is chosen, then the corresponding working memory should be added to this table, to store the 2-D DCT coefficients.

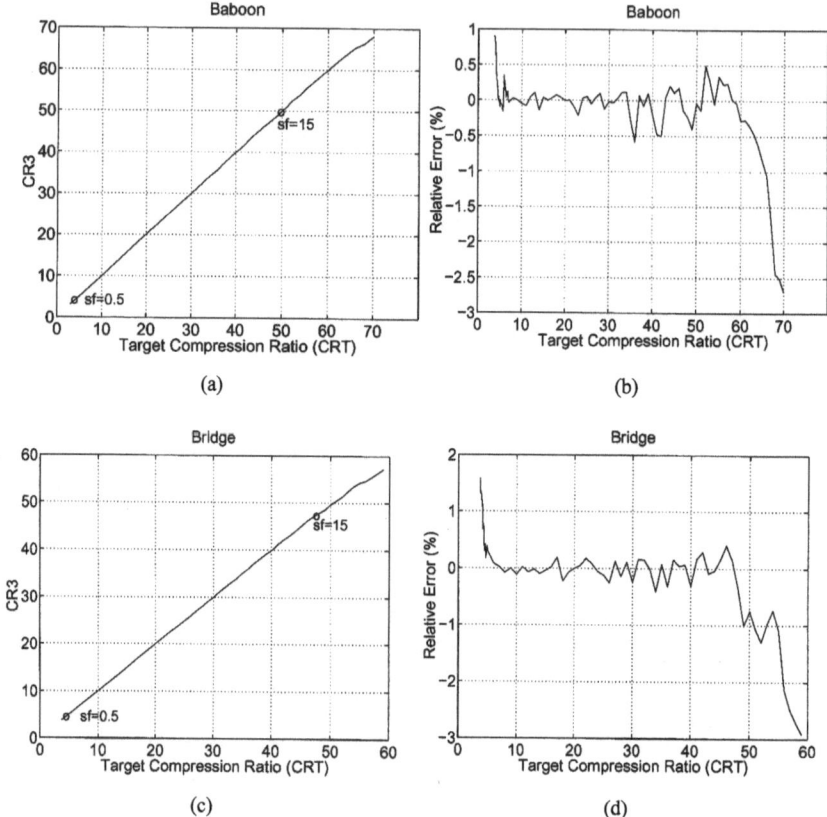

Figure 8.18.   Results for the images Baboon and Bridge (three-pass method).

9.2   Three-Pass Method

- *Particular requirements*: To find the slope m_i that best fits the CR-SF characteristic in all the regions defined in Table 8.4, a function $m_i = m_{i0} \cdot CR1 + b_{i0}$ must be evaluated. In order to store the different m_{i0}'s and b_{i0}'s, twelve units of permanent memory are required. Working memory, is required to temporarily store the values of CRT, CR1, CR2, SF2, and SF3.

  The number of operations will be computed with respect to the worst-case (from a computational complexity point of view), which happens when the final compression ratio lies in the region $0.5 \leq sf < 1.0$ (the region $10 < sf \leq 15$ is equally a worst case region).

  One comparison indicates that the target region is either region No. 3, 2 or 1. Given the value of CR1, the slope m_3 is calculated with the appropriate equation given in Table 8.4, requiring 1 addition and 1 multiplication.

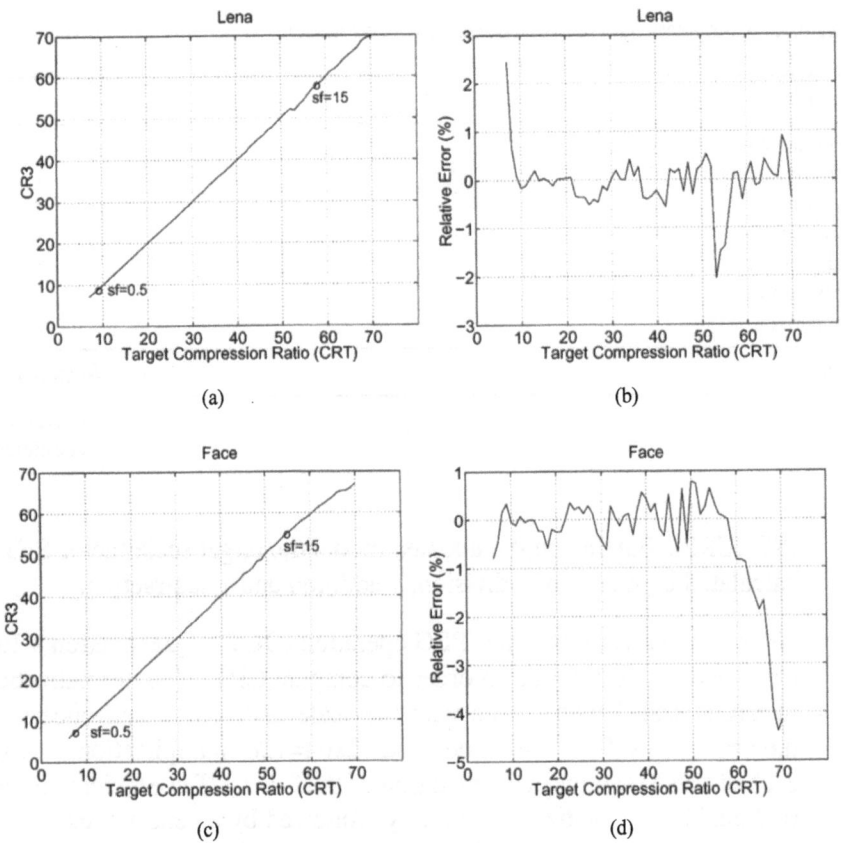

Figure 8.19. Results for the images Lena and Face (three-pass method).

The first SF2 (later named SF2′) is computed as $SF2 = 2 + (CR_T - CR1)/m_3$. One more comparison indicates that the target region is region No. 2. The computation of CR_{II} demands 1 division, 1 addition and 1 subtraction. In order to find the slope m_2, 1 addition and 1 multiplication are required. The computation of the second SF2 (later named SF2″, see Appendix) requires again 1 division, 1 addition and 1 subtraction. One additional comparison indicates that the target region is region No. 1. The computation of CR_I (i.e., the CR corresponding to a scale factor $sf = 1$) involves 1 division, 1 addition and 1 subtraction.

The final SF2 is calculated and used to compress the input image, which produces the value of CR2. A comparison indicates that CR_T is not approximated with the predefined precision and that the computation of CR3 is necessary. In order to do so, an intercept adjustment ($b_1 \rightarrow b_{1'}$) is necessary, whose new value is implicitly known by having the point

Table 8.5.   Computational complexity of the two-pass method.

Operation	Number
Quantization	1
Entropy Coding	1
Division	3
Multiplication	66
Addition	5
Subtraction	3
Comparison	2

Memory	Requirement
Permanent	6 coefficients
Working	3 coefficients

(SF2,CR2). Finally, on the updated model the target scale factor SF3 is calculated by means of 1 division, 1 addition and 1 subtraction.

- *General requirements*: One JPEG operation (DCT + Quantization + Entropy coding) is required in order to compute CR1. The normalization matrix is scaled twice, once with the scale factor SF2, and once with the scale factor SF3, which demands 2x64=128 multiplications. After each of these scalings, a quantization of the 2-D DCT coefficients is carried out. Each quantization is always followed by an entropy encoding. The first iteration produces the value of CR2, while the second produces the final bitstream representing the compressed image at an excellent approximation CR3, of the requested compression ratio.

Table 8.6 summarizes the total number of operations and memory requirements for the piecewise-linear approximation three-pass method. The number of operations given is per image, which for the quantization and entropy coding depends on the space resolution of the input image. The operations that correspond to the default JPEG compression are not included. The same comment that was made at the very end of Section 9.1, regarding the possible requirement of additional working memory, holds true for Table 8.6.

10.   About JPEG2000

JPEG2000 [17] is a recently introduced standard for still image compression. It presents a wide range of advanced encoding features, and it is particularly efficient for compressing images at low bit rates. JPEG2000 uses the wavelet transform as its core data decorrelation mechanism, which allows good spatial and quality scalability functionalities. It is important to point out that JPEG2000

Table 8.6. Computational complexity of the three-pass method.

Operation	Number
Quantization	2
Entropy Coding	2
Division	6
Multiplication	130
Addition	8
Subtraction	6
Comparison	4

Memory	Requirement
Permanent	12 coefficients
Working	5 coefficients

is not intended to replace JPEG but to complement it. At high bit rates, i.e., for applications requiring very good reconstruction image quality, the results produced by both methods are quite comparable. Yet, JPEG requires a significantly lower computational complexity. This feature makes JPEG a good choice in low-power, low-cost, portable-oriented hardware implementations, particularly in applications requiring very good picture quality and no other functionality but plain compression.

11. Conclusions

In this chapter two methods for controlling the compression ratio produced by JPEG have been presented. A study of the characteristic of the compression ratio versus the scale factor has been reported. The results of this study were used to build a linear and a piecewise-linear model of the CR-SF characteristic for any given image. The first reported method consisted in defining a single average-slope model; this first approach is then improved by using a more elaborated piecewise-linear model that requires one less iteration to produce the same results. Finally, a more general bit-rate control method was defined on an extended range for the scale factor. The validations of all these methods have been equally reported, showing the effectiveness of the bit-rate control techniques. The piecewise-linear technique has been successfully implemented in a consumer electronics industrial prototype.

Acknowledgments

The preparation of this document was supported by the Swiss Federal Office for Education and Science under Grant OFES C00.0105 (COST 276 project).

Appendix: Piecewise-Linear Approximation Algorithm

1 Enter the target compression ratio CR_T.

2 Compress the input image using a scale factor of 2, for the JPEG normalization array. Let us call CR1 the resulting compression ratio.

- If the absolute value of the relative error between CR1 and CR_T is less than 0.05 then the process stops here. Otherwise it will continue in step 3.

3 If ($CR_T >$ CR1) then go to step 4 otherwise go to step 11.

4 Insert the value of CR1 in the equation m : CR1 of region No. 4 in Table 8.4, in order to find the slope of the linear model for the region $2 < sf \leq 5$. Having the value m_4 and the point (2,CR1), the intercept b_4 is implicitly defined and so is the linear model $cr = m_4 \cdot sf + b_4$. Use this linear model to find the scale factor that corresponds to a compression ratio $cr = CR_T$. Let us call SF2 this computed scale factor.

- If (SF2 $>$ 5) then go to step 5 otherwise go to step 10.

5 Use the linear model $cr = m_4 \cdot sf + b_4$, in order to find the compression ratio CR_V that corresponds to a scale factor $sf = 5$. This operation defines the point (5,CR_V).

Insert the value of CR1 in the equation m : CR1 of region No. 5 in Table 8.4 in order to find the slope m_5 of the linear model for the region $5 < sf \leq 10$. Having the value m_5 and the point (5,CR_V), the intercept b_5 is implicitly defined and so is the linear model $cr = m_5 \cdot sf + b_5$. Use this linear model to find the scale factor that corresponds to a compression ratio $cr = CR_T$. Let us call SF2 this computed scale factor. The previous value of SF2 is overwritten, becoming SF2'.

- If (SF2 $>$ 10) then go to step 6 otherwise go to step 9.

6 Use the linear model $cr = m_5 \cdot sf + b_5$, in order to find the compression ratio CR_{VI} that corresponds to a scale factor $sf = 10$. This operation defines the point (10,CR_{VI}).

Insert the value of CR1 in the equation m : CR1 of region No. 6 in Table 8.4 in order to find the slope m_6 of the linear model for the region $10 < sf \leq 15$. Having the value m_6 and the point (10,CR_{VI}), the intercept b_6 is implicitly defined and so is the linear model $cr = m_6 \cdot sf + b_6$. Use this linear model to find the scale factor that corresponds to a compression ratio $cr = CR_T$. Let us call SF2 this computed scale factor. The previous value of SF2 is overwritten, becoming SF2''.

- If (SF2 $>$ 15) then go to step 7 otherwise go to step 8.

7 Signal that the quality of the compressed image, with the requested compression ratio, will be objectionable. If desired, the image is compressed using SF2 as scale factor.

- Stop.

8 Compress the input image using a normalization array scaled by the value of SF2. Let us call CR2 the resulting compression ratio.

- If the absolute value of the relative error between CR2 and CR_T is less than 0.05 then the process stops here. Otherwise, make $m_t = m_6$ and $b_t = b_6$ and continue in step 18.

9 Compress the input image using a normalization array scaled by the value of SF2. Let us call CR2 the resulting compression ratio.

- If the absolute value of the relative error between CR2 and CR_T is less than 0.05 then the process stops here. Otherwise, make $m_t = m_5$ and $b_t = b_5$ and continue in step 18.

10 Compress the input image using a normalization array scaled by the value of SF2. Let us call CR2 the resulting compression ratio.

- If the absolute value of the relative error between CR2 and CR_T is less than 0.05 then the process stops here. Otherwise, make $m_t = m_4$ and $b_t = b_4$ and continue in step 18.

11 Insert the value of CR1 in the equation $m : CR1$ of region No. 3 in Table 8.4, in order to find the slope of the linear model for the region $1.5 \leq sf < 2$. Having the value m_3 and the point (2,CR1), the intercept b_3 is implicitly defined and so is the linear model $cr = m_3 \cdot sf + b_3$. Use this linear model to find the scale factor that corresponds to a compression ratio $cr = CR_T$. Let us call SF2 this computed scale factor.

- If (SF2 < 1.5) then go to step 12 otherwise go to 15.

12 Use the linear model $cr = m_3 \cdot sf + b_3$, in order to find the compression ratio CR_{II} that corresponds to a scale factor $sf = 1.5$. This operation defines the point $(1.5, CR_{II})$.

Insert the value of CR1 in the equation $m : CR1$ of region No. 2 in Table 8.4 in order to find the slope m_2 of the linear model for the region $1 \leq sf < 1.5$. Having the value m_2 and the point $(1.5, CR_{II})$, the intercept b_2 is implicitly defined and so is the linear model $cr = m_2 \cdot sf + b_2$. Use this linear model to find the scale factor that corresponds to a compression ratio $cr = CR_T$. Let us call SF2 this computed scale factor. The previous value of SF2 is overwritten, becoming SF2$'$.

- If (SF2 < 1) then go to step 13 otherwise go to step 16.

13 Use the linear model $cr = m_2 \cdot sf + b_2$, in order to find the compression ratio CR_I that corresponds to a scale factor $sf = 1$. This operation defines the point $(1, CR_I)$.

Insert the value of CR1 in the equation $m : CR1$ of region No. 1 in Table 8.4 in order to find the slope m_1 of the linear model for the region $0.5 \leq sf < 1$. Having the value m_1 and the point $(1, CR_I)$, the intercept b_1 is implicitly defined and so is the linear model $cr = m_1 \cdot sf + b_1$. Use this linear model to find the scale factor that corresponds to a compression ratio $cr = CR_T$. Let us call SF2 this computed scale factor. The previous value of SF2 is overwritten, becoming SF2$''$.

- If (SF2 < 0.5) then go to step 14 otherwise go to step 17.

14 Compress the input image using a scale factor of 0.5.

- Stop.

15 Compress the input image using a normalization array scaled by the value of SF2. Let us call CR2 the resulting compression ratio.

- If the absolute value of the relative error between CR2 and CR_T is less than 0.05 then the process stops here. Otherwise, make $m_t = m_3$ and $b_t = b_3$ and continue in step 18.

16 Compress the input image using a normalization array scaled by the value of SF2. Let us call CR2 the resulting compression ratio.

- If the absolute value of the relative error between CR2 and CR_T is less than 0.05 then the process stops here. Otherwise, make $m_t = m_2$ and $b_t = b_2$ and continue in step 18.

17 Compress the input image using a normalization array scaled by the value of SF2. Let us call CR2 the resulting compression ratio.

- If the absolute value of the relative error between CR2 and CR_T is less than 0.05 then the process stops here. Otherwise, make $m_t = m_1$ and $b_t = b_1$ and continue in step 18.

18 In order to appropriately approximate CR_T, an intercept adjustment is necessary. The slope m_t and the point (SF2,CR2) implicitly define a more accurate model $cr = m_t \cdot sf + b_{t'}$. In this updated model, find the scale factor that corresponds to a compression ratio $cr = CR_T$. Let us call SF3 this computed scale factor.

19 Finally, using SF3 to scale the JPEG normalization array, compress the input image.

- Stop.

References

[1] J. Bracamonte, M. Ansorge, and F. Pellandini, "Procédé de contrôle du taux de compression d'images numériques," Patent pending PCT/CH98/00413.

[2] International Organization for Standardization, "ISO/JTC1/SC2/WG8 N800, initial draft for adaptive discrete cosine transform technique for still picture data compression standard, section 2.1.6 and annex E: page 49," May 1988.

[3] G. Park, "Image compression method for bit-fixation and the apparatus therefor," U.S. Patent 5,416,604, May 16 1995.

[4] J. Lee, Y. Park, and D. Yoon, "Variable length-adaptive image data compression method and apparatus," U.S. Patent 5,475,502, Dec. 12 1995.

[5] A. Razavi, R. Adar, I. Shenberg, R. Retter, and R. Friedlander, "VLSI implementation of an image compression algorithm with a new bit rate control capability," in *Proc. IEEE Int'l. Conf. on Acoustics, Speech, and Signal Processing (ICASSP)*, vol. 5, (San Francisco, CA, USA), pp. 669–672, March 1992.

[6] H. Lohscheller, "A subjectively adapted image communication system," *IEEE Trans. on Communications*, vol. 32, pp. 1316–1322, Dec. 1984.

[7] A. Watson, "Perceptual optimization of DCT color quantization matrices," in *Proc. IEEE Int'l. Conf. on Image Processing (ICIP)*, vol. I, (Austin, TX, USA), pp. 100–104, Nov. 1994.

[8] W.-H. Chen and W. Pratt, "Scene adaptive coder," *IEEE Trans. on Communications*, vol. 32, pp. 225–232, 1984.

[9] N. Jayant and P. Noll, *Digital Coding of Waveforms. Principles and Applications to Speech and Video*. Englewood Cliffs, NJ, USA: Prentice-Hall, 1984.

[10] D. Huffman, "A method for the construction of minimum redundancy codes," *Proc. of the IRE*, vol. 40, no. 10, pp. 1098–1101, 1952.

[11] G. Wallace, "The JPEG still picture compression standard," *IEEE Trans. on Consumer Electronics*, vol. 38, pp. xviii–xxxiv, Feb. 1992.

[12] ISO/IEC, "ITU-T recommendation T.81 digital compression and coding of continuous-tone still images," Sept. 1992.

[13] A. Habibi and P. Wintz, "Image coding by linear transformation and block quantization," *IEEE Trans. on Communications*, vol. 19, pp. 50–63, 1971.

[14] J. Zdepski, D. Raychaudhuri, and K. Joseph, "Statistically based buffer control policies for constant rate transmission of compressed digital video," *IEEE Trans. on Communications*, vol. 39, pp. 947–957, 1991.

[15] C. Cheng and A. Wong, "A self-governing rate buffer control strategy for pseudoconstant bit rate video coding," *IEEE Trans. on Image Processing*, vol. 2, pp. 50–59, 1993.

[16] W. Pennebaker and J. Mitchell, *JPEG Still Image Data Compression Standard*. New York, NY, USA: Van Nostrand Reinhold, 1993.

[17] C. Christopoulos, A. Skodras, and T. Ebrahimi, "The JPEG2000 still image coding system: An overview," *IEEE Trans. on Consumer Electronics*, vol. 46, pp. 1103–1127, Nov. 2000.

IV

VOICE INTERFACES AND PROCESSING

Chapter 9

MAN-MACHINE VOICE ENABLED INTERFACES

Andrzej Drygajlo
Swiss Federal Institute of Technology Lausanne,
School of Engineering,
1015 Lausanne,
Switzerland
andrzej.drygajlo@epfl.ch

Abstract   A speech interface, in a user's own language, is ideal because it is the most natural, flexible, efficient, and economical form of human communication. Spoken language interfaces to computers is a topic that has fascinated engineers and speech scientists alike for over five decades. For many, the ability to converse freely with a machine represents the ultimate challenge to our understanding of the production and perception processes involved in human speech communication. In addition to being a fascinating topic, spoken language interfaces are fast becoming a necessity. In the near future, interactive networks will provide easy access to a wealth of information and services that will fundamentally affect how people work, play and conduct their daily affairs. Advances in human language technology are needed for the average citizen to communicate with networks using natural communication skills and everyday devices, such as telephones. Without fundamental advances in user-centered interfaces, a large portion of society will be prevented from participating in the age of information, resulting in further stratification of society and tragic loss in human potential. While the principles of feedback to the user are the same for different applications of speech technology, the particular design decisions depend on the task, user, and context of use. In this chapter we present two implementations of such technology: 1) securized flexible vocabulary voice messaging system, 2) voice-enabled interface for interactive tour guide robots.

Keywords: Speech processing, man-machine communication, voice messaging system, human-robot interface.

J.F. Tasič et al. (eds.), Intelligent Integrated Media Communication Techniques, 305-336.

1.   Introduction

Human language technologies play a key role in the age of information. We have just passed through the turn of the century and we have already seen strong evidence that the key information technologies that are evolving are those that support multimedia communication and multimedia computing. The vision of the 1990s was ubiquitous, low-cost, easy-to-use communication and computation for everyone. Presently, one of the key technologies that is evolving and growing rapidly to support this vision is that of interactive voice processing. Computers now have the ability to talk, listen and perhaps, even understand. The field of voice processing covers a broad range of activities with the goal of enabling people to communicate with machines using natural voice communication skills. Although research in this domain has been carried out for several decades, it has been the confluence of cheap computation and algorithm improvements that has stimulated a wide range of uses for voice processing technology across the spectrum of information processing. Today, the benefits of information and services on computer networks are unavailable to those without access to computers or the skills to use them. As the importance of interactive networks increases in commerce and daily life, those who do not have access to computers or the skills to use them are further handicapped from becoming productive members of society. Advances in human language technology offer the promise of nearly universal access to on-line information and services [1]. Since almost everyone speaks and understands a language, the development of spoken language systems will allow the average person to interact with computers without special skills or training, using common devices such as telephone [2]. These systems will combine spoken language understanding and generation to allow people to interact with computers using speech to obtain information on virtually any topic, to conduct business and to communicate with each other more effectively. The field of voice processing encompasses five broad technology areas, including:

- speech compression, the process of compressing the information in a voice signal so as to either store it or transmit it economically over a channel;

- speech synthesis, the process of creating a synthetic replica of a voice signal so as to transmit a message from a machine to a person, with the purpose of conveying the information in the message;

- speech recognition, the process of extracting the message information in a voice signal so as to control the actions of a machine in response to spoken commands;

- speaker recognition, the process of either identifying or verifying a speaker by extracting individual voice characteristics, primarily for the purpose of restricting access to information, networks, or physical premises;

- spoken dialogue, the process of an appropriate control of the speech recognizer and the speech synthesizer, which permits intelligent, real-time interaction between humans and computers.

This chapter surveys the state of the art of man-machine voice communication. The goal of the survey is to provide an interested reader with an overview of the field: the main areas of work, the capabilities and limitations of current technology, and the technical challenges that must be overcome to realize the vision of graceful human computer interaction using natural voice communication skills. The structure of the chapter is based on a simplified version of many system architectures existing in these days. Most of the sections describe components of such an architecture or ways for deriving useful data structures. Parameters and models of these structures are often obtained by statistical analysis of experimental corpora. The acoustic and the language models are only an approximation of the reality and interact so that, very often, one compensates for the weakness of the other.

2. Human Factors in Man-Machine Communication

Speech is a preferred means for communication among humans. Because humans are designed to live in an air atmosphere, it was inevitable that they learn to convey information in the form of pressure waves supported by displacement of air molecules. But of the variety of acoustic information signals, speech is a very special kind. It is constrained in three important ways:

- by the physics of sound generation in the vocal system,

- by the properties of human hearing and perception, and

- the conventions of language.

These constraints have been central to man-man communication and remain of paramount importance for man-machine communication today. Man-man voice communication is the transfer of information from one person (speaker) to another (listener) via speech, which consists of four-dimensional acoustic signal coming from the mouth of a speaker and propagating through the air. The chain of events from conception of a message in the speaker's brain to the arrival of the message in the listener's brain is called the speech chain. The chain consists of a speech production mechanism located in the speaker, transmission through a medium such as air, and a speech perception process in a listener's auditory system (ears and brain). A speaker forms an idea to convey, converts that idea into a linguistic structure by choosing appropriate words or phrases to represent that idea, orders the words or phrases based on learned grammatical rules associated with the particular language, and finally adds any additional local or global characteristics such as intonation or stress to emphasize aspects important for overall meaning. Once this has taken place, the human brain produces a sequence of motor commands that move the various muscles of the vocal system to produce the desired sound pressure wave. The speech perception process begins when the listener collects the sound pressure wave at the outer ear, converts this into neurological pulses at the inner ear, and interprets these pulses in the auditory cortex of the brain to determine what idea was received. Voice communication from one person to another appears to be so easy and simple. Although speech technology has reached a point where it can be useful in certain applications, the prospect of a machine understanding speech with the same flexibility as humans do is still far away. The interest in using speech interface to machines stems from our desire to make machines easy to use. Using human performance as a benchmark for the machine tells us how far we are from that goal [3].

Along the way to giving machines human-like capability to speak and understand speech there remains much to be done about how the knowledge of structure and meaning in language can be incorporated into usable systems. Continuing improvement in the effectiveness and naturalness of human-machine voice communication systems depends on concepts and results from many fields, including microelectronics, computer science, digital signal processing, acoustics, auditory science, linguistics, phonetics, cognitive science, statistical modeling, and psychology.

The major issues in implementing man-machine voice communication systems are :

- natural speech input and output,

- speech recognition and understanding for voice input,

- speech synthesis and language generation for voice output,

- dialogue and usability factors related to how humans interact with machines.

With a microphone to pick up the human voice and a speaker to deliver a synthetic voice from the system to the human ear, the human can communicate with the system, which in turn can cause desired actions to occur. In order to do this, the voice communication system must take in the human voice input, determine what action is called for, and pass information to other systems. In some cases generation of the equivalent text (e.g. dictation) or other symbolic representation of the speech is all that is necessary. In other cases, such as in natural language dialogue with a machine, it may be necessary to extract the meaning of the utterance. Such a system can be used in many ways. In one class of applications, the human controls a machine by voice ; for example to complete a telephone call. Another class of applications involves access and control of information ; for example, the system might respond to a request for information by searching a database and then providing the answers to the human by synthetic voice, or it might even attempt to understand the voice input and speak a semantically equivalent utterance in another language. While this may be the most challenging part, the ability to recognize or understand speech and to produce synthetic voice output is still only one aspect of the problem. Another important aspect concerns usability by the human. The system must be designed to be easy to use, and it must be flexible enough to cope with the wide variability that is common in human speech and acoustic environment. Finally, the technology available for implementing such a system must be an overarching concern.

3. Voice Processing Systems

Man-machine voice communication can be seen as an exchange of information coded in a way suitable for transmission through the air. Two processes characterize such a processing : coding and decoding. Coding is the process of producing a representation of what has to be communicated. The content to be communicated is structured using words represented by sequences of symbols of an alphabet and belonging to a given lexicon. Phrases are made by concatenating words according to the rules of a grammar and associated in order to be consistent with a given semantics. These various types of constraints are knowledge sources with which a symbolic version of the message to be exchanged is built. The symbolic version undergoes further transformations to obtain a speech signal transmittable through the air. Decoding is the inverse process which uses knowledge sources to transform the message carried by a speech signal into different levels of symbolic representation. Decoding can produce word sequences or conceptual hypotheses. Unlike man-man communication, man-machine communication is expected to produce instances of computer data structures in a deterministic way. This means that a computer system has to produce the same representation for the same signal, every time this signal is processed. So, with current technology, speech interpretation by machine is performed by predesigned, predictable reactions to the data. The knowledge sources used by machines in the decoding process are only models of the ones used by humans for producing their messages. The decoding process has to deal with ambiguity due to distortions introduced by the transmission channel, the limits of the knowledge used, and often to intrinsic imprecision of the spoken message. The imprecision is due to the fact that the speaker's intention may not be that of producing exactly an instance of the data structures belonging to the knowledge of the decoder. To a certain extent, ambiguities can be reduced by exploiting message redundancy. In practice, knowledge is used to transform the input speech signal into more suitable sequences of vectors of parameters and to obtain from them various levels of symbolic representations. The first level of symbolic representation can be for example a phoneme, a syllable or a word. The choice of knowledge systems and the way they are used in a system determines the system architecture. System architecture design should be based on a number of performance indices. The most important of them for decoding are coverage and precision. Coverage means that the system has to be able to recognize all the word sequences it was conceived for. The requirement of precision means that knowledge systems and methods for their use should produce the lowest recognition or understanding error rates. Knowledge can be manually compiled or obtained by automatic learning from a corpus of data. Coverage and precision of manually derived knowledge are often limited. The best results so far have been obtained using component models having a simple, manually decided structure. Statistical parameters of these models are estimated by automatic training. Complex knowledge structures are obtained by composition of basic models. The requirements for speech synthesis in the coding process are quite different since it is acceptable to produce only those sentences that can be generated by a suitable grammar with just one or few voice types for the same sentence. In this case, the new dimensions of quality and naturalness become of fundamental importance. A common requirement for coding and decoding processes is acceptable computational complexity, both in terms of time and space. This has an impact on the memory requirements for the hardware system. Having responses close to real-time is a necessary condition. A functional architecture model for man-machine voice communication is shown in Fig. 9.1.

This model is valid if the voice only user interface is available, e.g. in telephone-based remote applications, or when voice is the main input channel because end users are not trained computer users (for example in database query by voice). The lowest layer handles input and output devices. The layer of speech recognition and synthesis represent a framework to build interactive applications, and is shared among different applications. The understanding component has to drive the dialogue with the user, assuming the functionalities of the dialogue control (or script) layer and interacting directly with the application functions layer which performs computations specific to the application.

4.   Natural Speech Input and Output

Speech sounds originate when air particles are induced to vibrate around their mean position with consequent changes in the air pressure with respect to

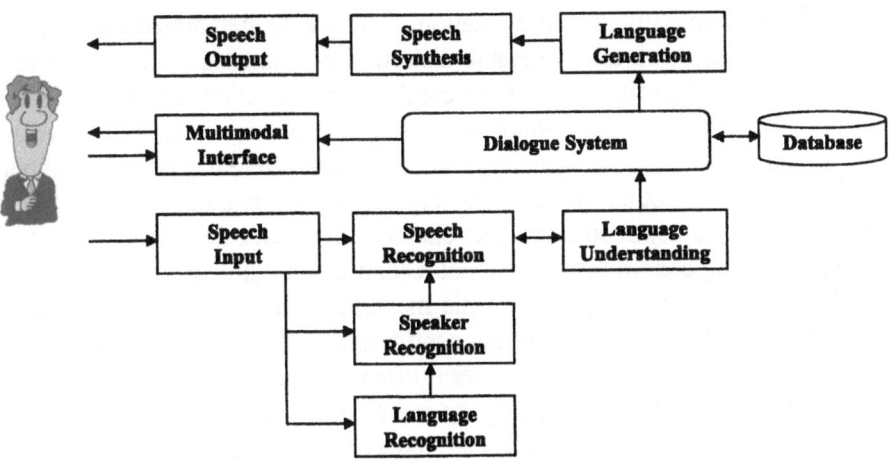

Figure 9.1. Man-machine voice communication system.

a static value. This results in a sound field in which variations of air density and pressure are function of time and space and propagate as acoustic waves. These variations are converted into electrical signals or are generated from electrical signals by transducers. Speech processing is performed by computers on numerical representations of these electrical signals. As transducers are not perfect, they introduce distortion to the information carried by the signals they transform. Furthermore, a microphone often captures a variety of signals present in the surrounding environment (e.g. speech and noise, or multiple conversation). In many cases, only one of the components is relevant for a given application and this important component has to be extracted from the mixture of sounds captured by the sensor. Performance of man-machine voice communication systems based on available technologies is influenced by a large number of factors, including type and use of microphones, while the quality and intelligibility of speech produced by synthesis systems is much less affected by the choice of the loudspeaker. For this reason, the emphasis of this section is on speech input.

4.1 Speech Signal Acquisition and Coding

Present acoustic-transduction technology makes available precise, robust, small microphones, e.g. electret microphones, often at fairly or very low cost. The type of microphone, the way in which it is used, and the environmental conditions in which it operates may have a fundamental impact on the performance that can be obtained in applications such as speech recognition. Speech acquisition involves a number of transformation steps, as information flows from the talker's mouth to a digital representation. Any poorly performed transformation step can cause an unrecoverable distortion of information. An important contribution to distortion is due to microphone transduction. In telephone applications, the microphone cannot be chosen to maximize recognition performance and the recognition system must account for different types of telephone handsets. But, for other applications (e.g. dictation, data entry, voice-driven interfaces, etc.) an optimal transducer choice can be made depending on microphone mounting conditions, directional response and other specifications. The ideal method for capturing an acoustic message is to use a transducer placed close to the emitting source. In this configuration, the captured power of the noise produced by other sound sources is often negligible when compared with the power of the signal produced by the source of interest. For speech recognition, available modeling techniques are based on the assumption that a single, clear speech message is conveyed by the acquired signal. When operating in a quiet environment, the use of a single microphone can allow high performance. However, this assumption is clearly not valid in many practical situations where the message to be processed is generated at some distance from the transducer and, therefore, is mixed with other sounds. In these conditions, it is advantageous to exploit the spatial selectivity of a microphone array to acquire the signal of the desired source. A microphone array consists of a set of acoustic sensors placed at different locations to spatially sample a sound pressure field. Using a microphone array it is possible to selectively pick up a speech message, while avoiding the undesirable effects due to distance, background noise, room reverberation and competitive sound sources. The directivity of a microphone array can be electronically controlled, without changing the sensor positions or placing the transducers very close to the speaker.

4.2   Speech Analysis and Feature Extraction

Speech input and output systems require the solution of problems of very different nature. A basic question is how to deal with the redundancy in the speech information, in order to derive a compact representation in terms of acoustic features. Every speech message can be seen as composed of a sequence of basic speech sounds. Every language has a set of basic units that represent the building blocks for the composition of every word or sentence. The simplest choice is to associate a basic speech unit with each phoneme of the language. A phoneme is an abstract linguistic unit, the smallest discriminant one in the phonology of a given language. Each language has its own set of phonemes that, generally, contains a few tens of elements. In principle, for each phoneme, there can be an infinite number of physical realizations, each of them resulting in different acoustic events. The properties of the corresponding speech waveforms are generally described by the amplitude evolution of the waveform or by the corresponding spectral content, obtained with the appropriate spectral analysis method. The speech signal can be considered as a non-stationary signal produced by articulators moving from one position to another with intrinsic mechanical time constants. Therefore, it is possible to define a time interval within which the signal can be considered time-invariant (i.e. stationary). On a segment of length comparable with that of the stationarity interval, standard spectral analysis methods can be applied. For speech recognition purpose, the main goal of the speech analysis and feature extraction step is the computation of a sequence of feature vectors that provide a compact representation of the given input signal. These representations aim to preserve the information needed to determine the phonetic identity of a portion of speech while being as impervious as possible to factors such as speaker differences, effects introduced by communications channels, and paralinguistic factors such as the emotional state of the speaker. They also aim to be as compact as possible. The continuous speech signal must first be converted into discrete samples at a rate between 6.6 kHz and 20 kHz. These discrete samples are then converted into a set of representative feature vectors. They are calculated on a frame-by-frame basis. Frame duration is defined as the number of samples over which a set of parameters is valid and the number of samples used to compute the frame parameters is known as the window duration. This short-term parameterization process is often referred to as the front-end of the speech recognition system. Typically, it compresses around 256 samples of speech window data down to between 10 and 20 parameters that are usually computed every 10 or 20 ms. In particular, computing the envelope of the short-term spectrum reduces the variability significantly by smoothing the detailed spectrum. The spectrum can be represented directly in terms of the signal's Fourier coefficients or as the set of power values at the outputs from a bank of filters. The envelope of the spectrum can be represented indirectly in terms of the parameters of an all-pole model, using Linear Predictive Coding (LPC), or in terms of the first dozen or so coefficients of the cepstrum - the inverse Fourier transform of the logarithm of the spectrum. One reason for computing the short-term spectrum is that the cochlea of the human ear performs a quasi-frequency analysis. The analysis in the cochlea takes place on a nonlinear frequency scale (known as the Mel scale). This scale is approximately linear up to about 1000 Hz and is approximately logarithmic thereafter. So, in the feature extraction, it is very common to perform a frequency warping of the frequency axis after the spectral computation. Researchers have experimented with many different types of features for use in speech recognition. Perhaps the most popular features used for speech recognition today are what are known as Mel-Frequency Cepstral Coefficients (MFCCs). These coefficients are obtained by taking the discrete cosine transform of the log spectrum after it is warped according to the Mel scale.

The static acoustic features are aimed at representing the power spectrum in a compact way and are derived from a short-term analysis. On the other hand, the dynamic features account for the time evolution of the power spectrum in a longest interval. The most common dynamic features used in speech recognition are based on the computation of time derivatives of the corresponding static features. Usually, between 8 and 16 cepstral coefficients, with their first- and second-order time derivatives, are used as feature vector in modern speech recognition systems. Spectral distortions due to various channels, such as a different microphone or telephone, can have enormous effects on the performance of speech recognition systems. To render recognition systems more robust to such distortions, many researchers perform some form of removal of the average spectrum. In the cepstral domain, spectral removal amounts to subtracting out the mean of the cepstra for a given telephone call. Other similarly simple methods of high-pass filtering of cepstral or log-cepstral trajectories have also been proposed for removing channel effects. In this case useful dynamic features can be obtained by applying the technique known as RelAtive SpecTrAl (RASTA) processing. The technique was first proposed and applied with Perceptually-based Linear Prediction (PLP) analysis of speech. For recognition systems that use statistical models, it is important to be able to estimate probability distributions of the computed feature vectors. In fact, the use of a large number of features for each frame yields a large number of statistical parameters to estimate and requires a great amount of data for system training. Because these distributions are defined over a high-dimensional space, it is often easier to use vector quantization to transform each feature vector to one of a relatively small number of template vectors, which together comprise what is called a codebook. A typical codebook would contain about 256 or 512 template vectors. Estimating probability distributions over this finite set of templates then results in a substantial reduction of storage requirements and computational complexity.

4.3   From Language Generation to Speech Synthesis Output

Speech synthesis is the process that allows the transformation of a string of phonetic and prosodic symbols into a synthetic speech signal [4]. The quality of the result is a function of the quality of the string, as well as of the quality of the generation process itself. Speech synthesis for voice communication systems is based on text-to-speech process which includes not only speech signal generation but also text processing. Another type of speech synthesis is based on concept-to-speech process. It differs from text-to-speech approach in that speech is generated from an abstract representation of semantic/pragmatic concepts rather than from text. Language generation deals with the generation of text and concepts.

The last step for speech output of the speech synthesis system is generation of the speech signal waveform, according to the segmental and prosodic parameters defined at earlier stages of speech synthesis. Speech signal generators, called synthesizers, can be classified into three categories: formant synthesizers, concatenative synthesizers, and articulatory synthesizers.

Formant synthesis is a descriptive acoustic-phonetic approach to synthesis. Speech generation is performed by modeling the main acoustic features of the speech signal. The basic acoustic model is the source-filter model. The filter, described by a small set of formants, represents articulation in speech. It models speech spectra that are representative of the position and movements of articulators. The source represents phonation. It models the glottal flow or noise excitation signals. Both source and filter are controlled by a set of phonetic rules (typically several hundred). High-quality rule-based formant synthesizers, including multilingual systems, have been marketed for many years.

Concatenative synthesis is based on speech signal processing of natural speech databases. The segmental database is built to reflect the major phonological features of a language. For instance, its set of phonemes is described in terms of diphone units, representing the phoneme-to-phoneme junctures. Other units are also used (diphones, syllables, words, etc.). The synthesizer concatenates speech segments, and performs some signal processing to smooth unit transitions and to match predefined prosodic schemes. Direct pitch-synchronous waveform processing is one of the most simple and popular concatenation synthesis algorithms. Other systems are based on multipulse linear prediction, or harmonic plus noise models. Several high-quality concatenative synthesizers, including multilingual systems, are marketed today.

Articulatory synthesizers are physical models based on the detailed description of the physiology of speech production and on the physics of sound generation in the vocal apparatus. Typical parameters are the position and kinematics of articulators. Then the sound radiated at the mouth is computed according to equations of physics. This type of synthesizer is rather far from applications and marketing because of its cost in terms of computation and the underlying theoretical and practical problems still unsolved.

5.   Speech Recognition

Speech recognition is the process of converting an acoustic signal, captured by a microphone or a telephone, to a set of words. The recognized words can be the final results, as for applications such as commands and control, data entry, and document preparation. They can also serve as the input to further linguistic processing in order to achieve speech understanding. Speech recognition is a very challenging problem in its own right, with a well-defined set of applications. However, many tasks that lend themselves to spoken input - e.g. making travel arrangements - are in fact exercises in interactive problem solving. The solution is often built up incrementally, with both the user and the computer playing active roles in the "conversation". Therefore, several language-based input and output technologies must be developed and integrated to reach this goal. Figure 9.1 shows the major components of a typical conversational system. The spoken input is first processed through the speech recognition component. The natural language component, working in concert with the recognizer, produces a meaning representation. For information retrieval applications illustrated in this figure, the meaning representation can be used to retrieve the appropriate information. If the information in the utterance is insufficient or ambiguous, the system may choose to query the user for clarification. Natural language generation and speech synthesis can be used to produce spoken responses that may serve to clarify the information. Throughout the process, discourse information is maintained and fed back to the speech recognition and language understanding components, so that sentences can be properly understood in context. The difficulties of speech recognition task can be described in terms of speech style, intra- and inter-speaker variabilities, vocabulary size, language constraints and adverse conditions. An isolated-word speech recognition requires that the speaker says words with short pauses in between, whereas a continuous speech recognition does not. Natural or spontaneous speech contains disfluencies, hesitations non-grammatical sentences, and is much more difficult to recognize than speech read from written text. A kind of generalization of an isolated-word recognition in which the speaker is not constrained to pronounce the words in isolation is often referred to as keyword spotting. In this case, one wants to detect keywords from unconstrained speech, while ignoring all other words or non-speech sounds. This leads to more user-friendly systems. Speaker dependent systems require speaker enrollment - a user must provide samples of his or her speech before using them, whereas other systems are said to be speaker-independent, in that no enrollment is necessary. For most applications (e.g. the public telephone network), only speaker-independent systems are useful. Speaker independence is usually achieved by using the same baseline statistical models that are trained on databases containing a large population of representative talkers. Recognition is generally more difficult when vocabularies are large or have many similar-sounding words. Adding more words increases the probability of confusion between words. When speech is produced in a sequence of words, language models or artificial grammars are used to restrict the combination of words. The simplest language model can be specified as a finite-state network, where the permissible words following each word are given explicitly. More general language models approximating natural language are specified in terms of a context-sensitive grammar. The task of continuous speech recognition is simplified by restricting the possible utterances. This is usually done by using syntactic information to reduce the complexity of the task and the ambiguity between words. The constraining power of a language model is usually measured by its perplexity, loosely defined as the geometric mean of the number of words that can follow a word at any decision point.

Finally, there are some external parameters that can affect speech recognition system performance, including the characteristics of the environmental noise and the type and the placement of the microphone. The past decade has witnessed significant progress in speech recognition technology. Substantial progress has been made in the basic technology, leading to the lowering of barriers to speaker independence, continuous speech, and large vocabularies. There are several factors that have contributed to this rapid progress. First, there is the coming of age of the Hidden Markov Model (HMM). HMM is powerful in that, with the availability of training data, the parameters of the model can be trained automatically to give optimal performance. Second, much effort has gone into the development of large speech corpora for system development, training, and testing. Some of these corpora are designed for acoustic phonetic research, while others are highly task specific. Nowadays, it is not uncommon to have tens of thousands of sentences available for system training and testing. Third, progress has been brought about by the establishment of standards for performance evaluation. Finally, advances in computer technology have also indirectly influenced our progress. The availability of fast computers with inexpensive mass storage capabilities has enabled researchers to run many large scale experiments in a short amount of time. This means that the elapsed time between an idea and its implementation and evaluation is greatly reduced. In fact, speech recognition systems with reasonable performance can now run in real time using high-end workstations without additional hardware - a feat unimaginable only a few years ago. In such systems the digitized speech signal is first transformed into a set of useful feature vectors at a fixed rate. These vectors are then used to search for the most likely word candidate, making use of constraints imposed by the acoustic, lexical, and language models. Throughout this process, training data are used to determine the values of the model parameters.

5.1 Acoustic Models

Acoustic models are stochastic models used with language models and other types of models to make decisions based on incomplete or uncertain knowledge. Given a sequence of feature vectors extracted from a speech waveform by the signal processing front end, the purpose of the acoustic models is that of computing the probability that a particular linguistic event, e.g. a word, has generated a sequence. These probabilities can then be combined with a priori probabilities given by the language model to find the sentence in the recognition language that is the most probable generator of the utterance. Acoustic models have to be flexible, accurate and efficient. Flexibility is required because the characteristics of the recognition task are often quite different from the training conditions, and because of the complexity of the language used in many speech recognition applications. Accuracy is needed because of the ambiguity which is inherent in the relation between speech sounds and linguistic contents. Finally, efficiency is important when a recognition system is used interactively, to avoid forcing the user to wait for the computer response to spoken input. Even if these problems have not been completely solved, hidden Markov models (HMMs) are a satisfactory solution, and most of the recent systems are based on their use. An HMM is a doubly stochastic model, in which the generation of the underlying phoneme string and the frame-by-frame, acoustic vectors are both represented probabilistically as Markov processes. Neural networks have also been used to estimate the frame scores; these scores are then integrated into HMM-based system architectures, in what has come to be known as hybrid systems.

In HMM approach each word is represented as a sequence of phonetic units, and each unit is represented via HMMs containing a predefined number of states. Each state is described by a probability distribution over the space of observations (parameters of acoustic vectors). If the observation vector takes values on a continuum range, the probability distribution is actually a density and the models are called Continuous Density HMMs (CDHMMs). If the observation vector is quantized over a given codebook, the distribution is discrete and the models are called Discrete Density HMMs (DDHMMs). CDHMMs are more precise than DDHMMs, but require more computation. Well known algorithms exist for learning the HMM distributions starting with a set of training data (e.g. the Baum-Welch forward-backward algorithm). The goal of the recognizer is to find the word sequence with the maximum posterior probability of the hypothesized Markov model λ (i.e. associated with a specific sequence of words) given an acoustic vector sequence $\mathbf{x}$. This involves maximizing the probability, $p(\lambda|\mathbf{x})$. The fundamental equation relevant to this process is a restatement of Bayes' rule as applied to speech recognition :

$$p(\lambda|\mathbf{x}) = \frac{p(\mathbf{x}|\lambda) \cdot p(\lambda)}{p(\mathbf{x})} \tag{9.1}$$

In Eq. 9.1 a likelihood $p(\mathbf{x}|\lambda)$ represents the contribution of the acoustic model, and a prior probability $p(\lambda)$ represents the contribution of the language model. Usually $p(\mathbf{x}|\lambda)$ and $p(\lambda)$ are trained separately on independent training corpora. By doing so, and assuming that $p(\mathbf{x})$ is independent of the parameters of the models, turns the speech recognition process into a maximum likelihood estimation (MLE) problem. The finding of the optimal word sequence according to Eq. 9.1, given the observation vectors and the acoustic models, may be carried out using different algorithms. The Viterbi algorithm is probably the most widely used; this can be extended with language modeling. One of the key issues in acoustic modeling consists of the choice of phonetic units. The most frequently used are phonemes, but one of the most successful approaches in continuous speech recognition employs the so called context dependent phone-like units. The rationale for this approach is that, because of an effect known as coarticulation, the same phoneme is realized differently depending on the phonemes that precede and follow it; representing each realization separately, therefore, allows us to achieve a higher recogniton accuracy. Simple phonemes (i.e. independent of context) are in the range of a few tens. However, context dependent units are several thousands.

5.2   Language Models

The use of linguistic structure (language model) in speech recognition systems is directed toward specifying allowable and/or likely sequences of words. The primary objectives of such a structuring are to: reduce perplexity, increase speed and accuracy, and enhance vocabulary flexibility. The most commonly used forms of structuring (grammars) are : finite-state grammars, N-gram models, and finite-state networks. The most frequently used grammar in speech recognition is the finite-state grammar. It reduces perplexity by delineating the words that are allowable at any point in the input. The deterministic nature of finite-state grammars makes them rigid. Only the word sequences that have been programmed into the grammar can be recognized. Once the user begins to say a sequence, it must be completed in one of the ways specified by the grammar. One of the most successful language models suitable for integration into the recognition process makes use of a special case of probabilistic finite-state automaton called N-grams. The role of N-grams is that of providing the a priori word sequence probability $P(\lambda)$. This goal is achieved by computing the probability of a sequence as the product of the probabilities of each word, conditioned on the occurrence of the previous $N - 1$ words. The underlying assumption is that only the last $N - 1$ words affect the occurrence of the next word. A large N provides a more detailed model, but the training material may be insufficient. In practice the cases $N = 3$ (trigrams) or $N = 2$ (bigrams) are used. An alternative to probabilistic N-grams is to use complete finite-state networks, which are compatible with the Viterbi decoding (also called beam search) or stack decoding to handle the challenges of large vocabulary systems.

5.3   Language Understanding

In man-machine communication, the ultimate goal of understanding is the execution of an action. Spoken language understanding involves two primary component technologies: speech recognition, and natural language processing. The integration of speech and natural language has great advantages. To natural language processing, speech recognition can bring prosodic information (information important for syntax and semantics but not well represented in text); natural language processing can bring to speech recognition additional knowledge sources (e.g., syntax and semantics). The result of understanding by a machine is represented by a data structure produced with an interpretation strategy using a semantic theory. For this purpose a semantic representation describes relationships at three different levels, namely the meaning of individual words (relative to a vocabulary and structure of concepts), the meaning of a group of words in a syntactic structure, and the meaning of a sentence in its context. Language understanding is a complex, multi-faceted task, and rarely appears in speech recognition systems. Although there have been significant recent gains in spoken language understanding, current technology is far from human-like: only systems in limited domains can be envisioned in the near term, and the portability of existing techniques is still rather limited.

6.   Dialogue Systems and Man-Machine Interaction

The need for a dialogue component in a system for man-machine interaction arises for several reasons. Often the user does not express his requirement with a single sentence, because that would be impractical; assistance is then expected from the system, so that the interaction may naturally flow in the course of several dialogue turns. Moreover, a dialogue manager should take care of identifying, and recovering from, speech recognition and understanding errors. While research on dialogue has progressed successfully with reference to text-based systems, the clearest practical benefits come from spoken dialogue systems. This is hardly surprising, since speech is the dominant modality for normal man-man dialogues. Spoken dialogue provides a valuable extension to the input/output capabilities of desktop systems, but here it serves an augmentative function, providing additional resources over and above more conventional input/output systems. However, the emerging perspective of accessing information through the World Wide Web by a large, international, heterogeneous population brings the problem of the quality of interaction to the forefront. This makes it necessary to consider natural language as an appropriate, if not the most, media for interaction. Spoken dialogue technology can be expected to succeed first in contexts where there is no viable alternative for communicating with computer systems, such as hand-held computers or "smart" portable telephones, which are too small to offer a usable physical interface. It is also the only modality available for communicating with systems over the telephone, where there is no (current) alternative to speech for complex or pleasant interactions. Developments in miniaturization are showing great promise, and will undoubtedly stimulate spoken dialogue research in new and challenging ways in years to come.

Interaction should be user-friendly (or convivial). Conviviality should be viewed and handled as a global feature of a system. Let us call a dialogue system, a more or less convivial interactive system that is capable of managing elaborate contextual exchanges with the user. The machine component of man-machine communication is an intelligent agent whose dialogue abilities are based on its primitive capacities of rational, cooperative behavior. Simple dialogue models reflect a dialogue strategy based on finite-state diagrams.

7. Voice User Interface

The telephony Voice User Interface (VUI) is what the basic voice processing technologies create - automated interaction over the telephone using essentially speech recognition to interpret the speech and synthesize the speech to respond. Just as the Graphical User Interface (GUI) changed the personal computers business and made it accessible to a much larger customer base, the telephony Voice User Interface (VUI) has the potential to change the way telephones are viewed and significantly broaden their use. Telephones have largely been used to connect to other people. The touch-tone pad is a very poor and limited user interface; nevertheless, the need for extensions in the use of telephones has led to its implementation in controlling telephone features (e.g., voice mail), in directing calls (e.g., auto attendants), and in customer transactions (e.g., retrieving account information). With a better voice interface, the range of automated interaction over the phone can expand tremendously. The telephone begins to be perceived as an assistant. With actual voice processing technology, the information, e.g., on Internet sites, can become accessible by phone, opening it to anyone with access to a telephone. On the other hand, the telephone is a quicker way than the Internet to obtain certain types of information or perform some transactions. It allows a "zero-weight" personal digital assistant, in that any telephone can allow access to personal and general information. It reduces or eliminates waiting time for the person phoning call centers. It makes information and services available 24 hours a day by telephone. It also reduces costs by automating repetitive tasks that now use a service representative as an interface between a person and a computer. When telephone subscribers pick up a telephone, they can have a personal assistant greet them rather than a dial tone. In the future, subscribers will eventually identify with their VUI, not their telephone company. Callers will feel secure because, for certain transactions, the VUI will verify that the caller is indeed the subscriber by the sound of their voice. Callers can accomplish their goals by saying what they mean.

8.   Security Considerations and Speaker Verification

Speaker recognition is the process of automatically recognizing who is speaking by using speaker-specific information included in speech waves. Speaker identity is correlated with the physiological and behavioral characteristics of the speaker. These characteristics exist both in the spectral envelope (vocal tract characteristics) and in the supra-segmental features (voice source characteristics and dynamic features spanning several segments). Generally, speaker recognition can be classified into speaker identification and speaker verification. Speaker identification is the process of determining from which of the registered speakers a given utterance comes. Speaker verification is the process of accepting or rejecting the identity claim of a speaker. Most of the applications in which voice is used to confirm the identity claim of a speaker are classified as speaker verification.

Speaker recognition methods can also be divided into text-dependent and text-independent methods. The former require the speaker to provide utterances of key words or sentences that are the same text for both training and recognition, whereas the latter do not rely on a specific text being spoken. Since the text-dependent method can directly exploit voice individuality associated with each phoneme or syllable, it generally achieves higher recognition performance than the text-independent method. However, there are several applications in which predetermined key words cannot be used. Therefore, text-independent methods have recently attracted more attention. Another advantage of text independent recognition is that it can be done sequentially, until a desired significance level is reached, without the annoyance of the speaker having to repeat the key words again and again. Both basic text-dependent and independent methods have a serious weakness. That is, these systems can easily be defeated, because someone who plays back the recorded voice of a registered speaker uttering key words or sentences into the microphone can be accepted as the registered speaker. To cope with this problem, some methods use a small set of words, such as digits, as key words, and each user is prompted to utter a given sequence of key words that is randomly chosen every time the system is used.

8.1   Text-Dependent Speaker Recognition

Text-dependent speaker recognition methods can be classified into dynamic time warping (DTW) or hidden Markov model (HMM) based methods. In DTW approach, each utterance is represented by a sequence of feature vectors. The timing variation of utterances of the same text is normalized by aligning the analyzed test feature vector sequence to the reference feature vector sequence using a DTW algorithm. The overall, accumulated distance between the test and reference utterances is used for recognition decision. The hidden Markov model (HMM) can efficiently model statistical variation in spectral features. Therefore, speaker recognition methods based on characterizing the utterances as sequences of subword units represented by HMMs were introduced as extensions of the DTW-based methods, and have achieved significantly better recognition accuracy.

8.2 Text-Independent Speaker Recognition

One of the most successful text-independent recognition methods is based on vector quantization (VQ). In this method, VQ codebooks consisting of a small number of representative feature vectors are used as an efficient means of characterizing speaker-specific features. Clustering the training feature vectors of each speaker generates a speaker-specific codebook. In the recognition stage, an input utterance is vector-quantized by using the codebook of each reference speaker; the VQ distortion accumulated over the entire input utterance is used for making the recognition determination. Temporal variation in speech signal parameters over the long term can be represented by stochastic Markovian transitions between states. Therefore, methods using an ergodic HMM, where all possible transitions between states are allowed, have been proposed. Speech segments are classified into one of the broad phonetic categories corresponding to the HMM states. After the classification, appropriate features are selected. In the training phase, reference templates are generated and verification thresholds are computed for each phonetic category. In the verification phase, after the phonetic categorization, a comparison with the reference template for each particular category provides a verification score for that category. The final verification score is a weighted linear combination of the scores from each category. This method was extended to the richer class of mixture autoregressive (AR) HMMs. It has been shown that a continuous ergodic HMM method is far superior to a discrete ergodic HMM method and that a continuous ergodic HMM method is as robust as a VQ-based method when enough training data is available. However, when little data is available, the VQ-based method is more robust than a continuous density HMM method. A method using statistical dynamic features has also been proposed. In this method, a multivariate auto-regression (MAR) model is applied to the time series of cepstral vectors and used to characterize speakers. It was reported that identification and verification rates were almost the same as obtained by an HMM-based method. Recently, a technique based on the maximum likelihood estimation of a Gaussian-mixture model (GMM) for speaker representation has been introduced [5]. Gaussian mixtures are noted for their robustness as a parametric statistical model and their ability to form smooth estimates of rather arbitrary underlying densities. This model can be regarded as a special case of a single-state continuous HMM. Furthermore, the VQ-based method corresponds to a degenerate case of GMM-based technique with a distortion measure being used as the observation probability.

In speaker verification, an identity claim is made by an unknown speaker and an utterance of this unknown speaker is compared with the model for the speaker whose identity is claimed. If the match is good enough, that is, above a threshold, the identity claim is accepted. The key features are the false acceptance rate and false rejection rate. The false acceptance rate is the fraction of attempts where an impostor is incorrectly accepted as valid. The false rejection rate is the number of times a valid speaker is classified as an impostor and denied access. To set the threshold at the desired level of customer rejection and impostor acceptance, it is necessary to know the distribution of customer and impostor scores. The equal-error rate is a commonly accepted overall measure of system performance. It corresponds to the threshold at which the false acceptance rate is equal to the false rejection rate. Adaptation of the reference model as well as the verification threshold for each speaker is indispensable to maintain high recognition accuracy for a long period. In order to compensate for the variations, two types of normalization techniques have been tried - one in the parameter domain, and the other in the distance/similarity domain.

8.3   Mode of Operation

Speaker verification enables access control of various services by voice in a way that is more secure than a password or account number alone. All systems require some sort of enrollment (training), where the user must provide a minimum amount of speech to create a baseline, a reference model (often called a "template"). These reference models must be associated with the user and stored in a database for later verification. The person who calls must claim an identity in some way, usually by giving an account number or PIN code. There are a wide variety of ways in which callers may be asked to prove their claimed identity, including: (1) User-chosen passwords: The user picks a phrase that he can remember. (2) User-specific passwords: Typically, users may be asked to say the last four digits of their social security number or their mother's maiden name as a password. Such passwords are somewhat less secure than a user-chosen password because the information can be obtained from other sources than the user. (3) All users say the same phrase: Some banks, for example, like the user to say the name of the service as a password. While this makes it easier for an impostor, it eliminates the problem of forgotten passwords and strengthens brand identification. (4) Verification from a recognized account number: If speech recognition is combined with speaker verification, the user can say the account number. This spoken number can be used to identify the account as well as to verify the account holder from the voice. This approach is the most convenient for the caller and avoids the necessity for the user to remember a password. On the other hand, it requires that the account number be spoken out loud, which may be of concern in some situations, such as a call from a public phone. (5) Prompted verification: The caller enrolls by saying more than one password or by saying digits. When calling in, the caller is asked to say one of the passwords or a string of arbitrary digits by the system. This defeats the line being tapped or the speech otherwise recorded, and that recorded speech being re-used by an impostor, since his recorded password may not be the one requested on the next call. This approach is generally used only when fraud prevention is a major issue, such as verifying internal access to a sensitive database or for tracking in law enforcement. (6) Background verification: Background verification does not assume specific words will be spoken. Instead, it measures general speech characteristics during a conversation or interaction with an automated system and voids the transaction or cancels the call if an impostor is detected. This technology usually requires a longer enrollment than word-specific technologies. In addition, some systems support an interaction that allows a variable number of times the caller is asked to repeat a password, with more repetitions if the calculated match to the reference model is poor. The variation could also involve asking for different information, depending on the match.

Most of the problems in speaker recognition arise from variability, including speaker-generated variability and variability in channel and recording conditions. As companies look to increase security to reduce fraud, biometrics is receiving particular attention. Companies are noting that speaker verification is inexpensive, less objectionable to the user than fingerprints or retinal scans, and in the field of comparable accuracy to other techniques. It is perfect for the telephone since no special equipment (such as a fingerprint scanner) is required. On the other hand, the telephone makes speaker verification harder; e.g. digital cellular transmission uses compression-decompression algorithms that do not preserve perfectly the personal voice features in the speech signal.

9. Interactive Voice Servers

The telephone is the most popular current platform for remote speech command and control. Speech also provides a simple, easy-to-use, uniform interface for securized call management and message processing operations and it is an important part of modern telephony applications, in particular interactive voice servers.

This section considers applications of automatic speech recognition and speaker verification techniques in developing efficient Voice Messaging Systems (VMSs)for Integrated Services Digital Network (ISDN) based communication.

Voice messaging, also known as voice mail, handles the access mechanisms to record and replay digitized messages and the distribution of these messages between users. Voice messaging provides the user with the opportunity to leave a digitally recorded speech message for someone who is not available but who has a remote securized telephone access to this message. The demand for this technology for different languages is increasing dramatically, due to the opportunity of savings in operating costs and the rise of mobile cellular telephony. There is also a definite need for voice messaging services to have flexible vocabulary, speaker independent recognition, speaker identity verification, and flexible, programmable dialogue structure.

9.1   Prototype Voice Messaging System (VMS)

The prototype demonstrator presented in this section was developed in the framework of cooperative project which involves two research institutes: the Signal Processing Laboratory (LTS) of the Swiss Federal Institute of Technology Lausanne (EPFL) and the Dalle Molle Institute for Perceptive Artificial Intelligence, Martigny (IDIAP), and three industrial partners: the Advanced Communication Services (aComm), SunMicrosystems (Switzerland) and the Swisscom [6, 7].

The prototype voice messaging system developed in this work offers two main functionalities: posting a message and consulting personal messages. The posting allows to choose the addressee and to record the message. The consulting allows to read the messages by the owner of the mail box. It includes a verification of the speaker identity by means of a vocal PIN-code.

The main VMS functions are: (1) *Command recognition* - to record a command, to recognize it, and to accomplish the associated action; (2) *User identity verification* - to record the voice of a user, and to verify his identity using three speaker verification methods in parallel; (3) *Message recording* - to record a message, and to store it. This message can be retrieved later by the owner of the mail box.

The main goal of the project was to develop a general "toolkit" which allows a universal VMS to be built by associating these three mentioned above functions. The basic hardware configuration of the prototype VMS demonstrator consists of a Sun workstation connected to SwissNet (the Integrated Services Digital Network (ISDN) in Switzerland) using ISDN BRI adapter card. The required software on the Sun workstation includes Solaris, SunISDN, XTL and a modular speech/speaker recognition system adapted to real-time processing. The languages used to program the demonstrator are C++ and C, and shell scripts (tcsh). The general structure of the prototype VMS is presented in Fig. 9.2.

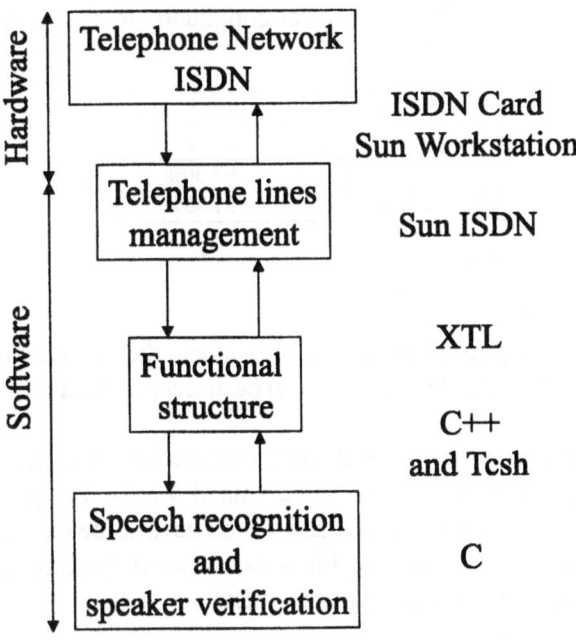

Figure 9.2.   General structure of the voice messaging system (VMS).

9.2   Speech Recognition

At the heart of automatic speech recognition system of the demonstrator lies a set of algorithms for recognition and training. In speech recognition the signal from a ISDN channel is processed using a Continuous Density Hidden Markov Model (CDHMM) technique where feature extraction and recognition using the Viterbi algorithm are adapted to a real-time execution. The approach selected for this work to model command words is the speaker independent flexible vocabulary approach. It offers the potential to build word models for any speaker using Swiss French and for any application vocabulary from a single set of trained phonetic sub-word units.

The major problem of a phonetic-based approach is the need for a large telephone quality database to train, once and for all, a set of speaker-independent and vocabulary independent phoneme models. This problem was solved using the Swiss French Polyphone database provided by the Swisscom. Standard three-state left-to-right phoneme HMMs with five Gaussian mixtures per state, characterized by 12 Mel Frequency Cepstral Coefficients (MFCC) plus energy and their first and second derivatives were used.

In order to validate the flexible vocabulary approach for VMS, we conducted a set of experiments to compare performances of HMMs using phonemes as units. The recognition rates obtained for command words are shown in Table 9.1.

Vocabulary type	Recognition rates
7 command words	98.38%
Digit recognition	97.30%
78 command words	91.95%

Table 9.1.   Recognition rates.

The reported recognition results clearly validate the flexible vocabulary approach, and show that phonetic models can be almost as efficient as whole word HMMs.

The VMS have been designed to work with command words. In practice, some speakers do not utter only the command words, but place them into phrases. Consequently, these command words have to be extracted from the phrase entered to the system. For this reason a word spotting algorithm was added to the demonstrator's recognition system.

9.3   Speaker Identity Verification

The VMS offers a securized access to personal messages. At the first connection, a user has to be registered. He is asked to repeat several times his Personal Identification Number (PIN) and to answer several questions. These utterances are the references used in the Speaker Identity Verification (SIV) process. The SIV is performed as follows: The speaker gives his PIN code. The PIN code is recognized using a HMM recognizer and its validity is tested. If the code is correct the utterance is used to perform a text dependent SIV using three methods based on: HMMs, DTW and second order statistical measures. All the three methods compare the scores obtained by the test utterance for personal models and for general models (speaker independent) called "world models". If the personal models obtain the best results the identity is validated. In the other case the identity is rejected. If the personal and the general models obtain an equivalent score, there is a doubt and more information is requested to take the decision.

9.4   Dialogue Structure

A problem one has to cope with when designing a VMS is to describe its functional structure in a compact, precise and readable way. Some graphical formalisms have been adapted to represent the states of the VMS. One of the possible states of the VMS is presented in Fig. 9.3. When a flexible vocabulary approach is used, the recognition system can be easily modified, upgraded or extended. The functional structure of the prototype demonstrator was constructed keeping in mind the flexibility of its structure. Two main parts have been developed. The first part is a service independent part that manages the telephone line, the recording, and the playing of messages.The second part is service dependent. It defines the functional structure of the VMS and controls the dialogue.

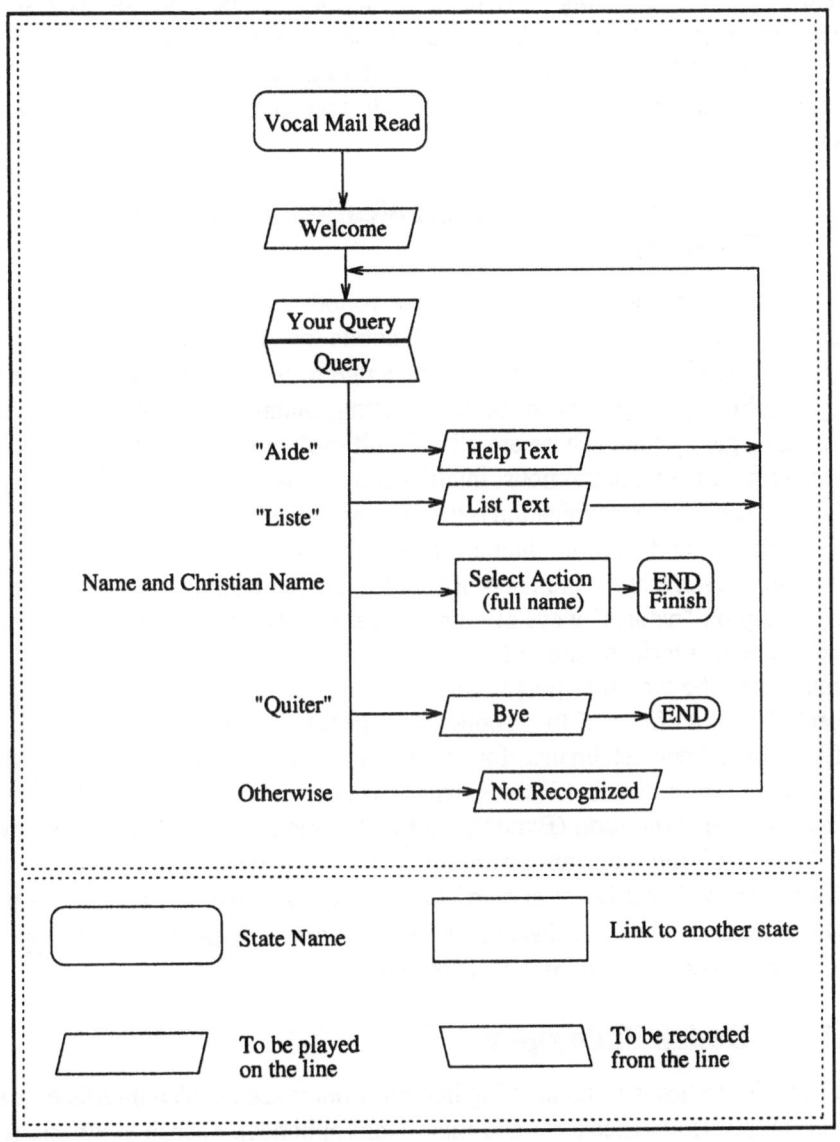

Figure 9.3. Graphical representation of one of the VMS states.

The service independent part is written in C++ programming language. It is fixed, and does not need to be updated when constructing a new VMS with changed dialogue. It consists of a loop with the following steps: play request for a command, record the answer, call the main shell script for the recognition and wait for the answer, play the result, go to the first step or quit the loop.

The service dependent part of the VMS is composed using shell scripts. According to the recognition result the state of the VMS can be changed. This part works in the following way: the main script is called from the fixed part; according to the state of the VMS, select the appropriate recognition script; this script does the recognition using the grammar and the vocabulary of this state; it chooses the answer and the next state according to what is recognized; it returns to the fixed part.

10.   Advanced Voice Communication Applications in Robotics

Human-centered and social, interactive robotics is a comparatively young direction in mobile robotic research, and mobile tour-guide robots intended for large-scale exhibitions pose a unique challenge to trans-disciplinary research. They require the integration of sensing, acting, planning and communicating within a single system. These are all difficult problems that must be solved if we are to have truly autonomous, intelligent systems.

Natural spoken communication is the most user-friendly means of interacting with machines, and from the human standpoint spoken interactions are easier than others, given that the human is not required to learn additional interactions, but can rely on "natural" ways of communication [8]. Although human-robot voice enabled interfaces are still in their infancy, they have the potential to revolutionize the way that people interact with tour-guide robots [9]. In this project, the joint efforts of the robotics and speech processing research groups of EPFL have been established for providing a voice enabled interface to the interactive tour-guide robot RoboX, which has been recently developed for the Swiss National Exhibition (Expo.02) at the Autonomous Systems Laboratory [10, 11]. Building such an experimental platform needs a lot of expertise and work on both hardware and software. The objective was to provide the project partners with the hardware devices, software tools and technical support necessary for real-life experiments (Fig. 9.4).

10.1   Design Philosophy

The major issues in implementing human-robot voice enabled interfaces are: speech output (loud-speakers) and input (microphones), speech synthesis for voice output, speech recognition and understanding for voice input, dialogue and usability factors related to how humans interact with tour-guide robots. They function by recognizing words, interpreting them to obtain a meaning in terms of an application, performing some action based on the meaning of what

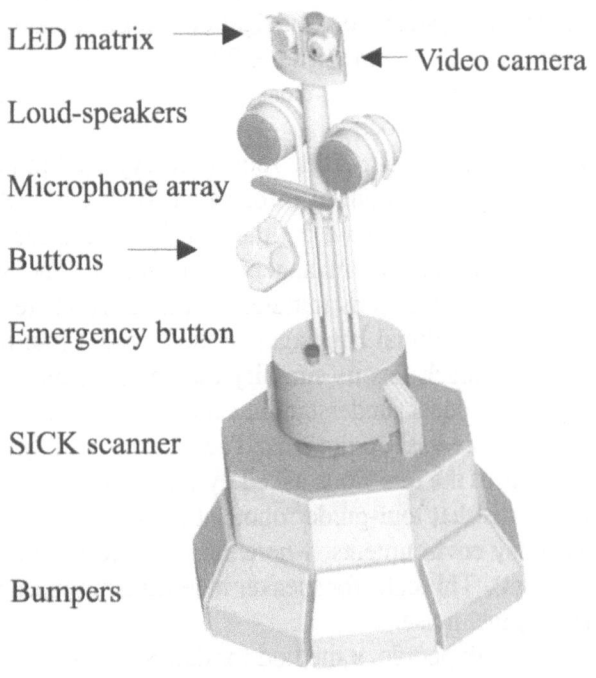

LED matrix
Video camera
Loud-speakers
Microphone array
Buttons
Emergency button
SICK scanner
Bumpers

Figure 9.4. The mobile tour-guide robot RoboX.

was said, and providing an appropriate spoken feedback to the user. Whether such a system is successful depends on the difficulty of each of these four steps for the particular application, as well as the technical limitations of the system. Robustness is an important requirement for successful deployment of such a technology (in particular speech acquisition and speech recognition) in real-life applications. For example, automatic speech recognition systems have to be robust to various types of ambient noise and out-of-vocabulary words. Automatic speech synthesis should not only sound naturally but also be adapted to an adverse acoustical environment. Lack of robustness in any of these dimensions makes such systems unsuitable for real-life applications.

While the issue of navigation of mobile robots is relatively well understood, the issue of voice enabled interaction in adverse environment remains largely open; despite the fact that interaction is a key ingredient of any successful application. Our aim was to create a system capable of collaborative human-robot interaction in a public environment. Studying specificities of autonomous, mobile tour-guide robots to be deployed at an exhibition led us to the following observations.

First, even without voice enabled interfaces, tour-guide robots are very complex, involving several subsystems (e.g. navigation, people tracking using laser scanner, vision) that need to communicate efficiently in real time. This calls for speech interaction techniques that are easy to specify and maintain, and that lead to robust and fast speech processing.

Second, the tasks that most tour-guide robots are expected to perform typically require only a limited amount of information from the visitors. These points argue in favor of a very limited but meaningful speech recognition vocabulary and for a simple dialogue management approach. The first solution adopted is based on yes/no questions initiated by the robot where visitors' responses can be in the four official languages of the Expo.02 (oui/non, ja/nein, si/no, yes/no). This approach lets us simplify the voice enabled interface by eliminating the specific speech understanding module and allows only eight words as multi-lingual universal commands. The meaning of these commands depends on the context of the questions asked by the robot.

A third observation is that tour-guide robots at the popular exhibition have to operate in very noisy environments, where they need to interact with many casual persons (visitors). This calls for speaker independent speech recognition and for robustness against noise.

To start interacting with people, a method for detecting them is needed. We have found that in the noisy and dynamically changing conditions of the robotics "exposition", a technique based on motion tracking using laser scanners, and on face detection with a color video camera gives the best results.

10.1.1   Speech synthesis.

In the noisy environment of the exhibition, the automatic speech synthesis system should generate speech signals that are highly intelligible and of an easily recognizable style; if possible, this style should correspond to the style of an excellent human guide. On the other hand, and to preserve the robot's specificity, the quality of its speech should not mimic perfectly the human speech, but such speech has to sound natural. Two main criteria that we have used to choose an appropriate method for automatic speech synthesis were intelligibility and naturalness. Therefore, a solution adopted for the speech synthesis event is a text-to-speech (TTS) system based on concatenation of diphones [4]. The intelligibility and naturalness of the synthesized speech highly depends on the quality of the segment database, grapheme-to-phoneme-translation and a prosodic driver for pitch and duration modification.

10.1.2   Speech recognition.

The first task of the speech recognition event is the acquisition of the useful part of the speech signal that avoids unnecessary overload for the recognition system. The adoption of limited in time (3 seconds) acquisition is motivated by the average length of yes/no answers. During this time the original acoustic signal is processed by the microphone array. The mobility of the tour-guide robot is very useful for this task since the robot, when using the people tracking system, can position his front in the direction of the closest visitor and this way can direct the microphone array. The pre-processing of signals of the array includes spatial filtering, de-reverberation and noise canceling. This pre-processing does not eliminate all the noise and out-of-vocabulary words. It provides sufficient quality and non-excessive quantity of data for further processing. At the heart of automatic speech recognition system of the robot lies a set of the state-of-the-art algorithms for training statistical models of words and then using them for the recognition task [7, 12].

10.1.3 Dialogue management. A particular problem, when designing a dialogue system is to describe its functional structure in a compact, precise and readable way. Some graphical state-based formalism has been adapted to represent the sequences of dialogues [13].

The design methodology proposed in [11] is conceived for developing voice enabled interfaces that are adapted to the nature of autonomous, mobile tour-guide robots with all their constraints, behavioral requirements of visitors and real-world noisy environments that the automatic speech recognition and synthesis systems have to cope with. In the approach presented, the development was focused on the potential user, from the very beginning of the design process through to the complete system.

10.2 Experimental Platform (hardware and software)

Speech is one of the input/output modalities within the multi-modal, multi-sensor interface of the robot and should naturally fit into functional layers of the whole system. On the other hand, from a functional and conceptual point of view, the addition of a voice enabled interface does not affect the overall system organization, implementation should take some specific constraints into account.

10.2.1 Hardware architecture. Multiple sensors and other input/output devices of the I/O layer are used by the robot to communicate with the external world, in particular with users. In this set of multi-modalities, loud-speakers and a microphone array represent the output and input of the voice enabled interface. Among input devices that have to cooperate closely with this interface, when verifying the presence of visitors, are two SICK laser scanners mounted at knee height and one color camera placed in the left eye of the robot. The blinking buttons help in choosing one of the four languages, and the robot's face, which consists of two eyes and two eyebrows can make the speech of the robot more expressive and comprehensive. Finally, a LED matrix display in the right eye of robot may suggest the "right" answer to the robot's questions (Fig. 9.5).

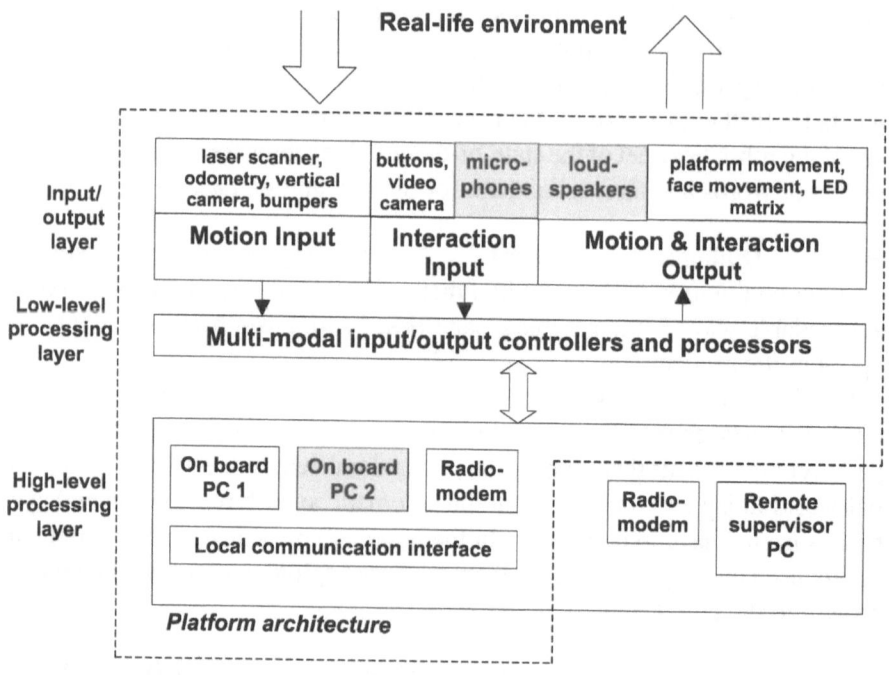

Figure 9.5. Architecture of the tour-guide robot RoboX.

The low-level processing layer contains hardware modules responsible for pre-processing of signals dedicated to input and output devices. The voice pre-processing is represented in this layer by the digital signal processor of the microphone array and the audio amplifier for the loud-speakers. The high-level processing layer consists of two on-board computers: one dedicated to all interaction tasks, including speech synthesis, speech recognition and dialogue management, and second one to navigation.

10.2.2   Software architecture.   The principal robot operations are controlled by one main program called sequencer, which executes a predefined sequence (task scenario) of events (scenario objects). The sequencer program is implemented in SOUL (Scenario Object Utility Language) designed at ASL to meet the requirements of the autonomous, interactive, mobile tour-guide robot [10].

11. Conclusion

In this chapter the concept of integration of generic spoken language technologies was presented. This concept was used to develop the methodology for designing and implementing human-machine voice enabled interfaces, in particular interfaces for voice messaging systems and tour-guide robots.

In the first specific case, the methodology for implementation of a workstation and ISDN platform oriented voice messaging system (VMS) was presented. The reported recognition results clearly validate the flexible vocabulary and speaker independent recognition approach for the telephone quality Swiss French. The proposed flexible software structure allows a user to modify easily the functional structure of the prototype VMS demonstrator.

In the second specific case, while the issue of navigation of mobile robots is relatively well understood, the issue of human-robot voice enabled interaction in adverse environment remains largely open. The design methodology proposed in this chapter is conceived for developing voice enabled interfaces that are adapted to the nature of autonomous, mobile tour-guide robots with all their constraints, behavioral requirements of visitors and real-world noisy environments that the automatic speech recognition and synthesis systems have to cope with. In the approach presented, the development was focused on the potential user, from the very beginning of the design process through to the complete system. We also demonstrated that this methodology allows for developing a variety of tour-guide scenarios, involving real-time state-based dialogue management on a single on-board computer responsible for the overall interaction of the robot and related modalities.

References

[1] D. Gibbon, R. Moore, and R. Winski, *Handbook of Standards and Resources for Spoken Language Systems*. Berlin: Mouton de Gruyter, 1997.

[2] D. Gibbon, I. Mertins, and R. Moore, *Handbook of Multimodal and Spoken Dialogue Systems*. Boston: Kluwer Academic Publishers, 2000.

[3] D. Gardner-Bonneau, *Human Factors and Voice Interactive Systems*. Boston: Kluwer Academic Publishers, 1999.

[4] T. Dutoit, *An Introduction to Text-to-Speech Synthesis*. Dordrecht: Kluwer Academic Publishers, 1997.

[5] D. A. Reynolds and R. C. Rose, "Robust text-independent speaker identification using Gaussian mixture models," *IEEE Trans. on Speech Audio Processing*, vol. 3, pp. 72–83, 1995.

[6] A. Drygajlo, J. L. Cochard, G. Chollet, O. Bornet, and P. Renevey, "Sun workstation and SwissNet platform for speech recognition and speaker

verification over the telephone," in *Proceedings of SIWORK'96*, (Zurich, Switzerland), pp. 1–4, May 13-14, 1996.

[7] P. Renevey and A. Drygajlo, "Securized flexible vocabulary voice messaging system on unix workstation with ISDN connection," in *Proceedings of 5th European Conference on Speech Communication and Technology (EUROSPEECH'97)*, (Rhodes, Greece), pp. 1615–1619, Sept. 22-25, 1997.

[8] X. Huang, A. Acero, and H. Hon, *Spoken Language Processing*. New Jersey: Prentice Hall, 2001.

[9] D. Spiliotopoulos, I. Androutsopoulos, and C. D. Spyropoulos, "Human-robot interaction based on spoken natural language dialogue," in *Proceedings of European Workshop on Service and Humanoid Robots (ServiceRob2001)*, (Santorini, Greece), 2001.

[10] B. Jensen, G. Froidevaux, X. Greppin, A. Lorette, M. Messier, G. Ramel, and R. Siegward, "The interactive autonomous mobile system RoboX," in *Proceedings of IEEE/RSJ International Conference on Intelligent Robots and Systems (IROS2002)*, (Lausanne, Switzerland), Sept. 30- Oct. 5, 2002.

[11] P. Prodanov, A. Drygajlo, G. Ramel, M. Meisser, and R. Siegwart, "Voice enabled interface for interactive tour-guide robots," in *Proceedings of IEEE/RSJ International Conference on Intelligent Robots and Systems (IROS2002)*, (Lausanne, Switzerland), Sept. 30 - Oct. 5, 2002.

[12] S. Young, J. Odell, D. Ollason, and P. Woodland, *The HTK Book*. Version 3.0, 2000.

[13] R. de Mori, ed., *Spoken Dialogues with Computers*. London: Academic Press, 1998.

Chapter 10

SPEECH CODING AND RECOGNITION IN NOISY ENVIRONMENTS FOR COMMUNICATION TERMINALS

Andrzej Drygajlo

Swiss Federal Institute of Technology Lausanne,

School of Engineering,

1015 Lausanne,

Switzerland

andrzej.drygajlo@epfl.ch

Abstract This chapter addresses the problem of speech processing, which is robust against noise for applications in communication terminals as front-ends to digital networks. By studying the limitations of auditory perception, particularly how it reduces the information rate of the speech signal through masking constraints, improvements may be made in the efficiency of: (1) speaker/speech recognition, (2) wide-band speech coding. In the first case, speech enhancement techniques derived from spectral subtraction are used not only for noise reduction, often unreliable, but also for the detection of missing (masked by noise or unreliable) features. We show that this detection technique can be combined with compensation techniques for missing features in the statistical models (Gaussian Mixture Models (GMMs) and Hidden Markov Models (HMMs)) to improve recognition results. In the second case, the spectral subtraction technique is used to design an integrated speech enhancement/coding system incorporating both ambient noise and quantization noise masking. The advantage of the method presented in this chapter over previous approaches is that perceptual enhancement and coding, usually implemented as a cascade of two separate systems are combined. This leads to a decreased computational load while controlling bit rate and maintaining acceptable speech intelligibility and quality.

Keywords: Speech enhancement, speech coding, speech recognition, speaker recognition.

J.F. Tasič et al. (eds.), Intelligent Integrated Media Communication Techniques, 337-357.

1.   Introduction

During the last decade one could observe an increasing interest in research in the domain of robust speech processing and speech enhancement in particular [1, 2, 3]. This interest has grown recently in an exponential manner due to the fact that automatic speech processing systems are becoming more and more used in new telecommunication applications in a variety of real-life environments. In many of these applications (e.g. digital mobile telephony, hands-free telephone systems, teleconferencing, interactive vocal servers, etc.) speech processing systems are confronted to adverse environments with high ambient noise levels. Therefore there is a strong need for powerful speech enhancement algorithms effectively combined with speech coding and automatic speech/speaker recognition systems capable of working at very low signal-to-noise ratios (SNRs).

1.1   Speech and Speaker Recognition

It is well documented that speech degraded by background noise renders the performance of many automatic speaker/speech recognition systems unacceptable. While current recognizers based on statistical modeling (Gaussian mixture models (GMMs) and hidden Markov models (HMMs)) give acceptable performance in the statistically averaged environment, present in the training database, their performance degrades rapidly when they are applied to other real-world conditions [2, 3]. Consequently, the research has to be concentrated in reducing the mismatch between training and testing conditions. It is impossible to train the models in all the possible adverse conditions. This is why it is necessary to design systems that adapt automatically to environmental changes. Moreover, when the speech used for automatic recognition tasks is degraded by intense background noise, some speech sounds are masked by noise, and in this case, classical pre-processing enhancement procedures such as spectral subtraction can be frustratingly ineffective. It is important for a system whose aim is to decrease the recognition error rate to take into account some properties of the human auditory system [4]. Auditory representations of clean speech contain much redundancy. Arguably, it is this redundancy which enables listeners to recognize speakers or speech in adverse conditions. Under the assumption that some time-frequency regions are too heavily masked to derive any reliable data, the auditory system faces the missing data problem. In automatic recognition terms, we face the missing features problem [5, 6]. This chapter describes our recent attempts to dynamically adapt the stochastic automatic speaker/speech recognition framework of GMMs and HMMs to handle the missing features problem with the help of the generalized spectral subtraction method [7, 8, 9]. In automatic recognition task there is no need to reconstruct the speech signal, and the missing feature components cannot only be estimated but also ignored by the spectral subtraction technique. In our case, the generalized spectral subtraction algorithm is used as a missing feature detector with a noise reduction, and not as a pure enhancement system. Recognition results are reported for challenging telephone-quality speaker and speech recognition tasks.

1.2 Coding

By studying the limitations of auditory perception through time-frequency masking constraints, improvements may be made in the efficiency of coding schemes for noisy speech [10, 11]. Indeed, sub-band speech coders calculate a noise masking threshold for the quantization stage similar to the one used for the adaptation of spectral subtraction parameters in perceptually tuned speech-enhancement algorithms. They can also use the same kind of time-spectral mapping to decompose the signal into decorrelated components. In this chapter, taking advantage of a common processing structure, we present a combined coding/enhancement approach to the design of a wide-band coder for noisy speech signals sampled at 16 kHz. Our approach merges fast orthogonal wavelet packet transform algorithms, which emulate an auditory critical band decomposition, with spectral masking models. This system controls bit rate while maintaining wideband speech quality. It also leads to a decreased computational load, compared with the use of separate coding and enhancement solutions.

2. Speech Enhancement Using Spectral Subtraction

Short time spectral amplitude estimation of clean speech by the spectral subtraction technique is the most commonly employed method in single channel speech enhancement systems [12]. The simplicity of the implementation and the speech quality improvement make it attractive.

2.1 Generalized Spectral Subtraction

The spectral subtraction technique and its derivatives make the assumption that the noise is additive in the magnitude spectral domain. Consequently, an estimate of the magnitude clean speech spectrum could be obtained directly by subtracting the magnitude noise spectrum estimated during speech pauses from the magnitude spectrum of the noisy signal. During non-speech periods, a speech activity detector provides an averaged estimate of magnitude noise spectrum $|\bar{N}(\omega)|$. The generalized spectral subtraction (GSS) algorithm is expressed by the following equations:

$$D_m(\omega) = |Y_m(\omega)|^\gamma - \alpha \cdot |\bar{N}(\omega)|^\gamma \qquad (10.1)$$

and

$$|\hat{S}_m(\omega)|^\gamma = \begin{cases} D_m(\omega) & \text{if } D_m(\omega) > \beta \cdot |\bar{N}(\omega)|^\gamma \\ \beta \cdot |\bar{N}(\omega)|^\gamma & \text{otherwise} \end{cases} \qquad (10.2)$$

where $\gamma = 1$ determines magnitude subtraction and $\gamma = 2$ power subtraction, $|Y_m(\omega)|$ and $|\hat{S}_m(\omega)|$ are the magnitudes of noisy and estimated clean speech, respectively, $\alpha \geq 1$ is the overestimation parameter and $0 < \beta \leq 1$ spectral flooring.

Given the estimate of the clean speech spectrum, the speech samples can be calculated by inverse discrete Fourier transform (IDFT). The reconstructed clean speech samples are approximated using the magnitude of estimated clean speech and the phase of noisy speech. Therefore, the reconstructed enhanced speech in the temporal domain is obtained by the following relationship:

$$\hat{s}(n) = \text{IDFT}[|\hat{S}_m(\omega)| \cdot e^{(j \arg Y_m(\omega))}] \qquad (10.3)$$

An important drawback of the spectral subtraction method resides in the fact that this method defines thresholds under which the speech is considered as masked by noise, and consequently, cannot be estimated. These thresholds are helpful in determining unreliable spectral features.

2.2   Missing Feature Detection

In the power subtraction case, according to Eq. 10.2, the short-term energy of the corrupted speech components lying below a threshold proportional to the noise energy is recognized to be deeply affected, and hence, the estimation of the clean speech is unreliable. In the frequency domain, the spectral components below the spectral floor are unreliable for the classifier and should be ignored. In such a case, they can be classified as missing features and the automatic speaker/speech recognition system can drop them from the recognition process:

$$|\hat{S}_m(\omega)|^\gamma = \begin{cases} D_m(\omega) & \text{if } D_m(\omega) > \beta \cdot |\bar{N}(\omega)|^\gamma \\ \text{Missing feature} & \text{otherwise} \end{cases} \qquad (10.4)$$

The components masked by noise can be detected in a flexible way by varying the parameters α and β. From Eq. 10.4, the detection threshold of masked speech is given as a function of the posterior signal-to-noise ratio ($\text{SNR}_{post}(\omega)$):

$$\frac{1}{\alpha + \beta} \cdot \frac{|Y_m(\omega)|^2}{|\bar{N}(\omega)|^2} = \frac{1}{\alpha + \beta} \cdot \text{SNR}_{post}(\omega) \leq 1. \qquad (10.5)$$

The parameters α and β can be estimated experimentally or calculated using additional criteria. One of these criteria is the "SNR criterion" defined for missing feature detection purposes in speech recognition [13]. This criterion treats data as unreliable when:

$$\text{SNR}_{prior}(\omega) = \frac{|\hat{S}_m(\omega)|^2}{|\bar{N}(\omega)|^2} \leq 1. \qquad (10.6)$$

Taking into account that speech enhancement is achieved by magnitude spectral subtraction, the SNR criterion is a special case of the detection rule of missing features in Eq. 10.5 with $\beta = 0$ and $\alpha = 3.41$. Under the assumption that noise is additive and stationary, the missing feature detection using generalized spectral subtraction can operate in critical sub-bands and various noisy conditions. Ignoring missing data means attempting to classify the feature components solely on the basis of the present information. In this case, the recognizers using multi-dimensional probability density functions (pdfs) such as GMMs and HMMs should be dynamically modified [14, 9].

In the sequel of this chapter, we introduce a simple missing feature model compensation (MFC) motivated by missing feature theory and the adaptation exhibited by humans. It is well known that locally weaker sound components do not contribute to the neuronal output of the human auditory system: they are masked and therefore can be considered missing for the purpose of further processing.

3. Statistical Modeling and Missing Feature Compensation in Recognition Systems

In this chapter, the performance of missing feature detection techniques is assessed using the marginal distribution of present features for likelihood computation. In spite of its simplicity, the method gives considerable decrease of error rate in the framework of GMM- and HMM-based speaker and speech recognition [15, 16, 9]. The likelihood in the case of complete set of features knowing the model λ is expressed as

$$p(\mathbf{x}|\lambda) = \sum_{i=1}^{M} p_i \Phi_i(\mathbf{x}, \mu_i, \Sigma_i) \tag{10.7}$$

where p_i is the mixture weight and Φ_i is a D-dimensional Gaussian density defined by the mean vector μ_i and the diagonal covariance matrix Σ_i. In the presence of missing features, this likelihood cannot be directly computed. In this case, it can be approximated as follows:

$$p(\mathbf{x}|\lambda) \approx p(\mathbf{x}_p|\lambda) \int_{\mathbf{x}_l}^{\mathbf{x}_u} p(\mathbf{x}_m|\mathbf{x}_p, \lambda) d\mathbf{x}_m \tag{10.8}$$

where $\mathbf{x}_l$ and $\mathbf{x}_u$ are the lower and upper limits of the missing feature space and $\mathbf{x}_p$ and $\mathbf{x}_m$ are the sub-vectors of present and missing features, respectively. Recently, it has been demonstrated that this approximation is in fact Bayesian optimal with respect to "zero/one" loss function. In the case where the knowledge about the missing features space is not provided and having diagonal covariance matrices for the statistical model, Eq. 10.8 can be replaced by

$$p(\mathbf{x}|\lambda) \approx \sum_{i=1}^{M} p_i \prod_{j \in present} \Phi_i(x_j, \mu_{ji}, \sigma_{ji}^2) \qquad (10.9)$$

where μ_{ji} is the mean, σ_{ji}^2 is the variance of the feature vector component x_j and p_i are the mixture weights corresponding to the mono-variate Gaussian densities $\Phi_i, i = 1, \cdots, M$ [6].

4.   Speaker Verification

During the last decade one could observe an increasing interest in research in the domain of robust speaker recognition, where robustness refers to the need to maintain good recognition accuracy even when the quality of the input speech is degraded. Generally, automatic speaker recognition can be classified into two main methods: speaker identification and speaker verification. Speaker identification is the process of determining from which of the registered speakers a given utterance comes. On the other hand, speaker verification is the process of accepting or rejecting the identity claim of a speaker. The verification decision, in the Bayes' sense, is based upon the likelihood ratio, the likelihood ratio of the conditional probability of the observed features of the utterance given the claimed identity ($\lambda = \lambda_c$) to the conditional probability of the observed features given the speaker is an imposter ($\lambda \neq \lambda_c$). A mathematical expression for the likelihood ratio in the logarithmic domain is

$$\log LR = \log p(\mathbf{x}|\lambda = \lambda_c) - \log p(\mathbf{x}|\lambda \neq \lambda_c) \qquad (10.10)$$

Automatic speaker verification systems are well suited to interactive vocal servers in telephony applications. As spoken language technologies are being transferred to real-life applications, the need for greater robustness in recognition technology is becoming increasingly apparent [17]. On the other hand, speaker verification is receiving a particular attention as inexpensive and less objectionable than other biometrics-based methods to be applied for secured access to terminals and teleservices. Therefore there is a strong need for powerful speaker verification systems capable for working at very low signal-to-noise ratios (SNRs).

The research work presented in this chapter is very well placed to fill in the gap in our knowledge concerning the integration of missing data theory in the context of robust speaker verification. This approach shows that speech enhancement techniques derived from spectral subtraction can be used not only for noise reduction, often unreliable, but also for unreliable or missing (masked by noise) feature detection. It also shows that this detection technique can be integrated with compensation techniques for missing features in the statistical Gaussian mixture models (GMMs) to improve speaker verification performance. In this chapter, we evaluate criteria for the detection of features that are unreliable due to masking by noise. Then, we investigate a set of compensation techniques, namely integration techniques, that deal with these unreliable features in the framework of speaker verification using the Gaussian mixture model classifiers [18, 19].

4.1   Detection of Unreliable Features

We implemented various speech enhancement techniques based on spectral subtraction (power spectral subtraction (SS) [14], generalized spectral subtraction (GSS) [7, 12], adaptive generalized spectral subtraction (AGSS) based on masking properties of the human auditory system [8], and minimum mean-square error (MMSE) spectral amplitude estimator of Ephraim and Malah [20]) in order to increase the performance of speaker verification in the presence of additive background noise.

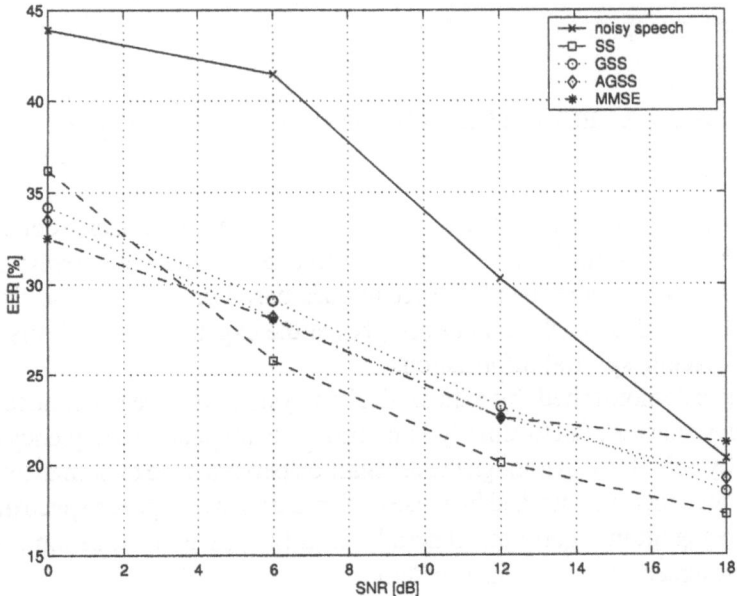

Figure 10.1.   EER at various SNRs. Critical band results for F-16 cockpit noise.

Generally, when comparing the performance of speaker verification systems, if the speech enhancement techniques are used as a front end to feature extraction (filter bank log-energies or Mel frequency cepstral coefficients (MFCCs)), the best performance is obtained by the MFCCs and the spectral subtraction (SS) algorithm. This is shown in Figs 10.1 and 10.2 where equal error rate (EER) is plotted for various SNRs. The other techniques, which over-estimate background noise, fail to give better results.

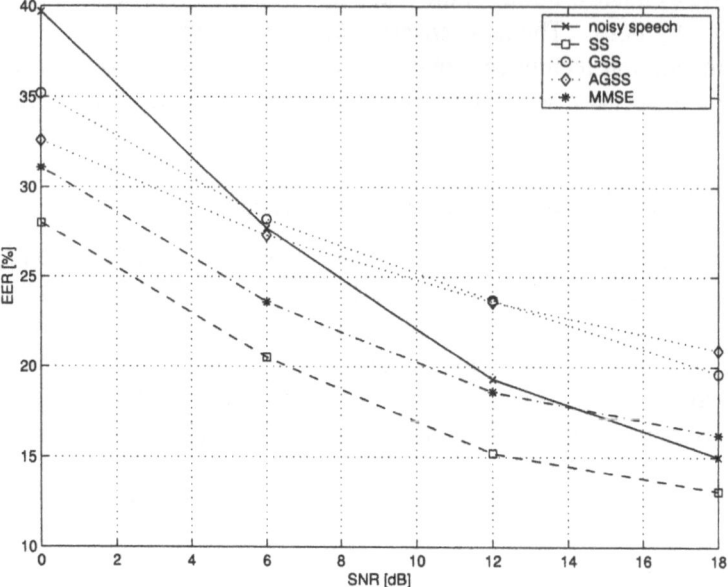

Figure 10.2.   EER at various SNRs. MFCCs results for F-16 cockpit noise.

Although the spectral subtraction technique reduces less background noise in the enhanced speech than the other techniques, it gives the lowest speech distortion. On the other hand, the techniques based on noise over-estimation substantially reduce background noise, but at the expense of speech distortion and non-reliable spectral information.

We have demonstrated, by a theoretical analysis of the speech enhancement techniques, that they fail to reliably estimate the clean speech in frequency bands masked by noise [16]. Although the recognition performance of human listeners is resistant to missing (masked by noise) information in the speech spectrum, the pattern recognition algorithms generally used in speaker verification systems are not designed to handle this problem.

We have also shown that the speech enhancement techniques based on spectral subtraction can be used to detect non-reliable features in time and frequency [14, 7, 8, 15]. The parameters that control the detection can be fixed a priori, as in the spectral subtraction (SS) algorithm and the MMSE spectral amplitude estimator method, or adapted, as in the adaptive generalized spectral subtraction (AGSS) method. We experimentally analyzed non-reliable feature detection by the speech enhancement algorithms for various noises and signal-to-noise ratios (SNRs) [21]. The power spectral subtraction (SS) algorithm gives the lowest detection rate whereas the generalized spectral subtraction (GSS) algorithm gives the highest. The MMSE spectral amplitude estimator provides a different detection rate. In speech frequency bands with higher energy, the MMSE detection curve is very similar to the power spectral subtraction detection curve. On the other hand, in higher frequency bands the MMSE detection curve approaches the detection curve of generalized spectral subtraction. The ability of the speech enhancement techniques to reduce the noise and to detect sub-bands masked by noise can be used in combination with missing feature compensation in statistical models to increase the noise robustness of automatic speaker verification systems [15, 21].

4.2   Integration Methods for Unreliable Feature Compensation

In this section, we present a solution to the problem of unreliable (missing) data in the task of speaker verification using some principles of the missing data theory [14, 7, 8, 15, 6, 22]. Assuming that speaker models were trained in clean speech conditions, we introduce the integration methods for missing feature compensation in the recognition phase.

The most well-known method of missing feature compensation is the *marginal distribution technique* [23, 6, 22]. It is a particular case of the integration method. The marginal distribution technique ignores non-reliable features in recognition by calculating the log-likelihood ratio using only the features present. Several tests were carried out on clean telephone speech databases degraded with artificially generated missing features [21]. Assuming that the missing features are known *a priori*, the experiments determined the upper limit of the verification performance. With the marginal distribution technique, we reported a significant increase in speaker verification performance. As an example, with 90% deleted features, we obtained approximately 21.41% of equal error rate (EER) whereas the classical method with no compensation gave about 50%.

We extended the use of missing feature compensation to applications where masking noises are present. For this task, we performed several experiments with noisy, telephone-quality speech [15, 24, 21]. We have shown that classical speech enhancement in a pre-processing stage performs worse than the combination of speech enhancement and missing feature compensation. The compensation of missing features using marginal distribution significantly decreased the EER. Considerable performance improvement was obtained by applying the marginal distribution technique after detection of missing features by the MMSE spectral amplitude estimator. The experiments carried out have shown that, in the presence of highly masking wide-band noise (i.e. white Gaussian noise at 0 dB SNR), the equal error rate was about 23% for approximately 90% of features detected as non-reliable. In this case, the combination of speech enhancement, missing feature detection and compensation did not give a large

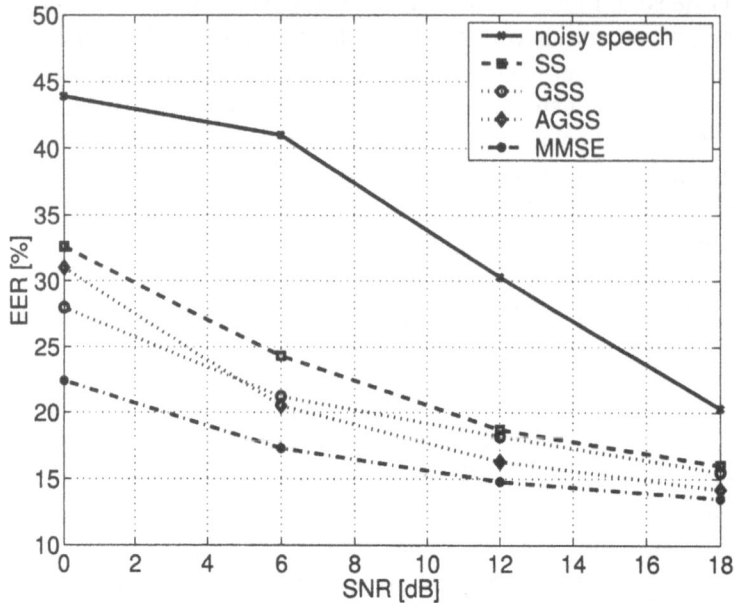

Figure 10.3.   EER at various SNRs using marginal distribution technique for F-16 cockpit noise and different missing feature detectors.

discrepancy of recognition error rates in comparison to the case of artificially generated missing features. A typical plot of EER as a function of the SNR for F-16 cockpit noise is presented in Fig. 10.3.

The use of speaker models with full covariance matrices significantly decreased the recognition error rate compared to the most often used reference system based on spectral subtraction (SS) and (MFCCs) (Fig. 10.4). In this case, the calculation of the marginal distribution, however, increased the computational cost due to matrix inversion.

By marginal distribution use, we assume that the feature components are not bounded and occupy an unlimited space $(-\infty, +\infty)$. Unlike the marginal distribution technique, the *bounded integration technique* takes some knowledge about missing feature space into account in the recognition process [15]. For experiments with this technique, we use speaker models with diagonal covariance matrices. The bounds of integration are chosen in the following way. As noise is additive in the spectral domain, the components masked by noise have an upper bound equal to the estimated noise energy. As the feature components are defined in the log domain, the lower bound of energies is equal to $-\infty$.

In Fig. 10.4, we compare bounded and unbounded integration techniques. Missing features are detected using the MMSE spectral amplitude estimator, since it gives the lowest EER. The bounded integration method noticeably im-

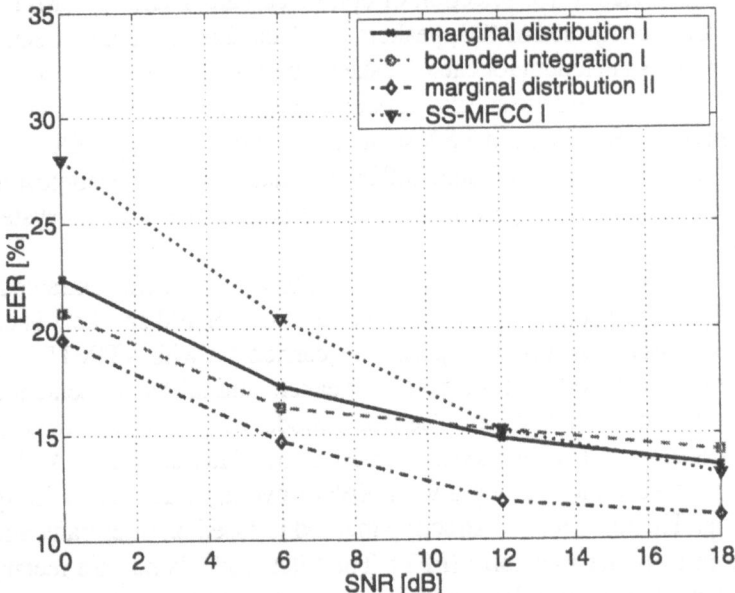

Figure 10.4. EER at various SNRs using marginal distribution and bounded integration techniques for F-16 cockpit noise. I denotes models with diagonal covariance matrices. II denotes models with full covariance matrices.

proves the performance at very low SNR conditions (SNR < 10 dB). However, including the knowledge about missing feature bounds in the marginalization technique does not reduce the EER at higher SNRs. These results show that adding any information about missing features can improve the performance of speaker verification in conditions where the number of missing features is high. As long as the use of the reliable sub-bands in the recognition process achieves sufficient information, the additional knowledge about missing features does not give any advantage compared to the marginal distribution technique.

4.3   Databases and Experiments

The speech database used in the experiments is extracted from the telephone quality NTIMIT corpus. This database is recorded after transmitting speech data with an 8-kHz bandwidth over local and long distance telephone lines. The sentences are uttered in only one session. The main degradations that could affect the recorded speech are the microphone handset and telephone line characteristics. In this chapter, the speech extracted from the NTIMIT corpus is called "clean" speech even though the average SNR is 36 dB. From NTIMIT, two subsets NTIMIT-T and NTIMIT-V are extracted to represent databases for test and validation, respectively. NTIMIT-T, a small database constituted of

22 speakers, is used in the selection of optimized parameters. NTIMIT-V is a larger database composed of 400 speakers. Both databases contain 10 sentences approximately 3 s in length for each speaker. These sentences are labeled "SA", "SI" and "SX". To build the speaker models, 8 sentences of about 24 s in length are chosen (i.e. 2 SA, 3 SI and 3 SX sentences). The remaining 2 SX sentences are used for the test phase. In total, 800 true speaker and 319200 impostor tests are provided to assess the performance of the speaker verification system for NTIMIT-V.

In experiments with NTIMIT-T, the world model has 32 pdfs extracted from the speech of 40 speakers. However, a world model of 100 speakers is chosen for likelihood normalization in experiments carried out with NTIMIT-V. These 100 speakers are different from the test speakers and all the 10 sentences are used for the model construction. The world model has 256 Gaussian pdfs. For both databases, the speaker models consist of 32 Gaussian pdfs. A Hanning window of 256 samples in length with 50% of overlap was chosen for speech windowing. The first feature extraction method is based on filter-bank analysis using the fast Fourier transform (FFT). The filter-banks bands are rectangular and do not overlap. In the telephony bandwidth, log-energies of fourteen critical bands are calculated according to the Bark scale. The second feature extraction method is based on cepstral analysis using 14 MFCCs. The windowed speech segments of 20 ms length are pre-emphasized before the computation of Mel scale filter-bank log-energies. The MFCCs and FFT-based filter-bank are both used to assess the performance of the speaker verification system. The MFCCs are used in models with diagonal covariance matrices while the filter-bank log-energies are evaluated with both speaker models having full and diagonal covariance matrices. All the experiments simulating noisy environments are carried out using four noises: white Gaussian noise, F-16 airplane cockpit noise, speech-like noise and factory noise extracted from the NOISEX-92 database (only results for F-16 cockpit noise are presented in this chapter).

5.   Speech Recognition

The Gaussian mixture model (GMM) can be regarded as a special case of a single-state continuous density hidden Markov model (HMM). Each state of the HMM is defined by emission and transition probabilities. For each state model λ of the HMM the probability to emit the feature vector x in the presence of missing features can be expressed as in Eq. 10.9.

The marginalization method consists of discarding the unreliable features from the recognition process. The marginal densities are used to compute the scores (log-likelihoods) of the models during the recognition. In experiments with *a priori* selection of unreliable features it can be shown that the spectro-temporal representation of the speech signal is redundant. If we select at random a certain percentage of the features and declare them as unreliable, we observe that 20% of the features are sufficient to obtain similar recognition rate than those obtained with the full set of features. Second, with a similar experiment we observe that the parameters with the highest energy (parameters whose energy is between 40% and 100% of the maximum energy in the same frequency band) are sufficient to obtain similar results than those obtained by the full set of features. These two experiments show that discarding some of the features from the recognition does not reduce the performance of the system. It is also possible to determine experimentally the best values for the parameters of the detector of unreliable features.

The bounded and weighted integration methods integrate the probability density functions of the HMMs over the possible values that the clean speech features can take [9, 25]. The standard approach, bounded integration, consists of considering that the possible values for the unreliable features is a uniform distribution between two bounds, generally zero and the value of the unreliable features. By studying the distribution of the reliable features, we have proposed to extend this integration method to the reliable features [9, 25]. The reliable features are not masked by the noise but their values are also distorted by the noise. Thus, we have proposed to evaluate the distribution of the possible values that the clean speech signal can take for the reliable features. We have introduced this weighted approach in the framework of the integration method. The proposed enhancement slightly improves the performances of the standard integration method for digit recognition task, and a significant enhancement has been obtained for medium size vocabulary recognition task.

5.1 Databases and Experiments

The recognition system was developed for digit recognition task using Hidden Markov Toolkit (HTK). The feature vector is provided by the 17 bands Bark filter bank analysis. Digit models were trained on TIDIGITS database downsampled to 8 kHz. 224 utterances from 152 speakers were extracted randomly from the database for recognition experiments. Lynx helicopter cockpit noise from the NOISEX-92 database was added to obtain test utterances. The SNR criterion was used to detect missing features. The results obtained were compared with baseline recognition performance. The assessment results show a significant increase in performance of the HMM based recognition system when using combinations of the spectral subtraction technique for speech enhancement of present (reliable) features and unreliable features detection with missing feature compensation in the HMMs. We can also observe that techniques which introduce less distortions to the enhanced signal such as simple power spectral subtraction ($\alpha = 1$ and $\beta = 0$) give better results.

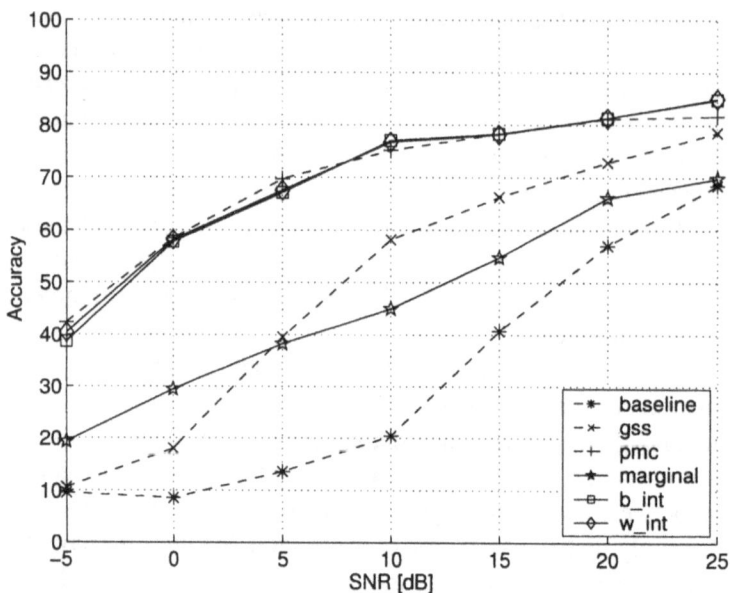

Figure 10.5.   Recognition results of the integration methods on the digits recognition task for Lynx helicopter noise.

We used two reference methods for the purpose of comparison of the results obtained by the proposed methods: (1) *General spectral subtraction (GSS)*; The following parameters were used: $\alpha = 1$ and $\beta = 0.2$ [12], (2) *Parallel model compensation (PMC)*; The models representing the Mel-log spectral parameters are compensated to cope with the additive noise [26]. The noise model used for the compensation is represented by the means and variances of the noise magnitude estimated during speech pauses.

The results are presented in Fig. 10.5 for the following methods:(1) *baseline:* baseline HMM recognition system, (2) *gss:* general spectral subtraction, (3) *pmc:* parallel model compensation, (4)*marginal:* marginalization, (5) *b_int:* bounded integration, (6) *w_int:* weighted integration.

6.   Combined Enhancement and Wideband Coding

When coding signals in noisy environments, enhancement is usually used as a pre-processing stage. The theory of the quantization of noisy sources states that the optimum quantization, in a minimum mean square error (MMSE) sense, of a Gaussian signal corrupted by additive Gaussian noise is composed of an optimum MMSE filter followed by a separate optimum MMSE quantizer [1]. This statement is maintained if the noisy signal undergoes an orthogonal transform. If only sub-band decomposition transform used in sub-band coding of noisy speech is orthogonal, enhancement and coding could be appropriately performed on the transform coefficients after the noisy signal has undergone only one transform operation.

On the other hand, orthogonal time-frequency transforms such as discrete wavelet packet transforms (DWPTs) have shown to be of particular value in the perceptually-tuned wide-band coding of speech signals. This approach, which we have recently introduced, integrates in an efficient way DWPT algorithms and psychoacoustic modeling [11]. The DWPT algorithm permits us to efficiently approximate the auditory critical band decomposition in the time and frequency domains. This allows us to make use of the temporal and spectral masking properties of the human auditory system to decrease the average bit rate of the encoder, while perceptually hiding the quantization error. The incorporation of a masking model in the DWPT coder leads to perceptually transparent coding of speech signals sampled at 16 kHz at average bit rate of 39.4 kbit/s [11]. Furthermore, if the amount of tolerable quantization noise is increasing beyond the masking threshold, such that it becomes more and more perceptible, the bit rate can be gradually reduced without drastically affecting the quality of the decoded speech signal. This result is mainly due to the DWPT fine temporal representation of higher frequency information, which preserves abrupt signal transitions.

This scheme can also be applied to the coding of noisy speech [10]. Indeed, subband speech coders calculate a noise-masking threshold for the quantization stage similar to the one used for the adaptation of spectral subtraction parameters in perceptually tuned speech enhancement algorithms.

The block diagram of the proposed coding-enhancement solution is presented in Fig. 10.6.

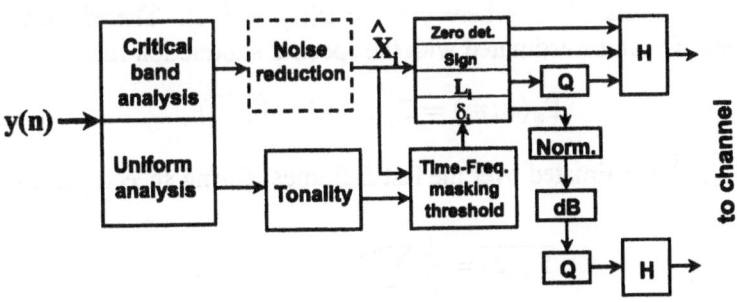

Figure 10.6. Proposed coding-enhancement system.

The noisy input signal $y(n)$ is decomposed by the DWPT. Then, the transform sequence Y_i is enhanced by a subtractive-type algorithm, to obtain the rough speech estimate $\tilde{X}_{i_l}^2$, which is further used to calculate a simultaneous masking threshold. Using this first rough masking threshold, a new subtraction rule

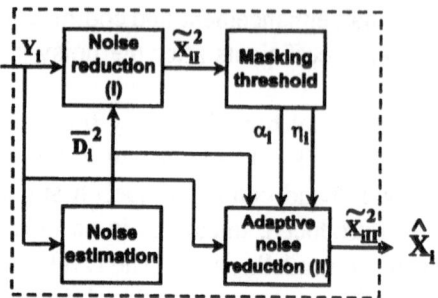

Figure 10.7.   Noise reduction mechanism.

is applied to compute $\tilde{X}^2_{i_{II}}$ which depends on masking. This is achieved by assuming that the high-energy frames of speech will partially mask the input noise (high masking threshold), reducing the need of a strong enhancement mechanism. On the other hand, frames containing less speech (low masking threshold) will undergo an overestimated subtraction. Finally, $\tilde{X}^2_{i_{II}}$ is quantized by re-computing its related masking threshold. Noise estimation is performed during speech pauses.

Such an approach saves the cost of the additional transformation necessary for the noise reduction operation. The optimum MMSE filter, also called Wiener filter, is often replaced by a spectral subtraction operation, employing the FFT. In the context of the DWPT, the spectral subtraction approach has to be justified.

If additive noise $d(n)$ is assumed to be uncorrelated with clean speech signal $x(n)$ and both are locally stationary sequences, over short time frames, it can be stated that $\Gamma_y(\omega, m) = \Gamma_x(\omega, m) + \Gamma_d(\omega, m)$; where $\Gamma_y(\omega, m)$, $\Gamma_x(\omega, m)$ and $\Gamma_d(\omega, m)$ are the short-time Power Spectra Densities (stPSD) in frame m for $y(n)$, $x(n)$ and $d(n)$, where $y(n) = x(n) + d(n)$. Therefore, the clean speech stPSD can be estimated with the spectral subtraction rule

$$\hat{\Gamma}_x(\omega, m) = \Gamma_y(\omega, m) - \overline{\Gamma_d(\omega)} , \tag{10.11}$$

where $\overline{\Gamma_d(\omega)}$ is estimated over the last L frames, during speech pauses, as

$$\overline{\Gamma_d(\omega)} = \frac{1}{L} \sum_{i=0}^{L-1} \Gamma_d(\omega, m-i) . \tag{10.12}$$

Using the DWPT, the approach is equivalent to the generalized Wiener filtering solution formulated for unitary transforms. Thus, for zero mean signals, the scalar filter that weights the transform coefficients can be expressed as

$$W_i = \frac{\sigma^2_{X_i}}{\sigma^2_{X_i} + \sigma^2_{D_i}} = 1 - \frac{\sigma^2_{D_i}}{\sigma^2_{Y_i}} \tag{10.13}$$

where the $\sigma_{[]_i}^2$ are the variances in the transform domain of the different signals. The MMSE variance of the estimated transformed speech signal in channel i is given by

$$\tilde{\sigma}_{X_i}^2 = W_i \sigma_{Y_i}^2 = \sigma_{Y_i}^2 - \sigma_{D_i}^2 . \qquad (10.14)$$

This last expression is similar to the spectral subtraction technique suggested in [12], which is computed using Eq. 10.11. Thus, PSD's in the Fourier domain have been replaced by variances in the DWPT domain. This last formulation is valid for stationary signals. If processing is done on a frame-by-frame basis, signals can be considered as stationary during the frame duration. As a consequence, the variances in the different transform channels can be replaced by the squared transform coefficients (i.e., $\sigma_{D_i}^2 \to D_i^2$, $\tilde{\sigma}_{X_i}^2 \to \tilde{X}_i^2$, ...). After these remarks we can rewrite (10.11) in the DWPT domain as

$$\tilde{X}_i^2 = Y_i^2 - \overline{D_i^2} . \qquad (10.15)$$

The basic spectral subtraction, as well as the DWPT subtraction presented above, suffer from residual noise with a perceptually annoying musical structure. To reduce this effect, a parametric-type approach, using an over-subtraction factor α_i and spectral flooring η_i can be implemented. In this method, α_i depends on the SNR of the frame, and η_i is a small constant value around 0.01. With the DWPT, this approach can be expressed as

$$\Delta(i) = Y^2(i) - \alpha_i \overline{D^2(i)} \qquad (10.16)$$

$$\hat{X}_i^2 = \begin{cases} \Delta(i) & \text{, if } \Delta(i) > \eta_i \overline{D^2(i)} \\ \eta_i \overline{D^2(i)} & \text{, otherwise .} \end{cases} \qquad (10.17)$$

A further improvement of this approach, to reduce residual noise, is based on the introduction of auditory masking properties and was proposed in [4]. In this last approach (using the FFT), both $\alpha_i(\omega, m)$ and $\eta_i(\omega, m)$ are made dependent on frequency and the current frame. Furthermore, they can be adapted based on the computation of a masking threshold. In this chapter, a similar approach is proposed for the enhancement scheme. Parameter α has been also made dependent on the masking threshold level of transform channel i and updated at each new frame. The perceptual rule that governs the variation of α_i is similar to the one that was proposed in [12], which renders α dependent on the SNR of the frame. With the help of the adapted α_i and Eq. 10.16, a refined spectral subtraction is performed to obtain $\tilde{X}_{i_{II}}^2$.

Prior to quantization, the coefficients of the transformed clean speech are estimated as

$$\hat{X}_i = \text{sign}\left(Y(i)\right) \sqrt{\tilde{X}_{i_{II}}^2} , \qquad (10.18)$$

where the sign of the estimated coefficient is the sign of the noisy signal. The encoding procedure has been explained in [11].

6.1 Performance Evaluation

The tests performed in this work make use of the French database BDSONS for the speech signals and the NOISEX-92 database for the perturbing noises. The speech samples employed are the phonetically-equilibrated "peq" phrases uttered by different female and male speakers. Noises were both real (car, airplaine, helicopter, ...) and synthetic (white noise, speech-like noise). All the signals are sampled at 16 kHz and quantized with 16 bits/sample.

The objective evaluation of speech signals to predict their subjective quality is an active research field. Most common methods are limited by the accuracy of the perceptual model they incorporate. Their results also depend on the similarity between their auditory model and the one implemented in the system under evaluation. For these reasons, the objective evaluation of the proposed coding system is based on the segmental SNR (SNR_{SEG}) measure [1].

The subjective evaluation of the coder and the enhancement-coding system was performed by trained subjects, during formal listening tests.

6.2 Discussion

The incorporation of a temporal masking model in the DWPT coder led to an average bit rate reduction of more than 5% with respect to considering only simultaneous masking [11]. Thus, perceptually transparent coding was achieved at 39.4 kbit/s. Furthermore, if the amount of tolerable quantization noise σ_{qi}^2 is increasing beyond the masking threshold, such that it becomes increasingly perceptible, the bit rate can be gradually reduced without drastically affecting the quality of the decoded speech signal. This result, which has been verified down to 15 kbit/s, is mainly due to the DWPT fine temporal representation of higher frequency information, which preserves abrupt signal transitions.

The objective quality of the encoded speech signals at 39.4 kbit/s is of SNR_{SEG}=20.02 dB. Speech signals encoded at this bit rate have been found to be subjectively equivalent to those encoded with the G.722 standard at 64 kbit/s. However, the processing delay of the proposed algorithm, of about 39.4 ms, is large compared with the 3 ms of G.722 and may cause echo problems in some telecommunication systems.

In this chapter, we presented two general techniques that allow us to combine in an efficient way several speech enhancement algorithms based on the spectral subtraction method with speaker/speech recognition and speech coding procedures. This paper shows new solutions to the problem of robust speaker verification and speech recognition in noisy environments in the context of missing data theory. It demonstrates that the detection and compensation of unreliable (masked by noise) features improve the performance of recognition systems in low SNRs. The new combination technique proposed for unreliable feature detection and compensation extends the capabilities of the spectral subtraction methods.

The second combination technique was proposed for merging speech enhancement and coding in the context of auditory modeling. The advantage of the method presented in this chapter over previous approaches is that perceptual enhancement and coding, usually implemented as a cascade of two separate systems, are combined. The computational complexity reduction is of the order of 50%, an amount that is mainly represented by the cost of the wavelet packet analysis and synthesis operations. A compromise among residual noise, distortion and bit rate is achieved by perceptually adaptive tuning of a parametric spectral subtraction rule. The integrated system, in noisy conditions, may lead to a slight increase of bit rate with respect to the DWPT coder alone in noise-free conditions. This is mainly due to a richer spectral content of speech caused by background noise and residual noise from enhancement. As this is a pioneering work in the field of the integration of speech enhancement with speech coding and speaker/speech recognition, this type of systems will certainly evolve in the future. An as yet unresolved problem is the optimization of the missing feature detection and compensation process in the recognition systems and of the bit allocation procedure simultaneously taking into account input noise, quantization noise and residual noise in the coding systems.

References

[1] J. R. Deller, J. G. Proakis, and J. H. L. Hansen, *Discrete-Time Processing of Speech Signals*. New York: Macmillan Publishing Company, 1993.

[2] G. Davis, ed., *Noise Reduction in Speech Applications*. CRC Press, Boca Raton, 2002.

[3] J.-C. Junqua and J.-P. Haton, *Robustness in Automatic Speech Recognition*. Boston: Kluwer Academic Publishers, 1996.

[4] N. Virag, "Single channel speech enhancement based on masking properties of the human auditory system," *IEEE Trans. on Speech and Audio Processing*, vol. 7, pp. 126–137, March 1999.

[5] M. Cooke, A. Morris, and P. Green, "Missing data techniques for robust speech recognition," in *ICASSP'97*, (Munich, Germany), pp. 863–866, April 1997.

[6] R. P. Lippman and B. A. Carlson, "Using missing feature theory to actively select features for robust speech recognition with interruptions, filtering, and noise," in *EUROSPEECH'97*, (Rhodes, Greece), pp. 37–40, Sept. 22-25, 1997.

[7] A. Drygajlo and M. El-Maliki, "Use of generalized spectral subtraction and missing feature compensation for robust speaker verification,"

in *Workshop on Speaker Recognition and its Commercial and Forensic Applications*, (Avignon, France), pp. 80–83, April 20-23, 1998.

[8] A. Drygajlo and M. El-Maliki, "Spectral subtraction and missing feature modeling for speaker verification," in *Signal Processing IX, Theories and Applications (EURASIP)*, (Rhodes, Greece), pp. 355–358, 1998.

[9] P. Renevey and A. Drygajlo, "Missing feature theory and probabilistic estimation of clean speech components for robust speech recognition," in *EUROSPEECH'99*, (Budapest, Hungary), pp. 2627–2630, Sept. 5-9, 1999.

[10] A. Drygajlo and B. Carnero, "Integrated speech enhancement and coding in time-frequency domain," in *ICASSP'97*, (Munich, Germany), pp. 1183–1186, April 1997.

[11] B. Carnero and A. Drygajlo, "Perceptual speech coding and enhancement using frame-synchronized fast wavelet packet transform algorithms," *IEEE Trans. Signal Processing*, vol. 47, pp. 1622–1635, June 1999.

[12] M. Berouti, R. Schwartz, and J. Makhoul, "Enhancement of speech corrupted by acoustic noise," in *IEEE Conf. on Acoust., Speech, Signal Processing*, (Washington, DC), pp. 208–211, April 1979.

[13] A. Vizinho, P. Green, M. Cooke, and L. Josifovski, "Missing data theory, spectral subtraction and signal-to-noise estimation for robust ASR: an integrated study," in *EUROSPEECH'99*, (Budapest, Hungary), pp. 2407–2410, Sept. 5-9, 1999.

[14] A. Drygajlo and M. El-Maliki, "Speaker verification in noisy environments with combined spectral subtraction and missing feature theory," in *ICASSP'98*, (Seattle, USA), pp. 121–124, May 12-15, 1998.

[15] M. El-Maliki and A. Drygajlo, "Missing features detection and handling for robust speaker verification," in *EUROSPEECH'99*, (Budapest, Hungary), pp. 975–978, Sept. 5-9, 1999.

[16] M. El-Maliki and A. Drygajlo, "Missing feature detection and compensation for GMM-based speaker verification in noise," in *COST 250 Workshop on Speaker Recognition in Telephony*, (Rome, Italy), November 10-12, 1999.

[17] J. Ortega-García and J. Gonzàlez-Rodríguez, "Overview of speech enhancement techniques for automatic speaker recognition," in *ICSLP'96*, (Philadelphia, USA), pp. 929–932, Oct. 1996.

[18] D. A. Reynolds, "Speaker identification and verification using Gaussian mixture speaker models," *Speech Communication*, vol. 17, pp. 91–108, 1995.

[19] D. A. Reynolds and R. C. Rose, "Robust text-independent speaker identification using Gaussian mixture models," *IEEE Trans. on Speech Audio
Processing*, vol. 3, pp. 72–83, 1995.

[20] Y. Ephraim and D. Malah, "Speech enhancement using a minimum mean-
square error short-time spectral amplitude estimator," *IEEE Trans. Acoustics, Speech, and Signal Proc.*, vol. 32, pp. 1109–1121, Dec. 1984.

[21] M. El-Maliki, *Speaker Verification with Missing Features in Noisy Environments.* Ph.d. thesis, EPFL, Lausanne, Switzerland, 2000.

[22] M. Cooke, P. Green, L. Josifovski, and A. Vizinho, "Robust automatic
speech recognition with missing and unreliable acoustic data," *Speech
Communication*, vol. 34, no. 3, pp. 267–285, 2001.

[23] M. Cooke, P. Green, and M. Crawford, "Handling missing data in speech
recognition," in *ICSLP-94*, (Yokohama, Japan), pp. 1555–1558, 1994.

[24] M. El-Maliki, P. Renevey, and A. Drygajlo, "Speaker verification for noisy
GSM quality speech," in *International COST 254 Workshop on Intelligent
Communication Technologies and Applications, with Emphasis on Mobile
Communications*, (Neuchâtel, Switzerland), pp. 303–306, May 5-7, 1999.

[25] P. Renevey and A. Drygajlo, "Estimation of unreliable data for robust
speech recognition," in *ICASSP'2000*, (Istanbul, Turkey), pp. 1731–1734,
June 2000.

[26] M. J. F. Gales and S. J. Young, "HMM recognition in noise using parallel
model combination," in *EUROSPEECH'93*, (Berlin, Germany), pp. 837–
840, Sept. 21-23, 1993.

V

DEDICATED HARDWARE INTERFACES

Chapter 11

LOW-POWER MICRO-CAMERAS FOR MOBILE MULTIMEDIA DEVICES

Steve Tanner, Michael Ansorge, Fausto Pellandini,

Institute of Microtechnology, University of Neuchâtel,

Rue A.-L. Breguet 2, 2000 Neuchâtel, Switzerland.

{steve.tanner, michael.ansorge, fausto.pellandini}@unine.ch

Nicolas Blanc

CSEM (Swiss Center for Electronics and Microtechnology S.A.),

Badenerstrasse 569, 8048 Zurich, Switzerland.

nicolas.blanc@csem.ch

Abstract CMOS-based image sensors offer the unique capability of integrating all electronic circuitry of a digital camera on a single chip. The resulting system, hereafter referred to as micro-camera because of its high level of integration, has a great potential in terms of low-power consumption, rendering a wide range of new applications in the field of portable multimedia devices possible. This chapter presents the concepts for the design of such low-power micro-cameras. In a first part, the photodetection principle and the basic CMOS sensor architectures are described. In a second part, the design of low-power Analog to Digital Converters suited for video applications is presented before shortly discussing on-chip control and pre-processing functionalities. Two examples of micro-camera realisations are then presented jointly to their specifications, architecture and performances. A summary followed by an outlook on future improvements and applications concludes the chapter.

Keywords: Image sensor, low-power design, CMOS, ADC, micro-camera, APS.

1. Introduction

Nowadays, electronic imaging is steadily gaining in importance as a support for the acquisition, processing and transmission of information. Recent advances in microelectronics have paved the way for new applications in the

J.F. Tasič et al. (eds.), Intelligent Integrated Media Communication Techniques, 361-401.

digital image-processing field. Deep sub-micron technologies render the implementation of very complex imaging systems in portable devices possible: MPEG players, visual-based identification and recognition systems, mobile phones of third generation (UMTS) with integrated video, etc. Consequently, low-power image sensors are strongly required. And, since the targeted markets include consumer products, cost-effective solutions are important, and are very often solved at the system level. The image sensor is therefore not seen as an isolated component, but encompasses more and more functions such as Analog to Digital converters (ADC), image correction, pre-processing algorithms and camera control.

This chapter develops some of the major issues for the design of low-power micro-cameras. In section 2, the CMOS micro-camera concept is presented. Section 3 describes the basic principles and architectures of CMOS image sensors, with their limitations and characterisation parameters. Low-power ADC design principles are recalled in section 4, whereas a design example is given in section 5. The optimisation of micro-camera architectures is then presented in section 6, on-chip digital circuitry being discussed in section 7, before describing two micro-camera realisations in section 8. Finally, section 9 is devoted to summarising the subject and presenting future perspectives in low-power CMOS image sensing.

2.   Micro-Cameras and their Applications

2.1   CMOS Imaging

From the early 1980's and for more than 15 years, charge-coupled devices (CCD) entirely dominated the market of solid-state image sensors. They offer excellent image quality and high resolution, but suffer from several limitations: high power consumption, incompatibility with mainstream microelectronic technologies (CMOS), image lag and smear, limited flexibility in the pixel addressing modes. In order to overcome those limitations, several research teams worked in the late 1980's on arrays of photodiodes addressed like random access memories that can be manufactured in a standard CMOS process. The early realisations, based on passive CMOS pixels, showed promising results [1]. In-pixel amplification was incorporated [2], leading to the concept of Active Pixel Sensor (APS) [3]. First commercial products based on CMOS image sensors appeared on the market in 1996-97.

The main technical advantages of the CMOS APS approach are a more flexible pixel operation (logarithmic mode, random access, etc.), reduced power consumption, and the ability to integrate on-chip additional functions and peripheral electronics, thus reducing components and packaging costs. CCDs are less noisy, they show a higher sensitivity and in general a better image cosmetic quality – low noise, high uniformity, low fixed pattern noise. They are how-

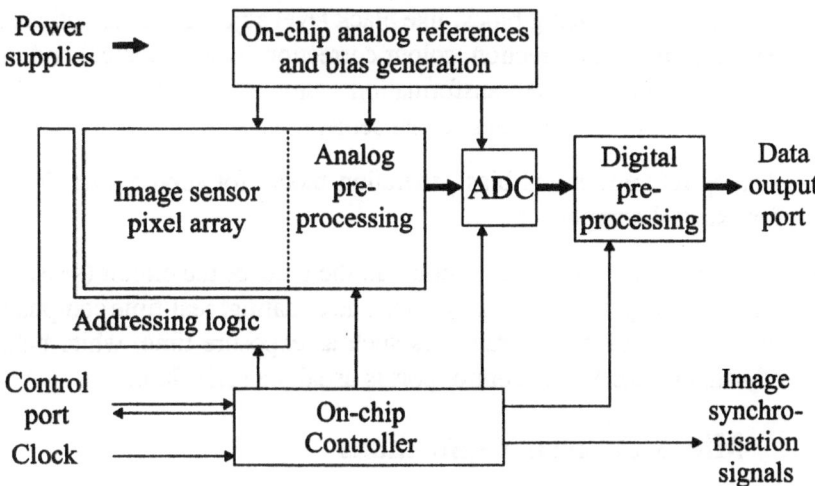

Figure 11.1. General architecture of a CMOS micro-camera.

ever more sensitive to an excess of generated charges (smear, blooming and lag artefacts). Finally, the key advantage of the CMOS approach is certainly to make use of standard CMOS process technologies, resulting in a low cost fabrication, in an access to many foundries, and in the use of standard tools and technological libraries (standard cells, simulation models, etc.).

2.2 The Single-Chip CMOS Micro-Camera

The CMOS technology allows to integrate, on a single chip, a complete system for image acquisition and pre-processing, concisely denoted hereafter by "micro-camera". The benefits of the single-chip approach are a reduced number of components (implying in turn a reduction in system dimensions and costs), a reduced power consumption, and an ease of system design and interfacing. The general architecture of a CMOS micro-camera is represented in Figure 11.1. The different blocks and their related functions are explained below:

- The pixel array, the associated addressing logic (horizontal and vertical) and the analog column processing stage form the image sensor block.

- Analog pre-processing block: this block directly follows the sensor, and performs tasks such as (correlated) double sampling for fixed pattern noise suppression, columns offset and/or gain correction, signal programmable amplification (for example with different gains for each colour channel), etc.

- Analog to Digital Converter (ADC).

- Digital pre-processing block, like black level adjustment, defective pixel and lens shading correction, colour correction (white balance) and interpolation, colour space transformation, gamma correction, spatial interpolation, filtering, sharpness enhancement, etc.

- Analog references and bias generation block, for suppressing the need for external components.

- On-chip controller block, piloting all the units of the circuit (sensor and ADC clocking, register initialisation and updating, real-time computation of image acquisition parameters such as exposure time, white balance adjustment, etc.). The control port is usually a serial link.

2.3   Micro-Camera Applications

CMOS micro-cameras are nowadays mainly used as a replacement component in a variety of imaging systems that were originally based on CCD image sensors. Such systems are:

- Web-cameras.

- Digital still cameras.

- Machine vision (using the random access of CMOS imagers).

- Surveillance and security.

Other applications are emerging, in which the CMOS sensor is not used simply as a replacement component for CCD, but as a micro-system that extensively uses the advantages of CMOS micro-cameras. Some of these applications are the digital video cameras integrated in portable devices, such as mobile phones of the third generation as evoked e.g. in [4], and palm-sized computers. Power consumption is in these systems often prevalent over higher image quality.

3.   CMOS Imaging Basis

3.1   Photo-Detection and Active Pixels

The photo-diode is the basic device used in both interline-transfer (IT) CCDs and most of the CMOS based image sensors for photo-detection. Alternatively, MOS capacitors are used as photosensing element notably in frame-transfer (FT) and full-frame (FF) CCDs, as well as CMOS photogate imagers [5]. A photodiode is a p-n junction formed at the contact of two different doped silicon regions, and naturally doted of an electric field (junction potential). This field is used to separate the electron-hole pairs that have been generated by the interaction of incident light (photons) with the semiconductor. Electrons and holes are collected at each side of the junction, and a photo-generated electric

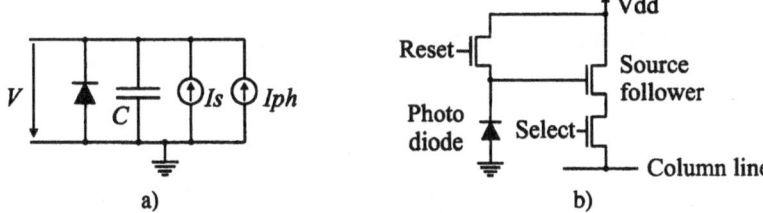

Figure 11.2. Reverse-biased photodiode: a) Small-signal equivalent circuit and b) Used in a basic APS pixel structure with three n-MOS transistors.

current is observed. For CMOS pixels, the photodiode is used in reverse-biased voltage mode for increased sensitivity to long-wavelength radiation and faster response time: a higher applied electric field increases both the width of the depletion region (hence the higher sensitivity) and the electron drift velocity. The increased depletion width moreover also reduces the depletion zone capacitance and sensor response time. The small signal equivalent circuit of the reverse-biased photodiode is given in Figure 11.2 a), where *Iph* denotes the photo-generated current, *Is* the dark current, and *C* the photodiode capacitance.

The reverse-biased photodiode can be operated in two main modes: The first is the integration mode, the most commonly used in APS image sensors. The photodiode capacitor is periodically charged through a transistor to a specific reverse bias voltage. Illumination causes the photodiode capacitor to discharge in proportion to the light intensity. The resulting almost linear response mode can be seen as a photo-generated charge integration mode. Linearity is limited mainly by the capacitor dependence of the reverse-bias voltage. The second mode is the current mode, in which the photodiode is not used as a capacitor, but only as a current source. Since the linear relationship between photo-current and incident light power is valid for at least 10 decades, very high dynamic range sensors can in principle be designed. Sensors with frequency output or non-linear response (e.g. sensors with logarithmic response) have already demonstrated a dynamic range in excess of 140 dB.

The innovation that allowed the emergence of CMOS sensors is the active pixel, whose main idea is to amplify/buffer the photodiode signal inside the pixel, improving significantly the sensor noise performances, and allowing to have an output signal with sufficient strength to drive the output readout bus of the pixel array. Without amplification, the readout procedure would be noisy and slow. Figure 11.2 b) depicts a standard APS pixel, with three n-doped MOS-FET transistors: 1/ a source-follower transistor (amplification), 2/ a reset transistor (reset of the photodiode capacitor) and 3/ a selection transistor for being able to read only one pixel at a time on the column line. The pixel operation includes three different phases:

1 Reset phase: the photodiode capacitor is reset to a constant reference voltage (usually Vdd) through the reset transistor.

2 Integration phase: during this phase, the internal floating node (see Figure 11.2 b) discharges with light.

3 Read phase: the pixel is selected with the selection transistor and the photodiode voltage can be read out.

Based on this first simple configuration, other electronics can be incorporated into the pixel, like photo-gate pixels [5], programmable current sources for offset pixel operation [6], or electronic shutters with on-pixel memory [7], etc.

3.2 APS Pixel Limitations

The light detection and amplification process occuring in an active pixel suffers from several fundamental limitations, described in this sub-section. A deeper description of these limitations can be found in [8].

Fill factor.   The fill factor is the ratio of the light sensitive area of the pixel over its total area. For a given pixel size, the more on-pixel electronics there are, the lower the fill factor and sensor's sensitivity. Usually, in APS image sensors, the fill factor is close to 30-50%, i.e similar to the fill-factor of IT CCDs that are commonly used in consumer applications.

Noise.   Noise sources limit eventually the performances of image sensors. One distinguishes between temporal (random) noise, and Fixed Pattern Noise (FPN). The readout noise includes dark current shot noise from thermally generated charge carriers, and electronic noise mainly from the in-pixel amplifier (1/f noise, thermal noise and reset noise) as well as ADC quantization noise. The ultimate limit is set by the photon shot noise.

Photon shot noise.   The photo-generation of carriers in the photodiode is a statistical process following a Poisson distribution. The maximum quantum-noise limited signal to noise ratio of the photodiode will then depend on the maximum number of electrons the photodiode capacitance can store (full-well charge). For example, if the photodiode capacitance is 15 fF, the charge-to-voltage conversion ratio is 10 μV per electron. A photo-generated charge of 100'000 electrons will result in a voltage swing of 1 V, with a corresponding noise of 316 electrons. The maximum SNR will then be 20 x log10 (100'000/316) = 50 dB [8].

Fixed pattern noise (FPN).   Due to process variations and device mismatches, each transistor has slightly different parameters (threshold voltage, etc.). As a consequence, each APS pixel will have a slightly different output level. This non-uniformity is constant over time and commonly referred to as "Fixed Pattern Noise". Its value is usually equal to a few mV RMS. The FPN can be removed by computing the difference between two output values achieved with and without illumination, an operation called "double sampling". Gain variations can be corrected off-chip as well, or by using unity gain amplifiers. Carefully designed sensors do not suffer from FPN anymore.

Reset noise. When the reset transistor closes, it induces a kTC noise in the pixel node, caused by the RC-filtered Johnson noise of the reset transistor channel resistor. The charge variance q^2 corresponding to this noise is given by [8]:

$$q^2 = kTC \qquad (k = 1.38e^{-23}[J \cdot K^{-1}], \ q = 1.602e^{-19}[C]) \qquad (11.1)$$

For instance, the RMS number of noise electrons in a photodiode with a capacitance of C = 100 fF, introduced by the reset operation at room temperature, is calculated to be 127 electrons [8]. This kTC noise is difficult to suppress because, unlike the FPN, it is different for each reset operation. A double sampling can also be used, but the same kTC noise must be read for both operations (CDS, Correlated Double Sampling), which is difficult with APS sensors because the two sampling operations are separated in time by the exposition itself. A special, patented structure was developed for non-destructive readout, called the photo-gate, and allowing CDS operation. It has been shown that the kTC noise can be reduced by a factor of 2 if a "soft reset" operation is made, consisting in using the reset transistor in limited saturation mode [9].

MOS-FET noise. The source follower transistor of the pixel is affected by two different noise sources:

- Flicker noise (or 1/f noise). This noise is caused by the charge carrier density fluctuation in the channel, and due to its 1/f power spectrum, it is dominant at low frequencies. Its influence can be neglected at the usual operating row readout frequency of image sensors (a few kHz) by using non-minimum-sized transistors.

- Johnson (resistor) noise in the channel with a white noise power spectrum (thermodynamic origin).

The Johnson noise in the channel can be referred back to the MOS-FET gate input and translated into an effective gate charge noise ΔQ. Its RMS value is, for a MOS-FET in saturation [8]:

$$\Delta Q = C_{GS} \sqrt{\frac{8kTB}{3g_m}} \qquad (11.2)$$

where C_{GS} is the equivalent gate capacitance (including the photodiode capacitance), B is the measurement bandwidth and g_m is the trans-conductance of the MOS-FET.

Shot noise or current noise.   This phenomenon of statistical nature describes the fact that a current is always associated with a statistical uncertainty of its exact value, due to the quantified nature of charge carriers. The shot noise variance of a current I is given by:

$$i^2 = 2qIB \qquad (q = 1.602e^{-19}[C]) \qquad (11.3)$$

where B is the measurement bandwidth. For instance, a current of 1 nA will show a current noise of 0.018 nA for a measurement bandwidth of 1 MHz.

Dark current.   Thermally generated charge carriers also set fundamental limits to the sensor performances. Also referred to as "dark current", these carriers spontaneously appear in the semiconductor and cannot be distinguished from photo-generated charge carriers. Their generation rate is strongly temperature and process dependent. For silicon operating under near to room temperature, this rate follows the same generation rate than the intrinsic concentration of mobile carriers ni: it typically doubles for each increase in temperature of 8-9 degrees Kelvin. As the integration time increases, the total number of thermal charge carriers increases proportionally. The dark current of the photodiode can thus be related to a noise expressed in electrons per second because it is proportional to the photodiode integration time. The average total number of thermally generated charges for a given temperature and integration time can be estimated, and subtracted from the actual measured signal. This reduces effectively the dark current average value, but not the associated dark current noise. Due to the stochastic nature of the thermal generation, the distribution of thermal electrons is described by a Poisson distribution. Assuming a total number of thermal electrons N_{th}, the corresponding noise is given by the standard deviation.

$$\Delta N_{th} = \sqrt{N_{th}} \qquad (11.4)$$

In Figure 11.3 are represented the noise contributions of a CMOS active pixel in function of the number of photo-generated electrons. The electronic noise (photon-independent) is dominant for low light levels. For higher intensities the photon shot noise becomes dominant, but since this noise is proportional to the square root of the number of electrons, the overall SNR increases and is maximum when the light intensity reaches its saturation value (full well).

3.3   Pixel Array

An Active Pixel Sensor includes the pixel array, a row addressing logic block, a column addressing block, an analog column processing block and an output

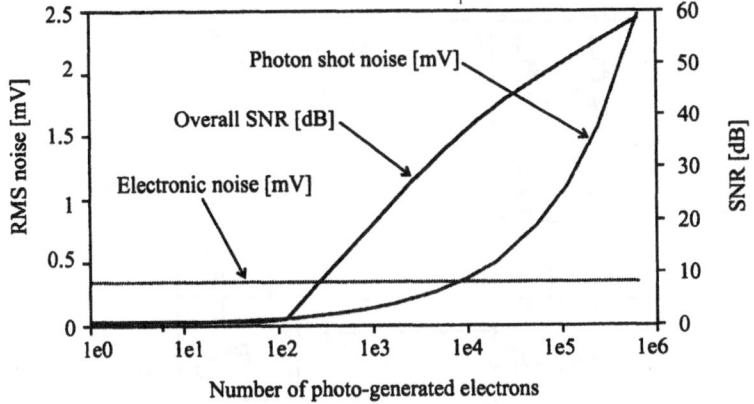

Figure 11.3. Overall noise and corresponding SNR in function of light intensity.

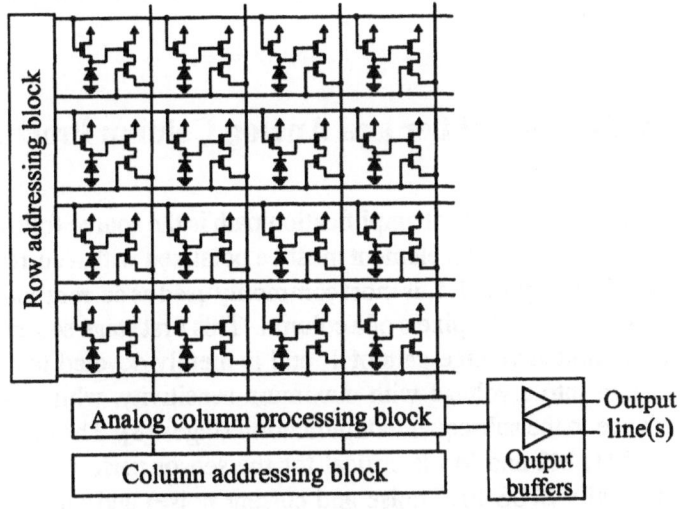

Figure 11.4. General architecture of an APS sensor.

buffer block (Figure 11.4). The pixels are addressed by the row addressing logic through horizontal lines. There is no restriction in the addressing scheme, so that a random access of the pixels is possible. The vertical direction is used for routing the output column lines and global signals to all the pixels (power supply, etc.). While the row addressing logic directly accesses the pixels, the column addressing logic addresses the analog column processing block, whose function will be explained below. Depending on the parallelism in the readout architecture, one or several columns can be amplified at the same time through the output buffer block.

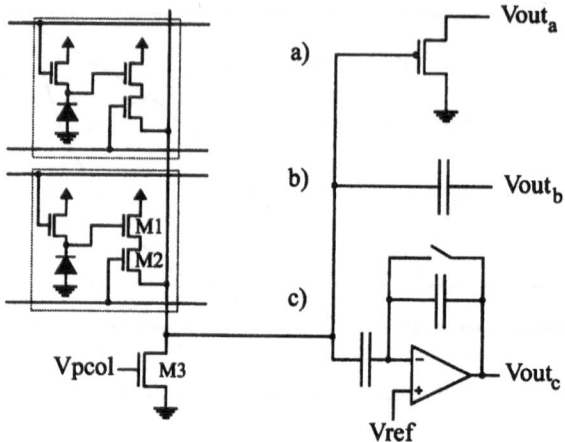

Figure 11.5.   Column line and pixel current source (left), and three examples of analog column processing stages (right) with a) source follower, b) double sampling capacitor, and c) double sampling stage with amplifier.

3.4   On-Pixel Amplifier and Analog Column Processing Block

The first element of the signal amplification path is the source follower of the pixel itself. This amplification element must be polarised with a current. A single transistor, M3 in Figure 11.5, is most commonly used as current source. This transistor is shared by all the pixels of a column. This first stage source follower amplifier has a limited voltage gain of 1, and is merely devoted to converting the photo-charge into a voltage with maximum sensitivity, while providing a lower impedance to the subsequent amplification stages [8]. The dimensioning of transistors M1, M2 and M3 is critical for the sensor performance in terms of noise (especially MOS-FET noise and current noise) and output dynamic. Once the source follower stage has converted the photo-charge into a voltage, the signal of each column readout line is fed into an analog column processing stage. There, several operations can be performed: noise suppression through a (correlated) double sampling stage, signal amplification, etc. Three examples of column analog processing stages are depicted in Figure 11.5 a) to 11.5 c). In this figure, *Vpcol* is the bias voltage for the column current source transistor M3, and *Vref* is an amplifier reference voltage [10].

3.5   Output Buffer

The signals coming out from the analog column processing stages are usually multiplexed into one (or several) output readout bus(ses) in order to form the sensor output signal. The selection of each column stage output is performed

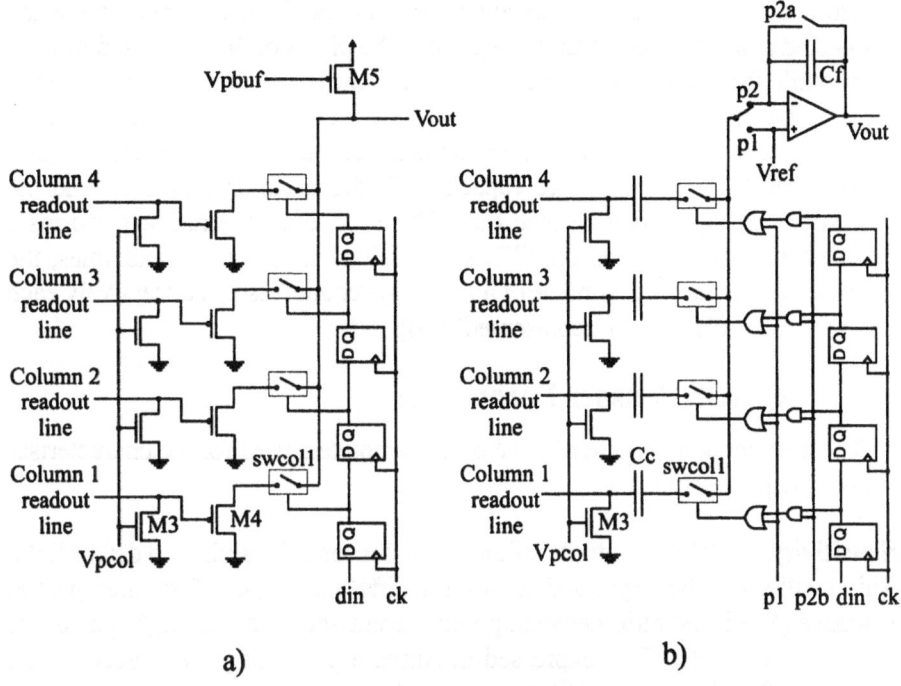

Figure 11.6. Two examples of column readout circuitry and output buffers.

through a switch by the column addressing logic. Finally, the signal is processed through the output buffer block that may perform different operations depending on the application, for instance:

- Buffering for driving other units or pads.

- Second stage of a double sampling unit.

- Programmable gain amplification.

- Non-linear (gamma correction) [11] or linear (RGB to Y-Cb-Cr) transformation.

Figure 11.6 shows two examples of output buffers with their corresponding analog column processing stage, column addressing logic and output bus. In Figure 11.6 a), the pixel output signals (column readout lines on the left) are amplified through a second stage made of a source follower transistor M4 (one for each column) and a current source M5 biased with the voltage *Vpbuf*. The column readout operation is performed through a shift register addressing successively, through column switches *swcol*, each column. This simple structure does not allow double sampling operation. Figure 11.6 b) represents a more complex output structure. During a first phase p1, the pixel readout values (after integration) can be sampled into column capacitors *Cc*, whose right plates are connected to a reference voltage *Vref*. Next, the pixels of the addressed row are reset while the right column capacitor plates are left floating. Consequently, the column capacitors are charged with the voltage difference between pixel read and reset values. By operating signals p2a and p2b several times, these charges can be transferred into capacitor *Cf* thanks to an amplifier, and the corresponding signal can be read out. This second configuration performs a double sampling operation for FPN suppression. In these two examples, the column addressing logic is made of a shift register addressing successively each column (four columns are represented) [10].

3.6   CMOS Sensors Parameters

The next paragraphs describe the basic parameters used for the characterisation of image sensors.

Sensitivity.   The sensitivity of an image sensor is its ability to detect light. This faculty can be expressed in several different ways. First, the spectral response (*SR*) is the ratio between photo current and incoming light power for a given wavelength. It is expressed in Ampere per Watt [A/W]. Second, the quantum efficiency η (or QE) is the ratio between the number of generated electrons and the number of incoming photons for a given wavelength. η and *SR* are linked by the following relation:

$$\eta = SR \cdot \frac{hc}{q\lambda} \qquad (h = 6.62e^{-34}[J \cdot s], \ c = 3e^{8}[m \cdot s^{-1}]) \qquad (11.5)$$

Alternatively, the input signal may be expressed in [lux], i.e. using photometric quantities instead of radiometric quantities. Photometric quantities are related to the human visual perception. These units depend however strongly on the spectral distribution of the illumination, which needs thus to be precisely defined. Moreover, these units cannot take into account the sensor properties in the near infrared range. It is therefore preferable to use radiometric units to directly compare the specifications of imagers proposed by various manufacturers. The quantum efficiency of electronic image sensors is mainly determined by the spectral response of the base silicon material according to the thickness and doping levels used for the various layers. The other thin film materials involved in the process, mainly aluminum, polysilicon, and the passivation layers, also have an impact on the QE. For both CCD and CMOS imagers, the QE is typically excellent in the visible range (above 50% in the range from 400 to 700 nm). It is limited by reflection and absorption in the top layers at the surface (metals, polysilicon). In particular the QE is limited in the UV (ultra-violet) range by the possible presence of polysilicon gates or recombination processes close to the Si-SiO interface. The decreasing absorption coefficient of silicon,

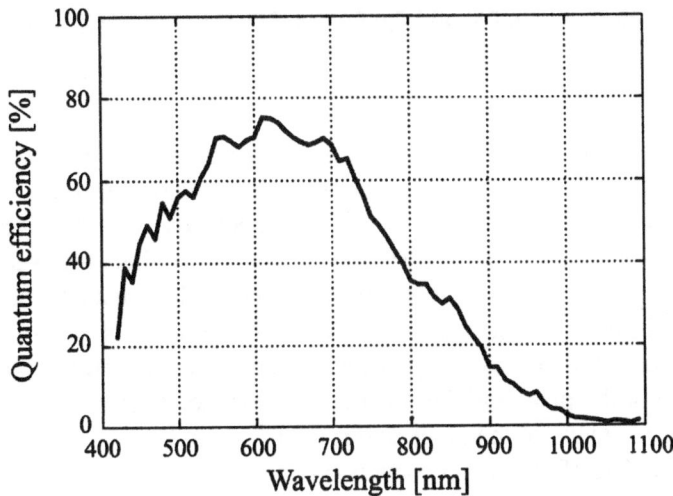

Figure 11.7. Measured quantum efficiency for a CMOS image sensor manufactured in a standard 0.5 micron technology.

which is essentially transparent above 1100 nm, limits the sensitivity of silicon in the IR (infra-red). The thin films used as passivation lead additionally to interference effects and possibly strong modulation of the QE as a function of wavelength. Figure 11.7 shows as an example the measured quantum efficiency for a CMOS image sensor manufactured in a standard 0.5 μm technology. It is worth noticing that nowadays optimized CMOS processes can achieve high quantum efficiencies comparable to the QE of backside illuminated CCDs.

Photodetection uniformity. Uniformity is an important parameter for image sensors. It requires the measurement of both the offset distribution (i.e. dark response or Fixed Pattern Noise) and gain distribution (also referred to as Photo-response non-uniformity, PRNU). Such measurements are ideally performed with an integrating sphere that allows generating a very uniform illumination, with typical variations of less than 0.5% over the complete field of view. Such an integrating sphere can be combined with various light sources, such as LEDs with selected wavelengths or even with a monochromator if a continuous tuning of the wavelength is desired.

Response and response linearity. The sensor response as function of the radiation power incident on the sensor is a further key parameter. Reference [12] describes an experimental set-up allowing the measurement of a sensor response of up to 12 decades of illumination levels, that is from 10^{-10} to 100 W/m^2. From such a measurement, the detection limit (noise floor), signal to noise ratio

(SNR), the saturation signal and dynamic range (DR) of image sensors can be determined.

Noise floor, SNR and dynamic range.   The noise floor of a sensor is measured in the darkness and with short integration times in order to minimize/avoid the effects of photon shot noise and leakage currents. The noise floor is computed either on the base of two successive images or on the base of many images. In the latter case, the standard deviation (rms value) of the output voltage for one pixel yields the sensor noise floor, usually expressed in millivolt. The noise can be converted into an RMS number of electrons through the charge conversion factor $Ceff$, which is the ratio between the photogenerated charge in the pixel and the pixel output voltage. $Ceff$ corresponds to the effective capacitance of the pixel, and is expressed in Farad. A typical sensing capacitance of 15 fF leads to a conversion gain of 10 microvolt per electron.

The signal to noise ratio SNR is given by the output signal voltage divided by the output signal RMS noise. Its value strongly depends on the measurement conditions (light intensity, integration time). For strong intensities, the SNR is limited by the photon shot noise, and for weak signals it is usually limited by the sensor electronic performances.

The dynamic range DR is defined by the ratio between saturation signal and noise floor. An excellent sensor with a linear response can reach a DR as high as 80 dB corresponding to 4 decades of light intensities, whereas a special design such as sensors with logarithmic output or frequency output can show a DR in excess of 140 dB.

Modulation transfer function (MTF).   The modulation transfer function describes the ability of a sensor to detect light variation in the spatial domain, and it is a quantitative measurement of the spatial resolution of an image sensor. Various techniques exist to determine the MTF of imagers. Very often standard test charts with sinus or grating patterns are used as for the measurements of optical elements. This approach requires however the use of an imaging optics, so that the MTF of the complete imaging system (i.e. optics and sensor) is actually obtained, but not the MTF of the sensor itself. The MTF for the optics needs to be known beforehand to compute the MTF of the sensor. Unfortunately, an accurate measurement of the sensor MTF is then basically limited to low spatial frequencies only: the MTF of an achromat for example is typically already reduced to 0.1 of its DC value at 150 lp/mm (line pairs / mm). A superior and simple method is based on a Michelson interferometer modified to use collimated light, also known as the Twyman-Green interferometer. In the image sensor plane a sinusoidal light pattern, whose frequency is defined by the inclination of one of the interferometer's mirrors, is created. The spatial frequency is directly proportional to the tilt angle of the mirror. Spatial frequencies from

0 up to 300 lp/mm can be reliably produced, sufficient to accurately qualify even sensors with extremely small pixel sizes (e.g. 2-3 micron). The MTF of electronic imagers is essentially defined by the pixel geometry (i.e. photodiode layout) and the cross-talk between pixels [13].

Blooming, image smear and lag.   When the sensor is over-illuminated in certain areas, the excess photo-generated charge spills into adjacent pixels, creating large saturated image areas. This effect is named blooming, and several techniques such as diffused overflow drain with control gates or implanted barriers have been developed in CCDs to soak these excess charge carriers. In APS CMOS sensors, an appropriate layout and the implementation of guard rings lead to sensors that are very robust to blooming. During read-out of Full-Frame CCDs and to a smaller extend Frame-Transfer CCDs, the photosites are continuously illuminated resulting in a smeared image. This effect does not appear in APS CMOS sensors due to their very different addressing and read-out scheme. Finally, any remanence in light intensity from an image to the next one is called lag. In CMOS sensors, lag is often due to an incomplete pixel reset operation, but it can be minimised [9].

Defective pixels.   Sensor arrays can contain defects. These are unavoidable in large area arrays and lead potentially to a low process yield. Normally, chip manufacturers offer large size image sensors in various grades, i.e. guaranteeing a maximum number of defects, such as single defects, column or row defects as well as clusters. Defective pixels are basically time-invariant and their effect on the image can be corrected for example by interpolation as long as their number is not too high, or they are not grouped in clusters. Defects are of different types. One can distinguish between [14]:

- Points defects, i.e. pixels whose output deviates significantly from adjacent pixels under similar illumination conditions.
- Dead pixels, showing very poor sensitivity.
- Hot point defects, i.e. pixels with very high dark current.

4.   Low-Power Video ADCs

4.1   Issues for Low-Power ADC Operation

The successful design of an ADC in terms of low power consumption involves classical methods used in low-power mixed-IC design, but mainly depends on an efficient optimisation made at the system and algorithmic levels, particularly regarding the choice of the conversion principle. The analog and digital parts of an ADC do not have always the same design constraints, especially concerning low-power implementation; for instance, decreasing the supply voltage is desired for the digital part, but degrades the SNR in the analog blocks. A compromise between the constraints related to analog and digital parts is most often to be found.

Digital low-power design.   The purely digital part of an ADC usually contributes to a small proportion of the total power consumption. This contribution tends nevertheless to increase with the resolution; high resolution ADCs need more control, decimation filters (for delta-sigma ADCs) or calibration logic in order to reach the required specifications. The following classic methods will be applied whenever possible for the design of the digital part:

- Minimising switching activity (by applying gated clocks, avoiding glitch activities, etc.).

- Suppression of power-consuming clock trees by using multi-phase clock systems.

- Reducing the voltage supply. An elegant solution consists in supplying the digital part of an ADC with a fraction of the voltage of the analog part (as an example, see [15]).

Analog low-power design.   The analog part of an ADC is usually the most power-consuming part; for instance, in the case of 8–12 bit range video switched-capacitor (SC) pipelined ADCs, most of the academic prototype circuits feature a power consumption for the analog part that lies between 50 and 90% of the overall ADC consumption. The main effort in the design of low-power ADCs must therefore be deployed in the analog domain. The following points are of concern:

- Active element design: most of the ADCs need active elements providing signal amplification, like operational amplifiers (OPAMPs) and operational transconductance amplifiers (OTAs). The simplest amplifier structures (simple 2-stage, class A, like the Miller OTA) are the most commonly used in low-power design because of their good compromise between performance and power consumption [16]. If a DC gain higher than 60-70 dB is required, other structures must be chosen, like the regulated folded-cascode amplifier [17], the replica-amp gain enhancement amplifier [18], the telescopic amplifier [19], gain-boosted techniques [20] and other approaches, at the expense of increased power consumption.

- Comparator design: dynamic comparators should be used whenever possible because they do not rely on active elements.

- Voltage supply reduction: for analog design, it is shown that a voltage supply reduction does not always lead to a power consumption reduction [21], [22] for several reasons: loss of maximal amplitudes (SNR degradation), limits of conduction in analog switches, low speed of MOS transistors, limited stack of transistors.

Architectural and algorithmic optimisation.   Usually, the most important gain in power consumption can be made at the architectural and algorithmic levels [23], [24]. The determining parameter in a low-power ADC design is the conversion principle, involving both analog and digital parts. Among all the possible approaches, the designer must select the most appropriate solution in order to fulfill the specifications, such as supply voltage, resolution, sampling rate, available die area, design complexity, environment (Power supply or substrate noise, etc.). In order to compensate for the loss of speed that is due to reduced voltage, a certain level of parallelism (or pipelining) can be chosen, which may have an impact on the conversion principle itself. The degree of parallelism is limited by economic reasons (die area), and, for analog blocks, by inter-channel matching problems.

Depending on the conversion principle, the complexity of the A/D converter can be balanced between the analog and the digital parts. For a low-power implementation, the analog complexity will usually tend to be minimised (reduced number of amplifiers, dynamic comparators, etc.) as far as the conversion principle allows it. Digital error compensation algorithms will usually be preferred to analog ones. Imperfections of analog components (mismatch, offset, etc.) must be solved at the system level. However, in some cases (folding architectures for instance), an apparently complex analog processing stage can dramatically reduce the overall power consumption [25]. Therefore, analog solutions should not be discarded without a meticulous study of their impact on the whole ADC system.

Switched-Capacitor (SC) solutions are usually preferred to Switched-Current (SI) because of their overall simplicity and efficiency [26]. Yet, the current-mode approach is a viable alternative for low-power implementation, and can lead to remarkable results [27].

4.2   Choice of the ADC Principle

The most commonly used architectures for video-rate ADCs are full-flash, folding, two-step (or sub-ranging) and pipeline, each of them having its own advantages and drawbacks. Among them, the pipeline approach offers good advantages for medium-rate ADCs, in particular regarding its ease of design and low-power potentiality. As an example, a parallel-pipeline ADC will be presented in detail in the next section, starting from the algorithmic study and ending with the circuit characterisation.

5.   Parallel-Pipelined ADC

5.1   Principle of the Pipelined RSD Conversion

The Redundant Signed Digit (RSD) converter, whose algorithm is depicted in Figure 11.8, was first published in [28], and is nowadays commonly used. The advantages of the RSD algorithm over a conventional binary decision algorithm are fully described in [28] and [29] and correspond mainly to :

- Tolerance to comparator offset (up to $\pm$ $Vref/4$).

- Tolerance to loop offset error: active element offset and switch charge injection result only in a digital offset and do not contribute to DNL (Differential Non-Linearity) or INL (Integral Non-Linearity).

- Allowance for a fully digital correction due to guaranteed non-missing decision levels.

- Active element saturation causes digital saturation (no distortion for low-level input signals).

The accuracy limitations of the RSD conversion depend on the way it is implemented, but they are mainly due to mismatch of the elements (for multiplication by 2 and subtraction of $Vref$), op-amp finite gain and settling time, noise and charge injection.

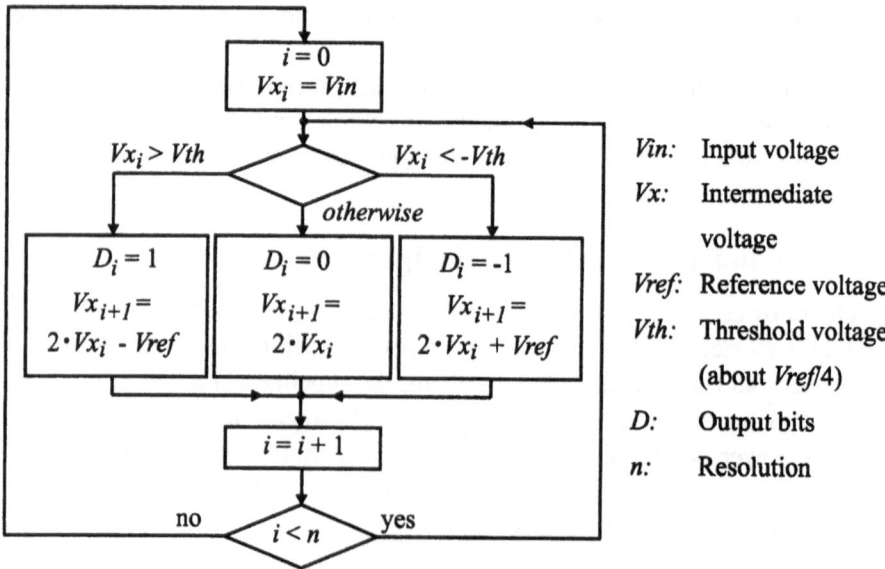

Figure 11.8.   Cyclic RSD converter algorithm.

5.2 Single-Ended Implementation

The RSD conversion algorithm, applied to a pipelined structure, results in the well-known "1.5 bit/stage" pipelined ADC. The ambiguous term "1.5-bit/stage" denotes the redundant nature of the RSD algorithm, giving a ternary intermediate result for a final binary code. Usually, a differential implementation is used because it offers a better immunity against noise, especially the noise in the power supply lines. However, it requires more die area, and the differential amplifier consumes more power than the equivalent single-ended one.

The proposed switched-capacitor implementation, shown in Figure 11.9, is a single-ended structure based on [30]. It uses three capacitors (*C1*, *C2* and *Cr*), one reference voltage *Vref* featuring a negative value, an OTA, two comparators (decision circuit) and switches. A conversion cycle is decomposed into two phases; in the first phase (sampling), both input voltage Vin and OTA offset voltage Voff are sampled into *C1* and *C2*. During the second phase (amplification), the OTA gain is used to perform a charge transfer operation between *C1*, *Cr* and *C2*. Depending on the ternary digital decision D_{i-1} of the previous stage, three cases can occur [10]:

- $D_{i-1} = 0$: The left plate of *Cr* is connected to ground during both phases 1 and 2. $Vout = 2 \cdot Vin$.

- $D_{i-1} = 1$: The left plate of *Cr* is connected to *Vref* during phase 1, and to ground during phase 2. $Vout = 2 \cdot Vin - Vref$.

- $D_{i-1} = -1$: The left plate of *Cr* is connected to ground during phase 1, and to *Vref* during phase 2. $Vout = 2 \cdot Vin + Vref$.

The resulting voltage *Vout* is used in the subsequent stages. The decision circuit is used to generate the conversion result D_i of the current stage.

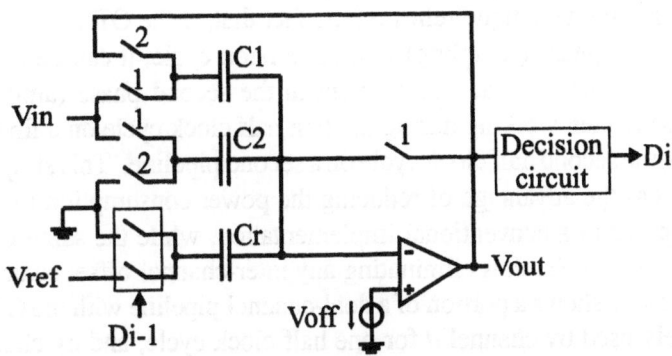

Figure 11.9. Schematic diagram of the proposed RSD pipeline stage.

5.3   System Equations and Simulations

The transfer equation of the proposed stage is:

$$Vout = \left(\frac{a}{a + 3 + Cp/C1}\right) \cdot$$
$$\left[\left(\frac{C1 + C2}{C1}\right) \cdot Vin + D_{i-1} \cdot \left(\frac{Cr}{C1}\right) \cdot Vref\right] \quad (11.6)$$

where D_{i-1} is the digital decision of the previous cycle, *Vref* is the reference voltage and *Cp* is the parasitic capacitor on the amplifier input node. Based on Equation 11.6, simulations were performed for a 10-bit pipelined ADC. The following errors were taken into account:

- Capacitor mismatch randomly distributed (-0.2% to +0.2%).

- OTA DC gain limited to 72, 69 and 66 dB, for OTA stages 1 to 3, respectively. The remaining stages (4 to 8) have a DC gain of 60 dB.

- Noise in the comparators corresponding to 100 mV.

- Parasitic capacitor amounting to about 20% of *C1*.

With these parameters, the obtained simulation results for the DNL and INL errors are plotted in Figure 11.10. Both DNL and INL are about 0.5 LSB, showing that a 10-bit resolution is achievable for this type of converter.

5.4   Active Element Sharing Technique

Various techniques have been applied in order to reduce the number of active elements in an ADC (duplication of the active element functionality, OTA sharing between two consecutive stages of a pipeline, etc.). The presented sharing principle has been proposed in [31], and consists of an OTA sharing scheme between two adjacent pipelines, leading to a dual-channel ADC architecture. The OTA sharing technique relies on the fact that, if the OTA is not requested during the first phase (sampling) of a conversion cycle, it can be used for another pipeline operating at that moment in the second phase (amplification). The OTAs are then working during the first half clock cycle on a first pipeline, and during the second half clock cycle on a second pipeline. This simple sharing technique has the advantage of reducing the power consumption by about 30-40% compared to a conventional implementation, while the same OTA offset affects both channels, thus eliminating any inter-channel offset errors.

Figure 11.11 shows a portion of a dual-channel pipeline with the OTAs being alternatively used by channel *a* for one half clock cycle, and by channel *b* for the second half clock cycle. The input sampling phases of channels *a* and *b* are then shifted by 180°.

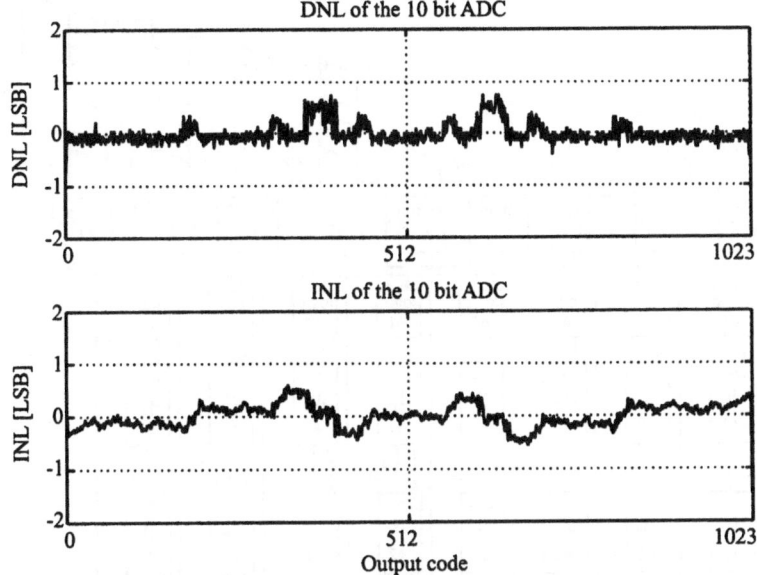

Figure 11.10. Simulated DNL and INL for a 10-bit pipelined RSD ADC.

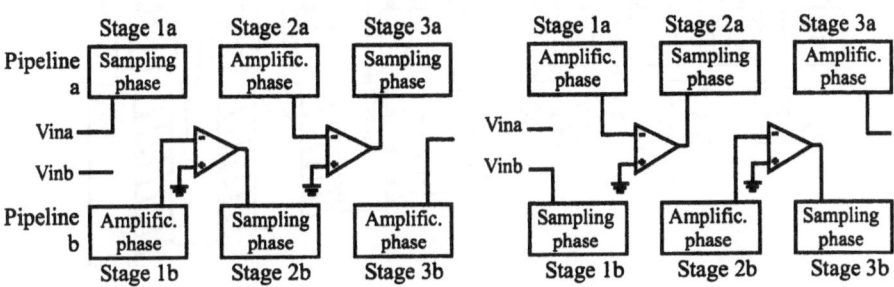

Figure 11.11. OTA sharing between two adjacent pipelines, depicted during first (left) and second (right) half clock cycles.

5.5 Pipeline Architecture

The proposed technique is applied to a 10-bit dual-channel pipelined ADC. Since the RSD conversion produces n bit for n-1 stages, whereas the last stage does not need an amplification, only eight OTAs are required for handling the two channels. The architecture of the dual-pipelined ADC is represented in Figure 11.12, with channel a appearing on the top and channel b on the bottom of the figure. The shared OTAs are located between the two channels. $\phi1$ and $\phi2$ represent the two clock signals of the system, featuring a mutual phase shift of 180°. $\phi1$ is clocking the odd-indexed stages of channel a and the even-indexed stages of channel b, while $\phi2$ is clocking the even-indexed stages of channel

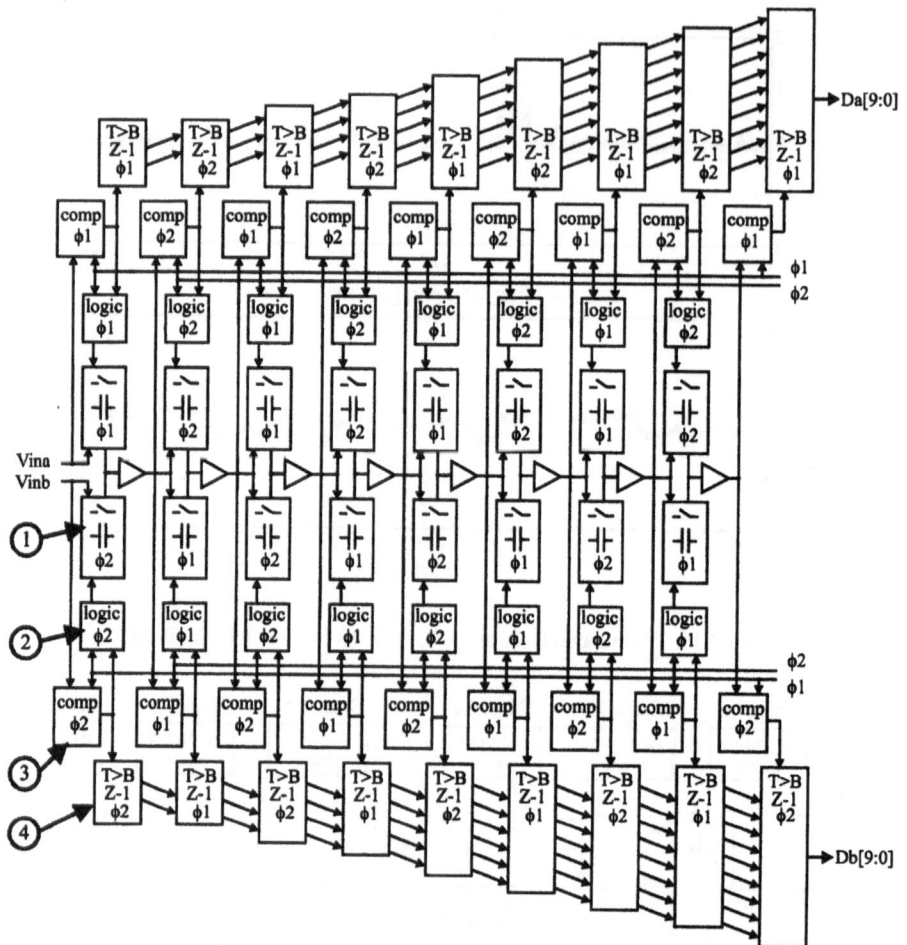

Figure 11.12.   10-bit dual-pipelined ADC architecture.

a and the odd-indexed stages of channel *b*. The structure of a channel stage is explained using the following indexes corresponding to the numbered labels appearing on the left-hand side of Figure 11.12: (1) capacitors and switches, (2) control logic for the switches, (3) two comparators and (4) digital pipeline (shift register) incorporating a ternary to binary conversion logic (T>B). The dynamic comparators presented in [17], offering low power consumption and reasonable speed performance, are used.

5.6 Parallel-Pipelined Realisation: Eight-Channel, 10-bit ADC

The parallel-pipelined ADC structure was integrated into an eight-channel, 10-bit ADC, whose sampling rate was targeted to be 1.25 Msample/s per channel. The corresponding settling time required for the amplifiers was about 300 ns. The amplifiers used correspond to two-stage Miller OTAs. Stage scaling was used in order to minimise power consumption. Table 11.1 provides the current consumption and capacitor values for all eight stages. To limit mismatch errors, the capacitors were chosen larger than what was requested by the sole kTC noise criterion.

Table 11.1. Capacitor values and current consumption of each stage.

Stage index	1	2	3	4	5-8
Capacitor (pF)	1.6	1.3	1.0	0.8	0.5
Current (μA)	164	144	112	84	64

The eight-channel pipelined RSD ADC was integrated into a 0.5-μm double-poly triple-metal CMOS technology from IMEC (Belgium), featuring an active area of 2.98 mm x 0.61 mm = 1.82 mm^2. The layout is depicted in Figure 11.13. Five circuits were tested and fully characterised. The extracted characteristics are based on the mean values of 5 x 8 = 40 different ADC parameters, and are summarised in Table 11.2.

Concerning inter-channel errors, only offset degrades the overall ADC performance, and must be corrected (digitally). Gain errors correspond to a limitation in SNR of 64 dB and can thus be neglected. Power consumption of the whole ADC corresponds to about 1.2 nJ/sample for a voltage supply of 3.0 V. Concerning frequency tests, it was not possible to fully characterise the chip at high frequencies due to the lack of Sample and Hold stage in the ADC input, this function being realised by the image sensor itself. The circuit was fully operating over a voltage supply range of ± 1.2 V to ± 1.6 V. All measurements

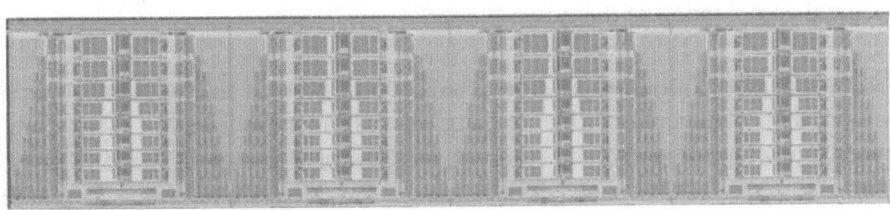

Figure 11.13. Layout of the eight-channel ADC.

are performed with a voltage supply of Vdd = ± 1.5 V at a temperature T = 20° C.

Table 11.2.   Measured parameters of the 8-channel ADC.

Each channel		
Nominal sampling frequency	1.25	MHz
Differential Non-Linearity (DNL)	± 0.5	LSB
Integral Non-Linearity (INL)	± 0.6	LSB
Signal-to-Noise Ratio (SNR) *	56	dB
Signal-to-Noise and Distortion Ratio (SNDR) *	55.5	dB
Effective Number of Bits (ENB)	8.9	bit
Power Supply Rejection Ratio (PSRR)	> 40	dB
All eight channels		
Analog input range	$\pm$ (Vdd - 0.2)	Volt
Nominal sampling rate	10	MHz
Power consumption (at nominal sampling rate)	12.4	mW
Inter-channel gain variance	74	dB
Inter-channel offset variance	45	dB

* Input signal frequency Fin = 28.07 kHz.

6.   Low-Power Image Sensor Architectures

The starting point for the design of a low-power digital image sensor is the optimisation of each individual block (sensor, ADC, etc.). For this purpose, the general rules summarised in sub-section 4.1 can be applied. However, those efforts may be fruitless if the combination of these blocks, at the architectural level, is not optimised. This section emphasizes some important issues in low-power image sensor design.

6.1   Sensor and ADC Interface

A good match between sensor and ADC performances is required for an optimal combination of these two blocks.

Sample & Hold function.   Since the readout time of a pixel is usually much shorter than its integration time, the pixel acts as a memory element. Therefore, when a pixel is addressed and read out, an additional Sample & Hold (S & H) function in the ADC is usually not required. This allows a reduction in power consumption. If the S & H function embedded in the pixel is not efficient (due to the fact that the required ADC acquisition time is too long compared to the

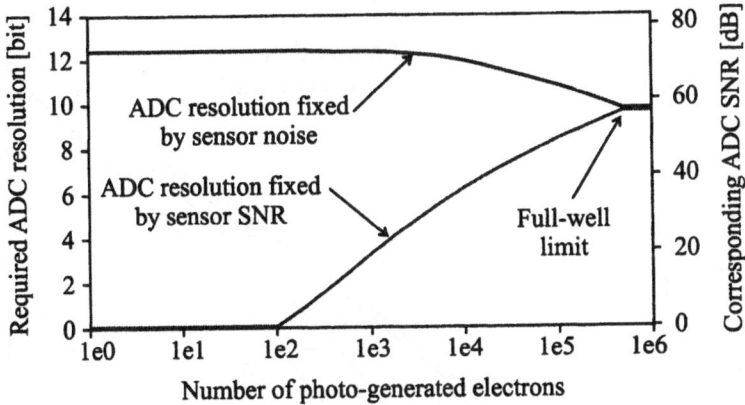

Figure 11.14. ADC resolution requirement in function of light intensity.

integration time), the S & H can be realised by the Double Sampling stage existing usually in the column analog processing stage.

ADC resolution. The ADC resolution should fit the sensor performance for optimum implementation. It would actually be ineffective to use a sensor with an SNR of 70 dB if the succeeding ADC has only 48 dB of SNR (equivalent to 8 bit). However, the performance matching of the sensor and the ADC is not so obvious because two contradictory parameters are to be matched:

- First, the ADC resolution (the smallest analog input step corresponding to an LSB) should match the noise of the sensor. Since this noise increases with incident light, the required ADC resolution decreases when the light increases.

- Second, the ADC SNR should match the output SNR of the sensor. Since the sensor SNR increases with light, the required ADC SNR (and consequently its resolution) will also increase with light.

These two requirements are represented in Figure 11.14 for a typical pixel equivalent capacitor (50 fF) and output swing (1.5 V). The noise floor is 0.3 mV.

Under full illumination conditions, referred to as the full-well limit, the sensor noise and SNR impose the same requirement for ADC resolution because both signal and noise are maximum. For a typical pixel, this requirement in resolution is about 9-10 bit. The minimal required ADC resolution is then fixed by the sensor SNR at full illumination. With this choice, the sensor and ADC performances are matched for high illumination levels. However, for low light levels (less than 10% of the dynamic), the noise decreases and a higher ADC resolution should be required. The first solution to this problem is to use

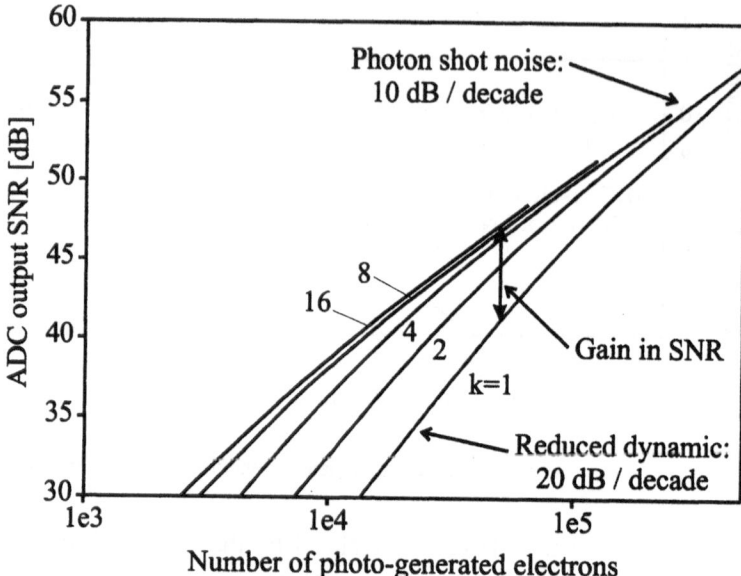

Figure 11.15. Overall SNR performance without pixel signal amplification (k=1) and with amplification (k=2, 4, 8, 16).

an over-dimensioned ADC (typically 12 bit of resolution). Unfortunately, the increase in power consumption is important. Another solution is to use so-called floating point ADCs [32] or relative precision ADCs [30]. They feature a limited resolution of 10 bit, but an input dynamic range of 13-14 bit.

Signal amplification before analog to digital conversion.   A third solution to the above mentioned problem consists in amplifying the sensor output signal in order to increase its dynamic so that it will fully exploit the ADC resolution. A programmable amplifier with adjustable gain is used and the gain is adapted to the light conditions. The effects of this amplification on the overall SNR performance were simulated and are shown in Figure 11.15, where the SNR after a 10-bit analog to digital conversion is plotted in function of light intensity.

The same pixel was used than for Figure 11.14, and the sensor electronic noise was set to 100 μV. The SNR without amplification (k = 1) is shown on the bottom right curve. For high illuminations, its slope reduces to 10 dB/decade due to the dominant photon shot noise. For lower illuminations, the slope amounts to 20 dB/decade due to the loss of output swing. In the other plots, the sensor output signal was amplified by gains of 2, 4, 8 and 16 so that its output dynamic matched the ADC input range. The amplifier noise was set to 100 μV and was then multiplied by the corresponding gain. In all plots the SNR for low light intensities is improved. As an example, the SNR is improved by 6 to

7 dB with a gain of 16 and light intensities of 1/20 of the full well; even with such low intensities, the SNR is still reasonable (46 dB).

6.2 Parallelism Optimisation

One of the major degrees of freedom in the design of a sensor architecture is the parallelism of the readout circuitry: the inherent parallel architecture of an image sensor is very well suited for exploiting a variable degree of parallelism in its subsequent readout electronics. Since parallelism is a common way to compensate for the loss of speed due to low-voltage operation, its application to sensor readout architectures (including ADCs) is well suited for a low-power sensor design. The degree of parallelism in digital sensor readout circuitry is given by the number of columns addressed at the same time during the scanning readout phase. Alternatively, this number corresponds also, in most cases, to the number of analog output lines connecting the sensor to the ADC block. The parallelism can vary from only one (no parallelism) to the number of columns appearing in the sensor (maximal parallelism). In this case, all the pixels of a row are read out and digitised at the same time.

The main advantages of parallelism in the readout architecture are:

- Thanks to a lower operating frequency, a voltage supply reduction is possible, leading to a lower power consumption.

- Relaxed speed constraints of analog components in the readout circuitry (switches, amplifiers, etc.) due to the reduced operating frequency and the reduced capacitive load of each analog path.

- Low-power consumption in the sub-sampling modes due to the possibility of switching off the unused signal paths.

However, the use of parallelism increases die area and induces new types of errors such as inter-channel mismatches (gain, offset, timing) [33] that must be carefully evaluated.

6.3 Pixel Structure and Operation

Another degree of freedom available for the design of sensor architectures is given by the selection of the pixel structure itself. An original pixel design, bringing new operation modes, has an impact on the overall sensor architecture, and can greatly improve some of its basic characteristics. As an example, we will consider the impact that a pixel with embedded analog memory has on the power consumption of an image sensor.

In traditional photography, the upper bound of the exposure time is usually limited to 1/50 s, corresponding to 20 ms. Beyond this value, the risk of getting blurred pictures is high. This limit also holds for electronic image sensors.

However, the readout time of an image has also to be taken into account in the latter case. The usual shutter mode with APS sensors is the line shutter mode, in which the exposure gets started and completed row-wise, as if a light aperture was sliding in front of the sensor and was immediately followed by the readout. If the subject is moving (translation, rotation, zoom), the image will get distorted under these conditions. The sensor readout time must then be as fast as possible and should be in the same order of magnitude as the maximum exposure time. The necessary readout frequency becomes very high for big sensors, which is critical for battery-powered, still picture image acquisitions devices, where the peak current consumption has to be minimised.

The requirement in high readout frequency can be relaxed if a special pixel with embedded analog local memory is used. The benefits are not only the possibility to use a global shutter mode (allowing a simultaneous exposure of all pixels, thus leading to an improved image quality), but also the ability to decrease the readout frequency, the information being temporarily stored in the pixel analog memory. An important reduction of the peak current is achievable this way.

6.4　Sensor Addressing Strategies

The last degree of freedom available in the design of low-power sensor architectures concerns the horizontal and vertical sensor addressing organization.

Sub-sampling addressing.　For sub-sampling modes, only a subset of the pixels is read out. Consequently, the required frequency and/or the hardware parallelism can be lowered, leading to a reduced power consumption. If the sensor architecture allows to switch off the unused readout paths, the power consumption of the sensor can remain proportional to the amount of pixels being read out.

Windowing addressing.　Another common addressing mode is encountered when only a given sensor area ("window") is read out. Only the rows and columns belonging to the window are addressed. The power consumption can be decreased if the unused column amplifiers are switched off.

Reduced duty cycle of active elements.　Power consumption can be further reduced if the active elements (current sources, amplifiers, etc.) are switched off whenever they are not used (this approach is similar to the "gated clock" principle applied for digital designs). In particular, the power consumption of the sensor can be lowered during the readout phase if the current sources for the in-pixel source follower (one for each column) are not switched on all the time.

7. On-Chip Control and Pre-Processing

As denoted in the introduction of this chapter, a micro-camera includes on-chip digital logic responsible for the tasks described below.

7.1 Clock and Control Signal Generation

This task is performed by a dedicated state machine that generates line and frame signals for the sensor, clocks for the ADC, and configuration signals for the system. The complexity of this unit varies depending on the sensor operation modes (windowing addressing, global shutter, sub-sampling addressing, etc.) but it is usually in the order of several thousands of gates.

7.2 Image Pre-Processing

A micro-camera may include diverse types of image pre-processing operations of varying complexity. A list of the most commonly used image pre-processing functions is discussed below.

Lens, sensor and ADC error correction. The image acquisition block, composed of the optical system, sensor array, and ADC, suffers from several fundamental limitations:

- For the lens: shading effect (decrease of intensity from the centre to the image borders), optical aberrations.

- For the sensor: pixel Fixed-Pattern Noise (offset and gain), column amplifier inter-channel errors (offset and gain), defective pixels, defective rows or columns, pixel and column amplifier non-linearity.

- For the ADC: inter-channel errors (offset and gain) in case of a parallel implemented architecture.

The correction of these errors involves a calibration, performed either when the circuit is fabricated, or when the component is mounted, or after every power-up, or finally in a regular manner during operation of the micro-camera. The correction hardware is dedicated.

Color interpolation, correction and space transform. In cost effective image sensors, color images are obtained with a single sensor in which the pixels are distributed into three groups, one for each basic color. Then, optical color filters are deposited on the pixel array, following a regular mosaic pattern, the most commonly used being the Bayer pattern, composed of two green, one red, and one blue pixel within a constitutive 2x2 pixels block. The resulting image is made of three intermixed and incomplete images, one green, one red and the last blue. An operation is therefore required for re-generating the missing information, for instance the red and blue information at the location of a green pixel.This operation is performed with an interpolation furnishing three full-resolution red, green and blue images. The hardware is complex and costly in computation, amounting up to 10 arithmetic operations per pixel.

Very often, the red-green-blue (RGB) color representation used with optical color filters must be converted into another space color representation through a matrix operation. A possible color correction (white balance) can be applied at the same time.

Image filtering.   A bilinear (or even non-linear) filtering can be applied for:

- Spatial image re-sampling (for zooming operation).

- Anti-aliasing filtering before down-sampling.

- Edge sharpness enhancement after color interpolation.

Image amplitude control.   The amplitude of the pixel as well as the transfer function of digital result versus light intensity can be adjusted for specific needs: gamma correction, smooth saturation of the pixels for high intensities, level equalisation for occupying the whole image dynamic, etc.

7.3   Image Acquisition Parameter Control

Some parameters have to be continuously adjusted in real-time, in response to the changing acquisition conditions. This task involves a dedicated hardware (state machine) or a programmable on-chip unit (microprocessor) for more sophisticated algorithms. The following parameters are concerned:

- Exposure time: single or multi-zone algorithms, linear or non-linear, hysteresis-based regulation, etc.

- Color correction coefficients computation (white balancing): depending on illumination source, the spectral composition of the white color changes and therefore the color correction coefficients have to be constantly adjusted.

- Autofocus algorithm (digital-based regulation).

- Flicker cancellation algorithm: the flicker of artificial light (50 or 60 Hz) can cause an inhomogenous image acquisition. The flicker frequency has to be detected and the exposure time limited to certain values.

All these algorithms use statistical parameters computed from the image (mean, histogram, etc.) by dedicated hardware.

7.4   Interfaces

The interfaces of the chip to the external world are important because they have an impact on the packaging cost and/or on the system dimensions. The interfaces are divided into the control port (usually serial because the throughput of information is low) and the image data port, usually parallel. Some applications require standard interfaces (USB, etc.) which may in their turn increase significantly the die size.

8.   Realisation Examples

This section presents two image sensors developed by CSEM (Swiss Center for Electronics and Microelectronics S.A.) and IMT (Institute of Microtechnology, University of Neuchâtel). They correspond to simple examples of low-power realisations.

8.1   First Realisation: APS256D

This first example [34] is a low-resolution (256 x 256 pixels), black and white digital image sensor fabricated with the ALP1LV (1 μm CMOS, 2-metal, 2-poly) technology from EM-Microelectronic Marin S.A., Switzerland. This circuit was specified to operate with a voltage supply of 2.7 to 3.3 V, and a maximum pixel rate of 4.2 MHz, corresponding to about 60 frames per second. The main focus was the low power consumption and the best SNR possible.

Architecture.   Due to the constraints of the technology and of a 15 μm pixel pitch, the simplest three-transistor, n+ p-substrate photodiodes pixel structure was chosen for the best possible fill factor (35% in this case). The operation modes of the pixel array were consequently reduced to the basic line shutter mode. The sensor readout architecture is the same as the one of Figure 11.6 a), with a parallelism of four readout paths. No FPN suppression is implemented, which has the advantage of a reduction in power consumption. In order to increase the sensor output dynamic, the voltage applied to the gate of the pixel reset transistors was generated with an on-chip voltage multiplier. The two bias voltages needed by the column and output buffer stages were also generated on-chip by a current reference.

The ADC is a four-channel, 10-bit parallel-pipelined RSD ADC as described in section 5. Programmable current references as well as voltage generators for the comparators were integrated on-chip. The on-chip controller is a full static synchronous state machine counting about 12600 transistors. It performs tasks such as sensor control, simple image statistics measurements (mean and number of saturated pixels), ADC inter-channel offset correction, black level adjustment, exposure time control (adaptive proportional regulation loop). A

 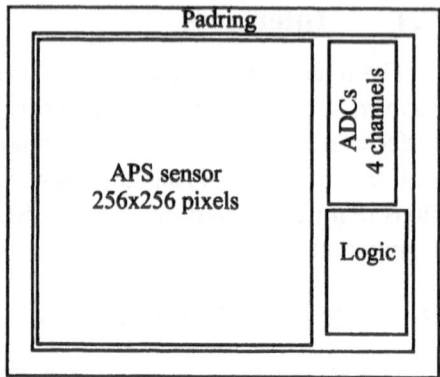

Figure 11.16. Photomicrography of the APS256D circuit (left) and floor-plan (right).

serial interface allows to externally access the eight 10-bit control registers. The interfaces of the chip include the control serial interface, the 10-bit output pixel port and frame synchronisation signals. The required external supply voltages are Vdd and Vdd/2. No other analogue voltages are required.

Circuit realisation.   The core-limited circuit layout of the APS256D is shown in Figure 11.16. The dominant block is the sensor including related logic (reset addressing logic on the top, read addressing logic on the bottom and column amplifiers and other units on the right). On the top right of the circuit are located the ADCs and on the bottom right the controller (standard cells). The overall chip size, including pad ring, is 6.4 x 4.8 mm, resulting in an area of 30.7 mm^2. The pads were placed on all four sides of the core in order to facilitate connections. The area occupied by the sensor, ADC, logic and pads are 61%, 8%, 6% and 15% respectively.

Measurement results.   The main measured sensor parameters are summarised in Table 11.3 [34]. The ADC [35] shows an SNDR (signal to noise and distortion ratio) of 56.5 dB, a DNL of ±0.5 LSB and an INL of ±0.9 LSB. The power consumption of the whole chip is 12.3 mW with a working frequency of 4 MHz and a voltage supply of 3.0 V. Contributions to the power consumption are 30% for the sensor, 30% for the ADCs, 15% for the logic and 25% for the pads.

8.2   Second Realisation: VGACAM

This second circuit [36] is a medium resolution (648 x 488 pixels) color image sensor specially developed for very low-power digital still cameras. Integrated into a 0.5-μm double-poly, triple-metal CMOS technology from IMEC (Belgium), this circuit was specified to operate with a voltage supply of 2.7 to

Table 11.3. Measured parameters of the APS256D circuit.

Parameter	Value	Units
Saturated output signal swing	1.2	Volt
Computed full well capacity	1'000'000	electrons
Conversion gain	1.5	μV/e
Fixed Pattern Noise	8.5	mV RMS
Noise floor	0.4	mV RMS
Dynamic range	70	dB
Sensitivity @ 626 nm	0.84	V/lux·s
Dark current	150	mV/s

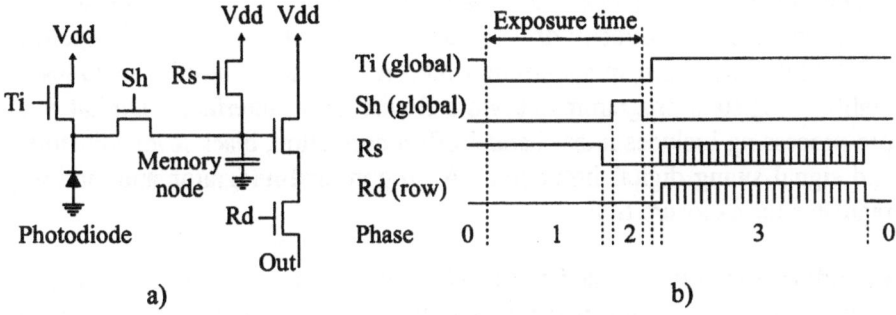

Figure 11.17. a) Schematic diagram of the VGACAM pixel and b) Related global shutter timing.

3.3 V, and a maximum pixel rate of 10 MHz, representing about 30 frames per second. It is equipped with a special 5-transistor pixel with an integrated analogue storage node, allowing global shutter operation. This feature is useful either for the acquisition of fast moving objects, or for allowing a reduced operation frequency while keeping a blur-free image (see sub-section 6.3).

Sensor architecture. The pixel schematic diagram is represented in Figure 11.17 a). Compared to the basic APS structure, it includes two more transistors. The first is the shutter (Sh) transistor, whose function is to disconnect the photodiode from the source follower amplifier transistor gate (memory node) at the end of the exposure time. The second is the timer (Ti) transistor enabling the resetting of the photodiode independently regarding the memory node. The reset (Rs) transistor is hence used for resetting the memory node.

The pixel can be operated in line shutter or global shutter mode, depending on the timing of the control signals. For the global shutter mode, the corresponding signals are given in Figure 11.17 b). In phase 0, the photodiode and memory nodes of all the pixels are reset. The exposition phase (1) begins when the photodiode node is released, and starts to integrate photo-generated charges. Then, at the end of phase 1, the memory nodes are released and a pulse is generated on all the shutter transistors (2), allowing charge transfer from the photodiode to the memory node. The exposure time stops with the end of phase 2. At this moment, all the pixel signals have been stored into the memory node, and the photodiode is not used any more and can consequently be reset. The memory nodes are then read out and reset row by row (phase 3) like in line shutter mode. For the line shutter mode, the shutter and timer transistors are set to constant values of 1 (switched on) and 0 (switched off) respectively, and no global signal is used.

The readout structure of the sensor is the same than the one represented in Figure 11.6 b), with a parallelism of eight readout paths. This readout circuit supports FPN suppression, as well as output offset adjustment. The eight output buffers directly drive a bank of eight-channels, 10-bit pipelined RSD ADCs presented in sub-section 5.6.The on-chip controller includes 32 eight-bit registers programmable via a 3-wire serial interface. Digital pixel pre-processing includes inter-channel offset correction, black level adjustment and signal swing digital limitation. A proportional-integrator automatically regulates the exposure time.

Circuit realisation.   The floor plan is optimised for minimum die size in the vertical direction. For this purpose, only two pad rows are used, one on the left side and the other on the right side of the chip. The circuit photomicrography is shown in Figure 11.18. The sensor block occupies the left part of the chip, while the ADCs and the on-chip controller are located on the top-right and on the bottom-right parts, respectively. The overall chip size, including padring, is 9.28 x 5.94 mm, resulting in an area of 55.1 mm^2. The area occupied by the sensor, ADC, logic and pads represent 76%, 3%, 3% and 11% respectively. Routing channels occupy the remaining area (7%). The pixels have a fill factor of 35% and a pitch of 10.5 μm.

Measurement results.   The main measured sensor parameters are summarised in Table 11.4. These measurements show that the FPN suppression stage provides poorer results than expected, and that the on-chip electronic noise is important. All the experimental results of the ADC block are presented in sub-section 5.6. Overall power consumption is 63 mW with a frame rate of 30 frames/s and a voltage supply of 3.0 V. The contributions of the blocks to the power consumption are 55% for the sensor, 20% for the ADCs, 3% for the logic and 22% for the pads.

Global shutter performance.   The global shutter was tested at a frequency of 4 MHz with the acquisition of a rotating fan, spinning at about 200-300 rpm.

Figure 11.18. Photomicrography of the VGACAM circuit.

The left side of Figure 11.19 shows an image acquisition in line shutter mode, with an exposure time of 1 ms. The image is strongly distorted due to the important scanning time of 83 ms. On the right, the fan was acquired in global shutter mode with identical exposure time and rotation speed. No distortion is observed; the efficiency of the global shutter is demonstrated. A more detailed analysis shows that the noise of the image in global shutter mode is higher than in line shutter mode (dispersion of leakage current between the pixels) and many more white pixels are observed.

9.   Summary and Perspectives

In this chapter were presented the basic techniques for the design of low-power CMOS digital image sensors suited for mobile imaging applications. Aspects such as photodetection limitations inherent to CMOS sensors, low-power techniques for video ADCs, and architectural issues of digital image

Table 11.4. Measured parameters of the VGACAM circuit.

Parameter	Value	Units
Saturated output signal swing	1.86	Volt
Computed full well capacity	266'000	electrons
Conversion gain	7	$\mu V/e$
Fixed Pattern Noise	10	mV RMS
Noise floor	<1.6	mV RMS
Dynamic range	>60	dB
Sensitivity @ 626 nm	2.3	V/lux·s
Dark current	118	mV/s

Figure 11.19. Acquisition of a rotating fan with line shutter (left) and global shutter (right) modes. Fan speed is about 200-300 rpm.

sensors, were covered. Two realisation examples were described, featuring a power consumption that is lower than commercial products presently available on the market.

However, there is still a great potential for improving the performances of such circuits. First, the power consumption can be further reduced by refining the techniques presented in this chapter, and by decreasing the voltage supply. A reduction in power consumption by an order of magnitude is expected with the use of voltage supplies as low as 1-1.2 V [37]. However, such a low voltage represents a design challenge for maintaining high sensor performances in terms of sensitivity, dynamic range and noise figure.

Second, the image sensor performances will benefit from the continuous refinements of digital CMOS processes. If today's electronic CMOS imagers are mostly based on 0.5 and 0.35 μm technologies, recent sensors manufactured in 0.25 and even 0.18 μm technologies have been reported. A smaller feature size clearly allows shrinking the pixel size down to 3-5 μm (and so will the

Table 11.5.   Expected performances of next generation CMOS image sensors.

Parameter	Value	Units
Technology	0.18	μm
Voltage supply	<1.2	Volt
Pixel pitch	5	μm
Fill factor	>50	%
Sensitivity	0.1	lux
Noise floor and FPN	<0.1	mV RMS
Dynamic range	>80	dB
ADC resolution	12	bit
Power consumption (sensor and ADCs)	<0.2	mW/MHz

overall sensor size and costs), corresponding to the practical limits set by optical diffraction. Alternatively, the fill factor or pixel functionality can be improved for a given pixel size. However, not all opto-electronic properties of imaging devices will improve with decreasing minimum feature size. In particular, the quantum efficiency, especially in the long wavelength range, is diminishing due to the use of thinner layers with higher doping levels. Also, increased dark currents, reduced full well capacities and decreased voltage supplies represent as many significant challenges to master, so as to achieve performances lying beyond those of nowadays CMOS and CCD imagers.

Third, deep sub-micron technologies will clearly render massive on-chip digital signal processing possible, hence reinforcing the role of the micro-camera as a sub-system and not only as a component. Several leading foundries have already optimised their deep sub-micron technologies for optical sensors along this direction. They are also offering additional backend processes supporting micro-lenses arrays, color filters, and dedicated packages, as required for producing cost effective digital color cameras. Thanks to such advances, a VGA color camera with on-chip MPEG real-time image compression will, in the coming years, consume less than 100 mW and occupy less than 50 mm^2 of silicon. Table 11.5 lists the performances of CMOS image sensors expected for the very next years.

Technically, the micro-camera technology has proven to fulfill the requirements of mobile video devices for a broad range of applications, such as those listed, for instance, in [4]. However, beyond technical feasibility, other key issues have to be examined to ensure the success of these applications, such as the ease-of-use aspect, and the "added function over added cost" ratio to ensure user acceptance. Important reflections on the user communication improvement by video information, on the provision of smart video-based user interfacing and personalization facilities, and on the identification of new functionalities enabled by embedded micro-cameras, are to be made. It seems that the success of micro-cameras in the consumer electronics arena will also depend on the combination possibilities between video information and complementary media, and on the intelligent exploitation of the added functionalities provided by the camera and related processing algorithms. The potentialities offered by micro-cameras are however very large, so that this field can be considered as one of the most active in nowadays semiconductor industry.

Acknowledgments

The authors express their gratitude to Asulab S.A., SBT Cerberus Division A.G. and EM-Microelectronics Marin S.A. (Switzerland) for the appreciated collaboration in the frame of the research activities presented in this chapter.

Glossary

ADC	Analog to digital converter.
APS	Active pixel sensor.
CCD	Charge-coupled device.
CDS/DS	Correlated double sampling / double sampling.
CMOS	Complemetary metal-oxyde semiconductor.
DNL/INL	Differential/Integral non linearity.
DR	Dynamic range.
FPN	Fixed pattern noise.
kTC	Indicates the thermal noise sampled by a capacitor.
LSB	Least significant bit.
MOS-FET	Field-effect transistor in metal-oxyde semiconductor (MOS).
OTA	Operational transconductance amplifier.
RMS	Root mean square.
RSD	Redundant signed digit.
SNR	Signal to noise ratio.

References

[1] D. Renshaw, P. B. Denyer, G. Wang, and M. Y. Lu, "ASIC image sensors," in *Proc. IEEE ISCAS'90*, (New Orleans, USA), pp. 3038–3041, May 1990.

[2] O. Yadid-Pecht, R. Ginosar, and Y. S. Diamand, "A random access photodiode array for intelligent image capture," *IEEE Trans. on Electron Devices*, vol. 38, pp. 1772–1780, 1991.

[3] E. R. Fossum, "Active pixel sensors - are CCDs dinosaurs?," in *CCDs and Optical Sensors III*, vol. 1900, pp. 2–14, 1993.

[4] M. Ansorge, F. Pellandini, S. Tanner, J. Bracamonte, P. Stadelmann, J.-L. Nagel, P. Seitz, N. Blanc, and C. Piguet, "Very low-power image acquisiton and processing for mobile communication devices," in *Proc. IEEE Int'l. Symp. on Signals, Circuits and Systems*, (Iasi, Romania), pp. 289–296, July 10-11, 2001.

[5] S. Mendis, S. E. Kemeny, and E. R. Fossum, "CMOS active pixel image sensor," *IEEE Trans. on Electron Devices*, vol. 41, pp. 452–453, March 1994.

[6] O. Vietze and P. Seitz, "Image sensing with programmable offset pixels for increased dynamic range of more than 150 dB," in *Proc. SPIE, Solid-State Sensor Arrays and CCD Cameras*, vol. 2654, (San Jose, USA), pp. 93–98, 31 Jan. - 2 Feb., 1996.

[7] C. H. Aw and B. A. Wooley, "A 128 x 128 pixel standard CMOS image sensor with electronic shutter," *IEEE JSSC*, vol. 31, pp. 1922–1930, December 1996.

[8] P. Seitz, *Handbook of Computer Visions and Applications*, ch. Solid state image sensing, pp. 165–222. Academic Press, 2000.

[9] B. Pain, G. Yang, and C. Ho, "A single-chip programmable digital CMOS imager with enhanced low-light detection capability," in *13th Int. Conf. on VLSI Design*, (Calcutta, India), January 4-7, 2000.

[10] S. Tanner, *Low-Power Architectures for Single-Chip Digital Image Sensors*. Ph. d. thesis, University of Neuchâtel, Switzerland, November 2000.

[11] G. Wang, D. Renshaw, P. B. Denyer, and M. Lu, "CMOS video cameras," in *Proc. IEEE EURO ASIC'91 Conference*, (Paris, France), pp. 100–103, May 1991.

[12] M. Willemin, N. Blanc, G. K. Lang, S. Lauxtermann, P. Schwider, P. Seitz, and M. Waeny, "Optical characterization methods for solid-state image sensors," *Optics and Lasers in Engineering*, vol. 36, pp. 185–194, 2001.

[13] A. J. Theuwissen, *Solid-State Imaging with Charge-Coupled Devices*. Kluwer Academic Publisher, 1995.

[14] G. C. Holst, *CCD Arrays cameras and displays*. JCD Publ., 1996.

[15] L. Grisoni, A. Heubi, P. Balsiger, and F. Pellandini, "Micro-power 14-bit ADC: 45 μW at $\pm$1.3 V and 16 kSample/s," in *Proc. of ISIC'97*, (Singapore), pp. 39–42, Sept. 10-12, 1997.

[16] H. S. Lee, "CMOS operational amplifiers for low-power, low-voltage circuits," in *Workshop on Low-Power/Low-Voltage IC Design*, (EPFL, Lausanne, Switzerland), June 27 - July 1, 1994.

[17] T. B. Cho and P. R. Gray, "A 10-bit, 20-MS/s, 35-mW pipelined A/D converter," *IEEE JSSC*, vol. 30, pp. 166–172, March 1995.

[18] P. C. Yu and H. S. Lee, "A high-swing 2-V CMOS operational amplifier with replica-amp gain enhancement," *IEEE JSSC*, vol. 28, pp. 1265–1272, December 1993.

[19] E. Sackinger and W. Guggenbuhl, "A high-swing, high-impedance MOS cascode circuit," *IEEE JSSC*, vol. 25, pp. 289–298, February 1990.

[20] K. Bult and G. J. Geelen, "A fast-settling CMOS op-amp for S-C circuits with 90-dB DC gain," *IEEE JSSC*, vol. 25, pp. 1379–1384, December 1990.

[21] E. Vittoz, "Limits to low-power, low-voltage analog circuit design," in *Workshop on Low-Power/Low-Voltage IC Design*, (EPFL, Lausanne, Switzerland), June 27 - July 1, 1994.

[22] V. Peluso, M. S. J. Steyaert, and W. Sansen, "A 900-mV low-power DS A/D converter with 77-dB dynamic range," *IEEE JSSC*, vol. 33, pp. 1887–1897, December 1998.

[23] A. P. Chandrakasan, S. Sheng, and R. W. Brodersen, "Low-power CMOS digital design," *IEEE JSSC*, vol. 27, pp. 473–484, April 1992.

[24] J. Rabaey, "Low-power design at architectural and system level," in *Workshop on Low-Power/Low-Voltage IC Design*, (EPFL, Lausanne, Switzerland), June 27 - July 1, 1994.

[25] P. Vorenkamp and R. Roovers, "A 12-b, 60-MS/s cascaded folding and interpolating ADC," *IEEE JSSC*, vol. 32, pp. 1876–1886, December 1997.

[26] V. Valencic, "Micropower CMOS data converters," in *Workshop on Integrated A/D & D/A Converters*, (EPFL, Lausanne, Switzerland), July 5-9, 1993.

[27] M. Yotsuyanagi, H. Hasegawa, and K. Sone, "A 2 V, 10 b, 20 MSamples/s, mixed-mode subranging CMOS A/D converter," *IEEE JSSC*, vol. 30, pp. 1533–1537, December 1995.

[28] B. Ginetti, A. Vandemeulebroecke, and P. Jespers, "RSD cyclic analog-to-digital converter," in *Symp. VLSI Circuits Dig. Tech. Papers*, pp. 125–126, 1988.

[29] B. Ginetti, P. Jespers, and A. Vandemeulebroecke, "A CMOS 13-bit cyclic A/D converter," *IEEE JSSC*, vol. 27, pp. 957–965, July 1992.

[30] A. Heubi, P. Balsiger, and F. Pellandini, "Micro power 13-bit cyclic RSD A/D converter," in *Proc. ISLPED'96*, (Monterey CA, USA), pp. 253–257, Aug. 12-14, 1996.

[31] K. Nagaraj, "A 250-mW, 8-b, 52-Msamples/s parallel-pipelined A/D converter with reduced number of amplifiers," *IEEE JSSC*, vol. 32, pp. 312–320, March 1997.

[32] L. Grisoni, *Low-Power Floating Point A/D Converters for Audio Signals*. Ph. d. thesis, University of Neuchâtel, Switzerland, 1998.

[33] W. C. Black and D. A. Hodges, "Time interleaved converter arrays," *IEEE JSSC*, vol. 15, pp. 1022–1029, December 1980.

[34] N. Blanc, S. Lauxtermann, P. Seitz, S. Tanner, A. Heubi, M. Ansorge, F. Pellandini, and E. Doering, "Digital low-power camera on a chip," in *Proc. of Int'l. COST 254 Workshop on Intelligent Communication Technologies and Applications with Emphasis on Mobile Communications*, (Neuchâtel, Switzerland), pp. 80–83, May 5-7, 1999.

[35] S. Tanner, A. Heubi, M. Ansorge, and F. Pellandini, "A 10-bit low-power dual-channel ADC with active element sharing," in *Proc. ISIC99*, (Singapore), pp. 529–532, September 8-10, 1999.

[36] S. Tanner, S. Lauxtermann, M. Waeny, M. Willemin, N. Blanc, J. Grupp, R. Dinger, E. Doering, M. Ansorge, P. Seitz, and F. Pellandini, "Low-power digital image sensor for still picture image acquisition," in *Proc. of*

SPIE Electronic Imaging 2001, vol. 4306, (San Jose, USA), pp. 358–365, January 21-26, 2001.

[37] K. B. Cho, A. Krymski, and E. R. Fossum, "A 1.2V micropower CMOS active pixel image sensor for portable applications," in *IEEE ISSCC*, (San Francisco, USA), February 7-10, 2000.

VI

DISTRIBUTED MEDIA

Chapter 12

RETROSPECTIVE OF THE DISTRIBUTED MEDIA SERVER TECHNOLOGY

Karolj Skala, Zvonimir Zelenika, Ivan Nikolić, and Tibor Skala

Rudjer Boskovic Institute
Bijenicka c. 54
10000 Zagreb, Croatia
skala@irb.hr

Abstract   The last five years the Web as a media server environment is moving quickly beyond static text and images, and even beyond mono-streaming and simple "push" as new generations of media are emerging and delivery the teleimmersion on the grid distributed systems. The media servers use the term webcasting, which includes intelligent push services and multi-streaming audio and video.

This article introduces the early generation of Virtual Reality Webcasting (VRW) by defining new media models while also identifying the new media architecture being developed to support these new, more complex forms of on-line multimedia and the new concept of teleimmersion technologies in the grid systems.

This retrospective article, describes and presents ideas in Virtual Reality Modeling Language (VRML) technology that are mostly focused on integrating the VRML with other technologies applied in Internet services (e.g. data streaming and push techniques, multicasting and multimedia). The Virtual Reality Webcasting technology is announced as a new web communication paradigm. Our example is a project involving a Virtual Reality Webcasting of the University Sailing Regatta, the first attempt at a real-time VR webcast of such an event in 1997. Furthermore we explore the great development of streaming multimedia and its implication for future grid systems, as the ultimate synthesis of media and networking technologies.

Keywords: Virtual reality, webcasting.

1.   Initial Scope

The project was intended to cover a sporting event in real time using the Internet global computer network as the principal medium. Such transmission

J.F. Tasič et al. (eds.), Intelligent Integrated Media Communication Techniques, 405-418.
© 2003 *Kluwer Academic Publishers.*

would allow hypermedia information servicing. This concept includes the option of handling several parallel video images and a three-dimensional model of the event site refreshed in real time, indicating the positions of the objects (competitors). This included 3D defined pictures and texts presented in diagrams, tables etc.

The goal was to accomplish the first attempt to webcast an event in Virtual Reality Webcasting (VRW). Our approach made the artificial world very realistic (event presentation) and globally interactive [1]. The advantage of such an approach over a videoconference or real-time audio-based webcast is a narrower bandwidth. Since we covered a rather large area, a 20 nautic miles (NM) event field, it would have been impossible to cover the entire event and area with digital cameras and then stream such data to the network. Because the only working VRML multi-user server system was inaccessible at the time, we had to make certain compromises in real-time world refreshing and data streaming. We call our solution "real-time virtual reality webcasting" to be discussed in detail.

2.   Introduction

In the early phase of video WAN servers, there were only a few high-level webcast video players available. Before we look into details of Virtual Reality Webcasting, we will look at some of the applications where video servers were in place and operations as of the middle of 1996. We will look at 1996 because it was about this time when the first-generation broadcast class of video servers, shown for the first time around 1994. It is also about this time that broadband delivery was declared almost dead and that the real market was shifting away from exclusively unique system configuration.

The larger projects undertaken in 1996 was Sweden's TV4. The video system was built with compressed digital servers employing motion-JPEG on Tektronix Profile platforms. A digital network employing a wide range of digital video production, distribution, and routing equipment was also part of this highly sophisticated system, and the enhanced MPEG 2 4:2:2P@ML equipment became available. The MPEG 2 servers have the potential to grow into other broadcast applications such as time-delay for programming, news story playback, and other general station operations. The new networked architectures and web programming created as the technology and media applications improve. Entire systems built around the local and the wide area network concepts. The development of those systems on other than local networks is expected to be a long time away. The VRML was conceived at the first annual WWW Conference held in Geneva, Switzerland during the spring of 1994 at the initiative of Tim Berners Lee, the creator of WWW and HTML. The idea was to create a standard procedure for describing virtual 3D environments that would be network and communication oriented. By May of 1995, the VRML 1.0 standard was created and made public by the VRML Authoring Group (VAG). The first standard was intended to be temporary and introductory since it defined a procedure for generating and describing static 3D environments of limited multimedia content [2], with rather simplified and limited network and interactive capabilities.

The VRML 2.0 [3], released in 1996 (based on the Silicon Graphics Movie Worlds Standard), greatly improved the multimedia nature of VRML worlds (embedded video or space-positioned and oriented sound), interactivity (proximity and other sensoring tools), movement (using position interpolator tools, that are a very neat way of avoiding scripting (VRML script or Java script) for such low-level functions). It also brought us the open approach: VRML 2.0 worlds can be closely integrated, controlled by or control other events trough Java and VRML scripting and Java language applications.

What it lacked are multi-user capabilities, that have been sought since the very conception of the idea of network-oriented Virtual Reality. However, due to the open architecture of the standard, using some extensions to the original VRML 2.0, some companies included in VRML standardizing and defining have created their own concepts for multi-user interaction within the same 3D environment. Among the first works in this area are the VirtualLife Network by MiraLab (Geneva, Switzerland), which was extremely high-end oriented, and some projects (NFS Net) by the US army [4] (that will not be discussed since they do not use the standard or extended VRML as their base).

3. Webcasting Technology

Audio and video file transfers have always been a time-consuming process due to their large size. Advances in digital compression, fueled by HDTV and multichannel broadcast, satellite and cable research, Virtual Reality technologies, and CD-ROM multimedia efforts, are now finding their way into the much more band-restrictive Internet environment. Ultra-high-speed estimation of motion elements in the video frame and the replacement of this information with much shorter motion-prediction codes have allowed for the massive reduction of redundant information. Sending only such codes to downstream processors capable of reconstructing the original signal has reduced the size of a transmitted video file more than a thousand-fold. These smaller files are sent, not as entire packages that have to be decoded only after all the data is received, but as a "stream" of information that can be displayed as if it were being read off a virtual hard disc.

Four technologies underlie the rapid growth of webcasting at that time:

- the development of Internet protocols (such as IP Multicast and UDP) that facilitate faster delivery of time-sensitive packets, and the diffusion of routers and the expansion of network backbone capacity that allow those protocols to operate;

- continuing advances in compression technology (developed for MPEG-2) which allow for a radical reduction in the quantity of information;

- a new class of multimedia PCs – "receiver" devices with enough processing power and memory to handle audio and video decompression "on the fly";

- high-speed modems capable of squeezing more data over ordinary phone circuits.

The constant growth of the World Wide Web continues to bring new challenges to public broadcasters. Just as stations are beginning to recognize that cyberspace is both a competitor for leisure time and an environment for building new "customer" loyalty, a new web phenomenon comes along that strikes at the heart of the radio and television medium as an interactive audio, video and VR webcasting. Faster modems, advances in compression technology and the proliferation of multimedia PCs have brought the next wave of challenges: Internet radio and net-TV. Public broadcasters are inherently in a good position to take advantage of these technologies. While real-time "intercasting" one's broadcast signal over the net offers a tempting opportunity to globally expand one's service territory, net congestion and economic factors will limit what today appears to be a free ride. In our opinion, the real opportunity presented by webcast technology is the ability to develop multiple on-demand program services based upon customer interest and customer need. Today, much of the Internet looks like the early days of radio: entry costs are low (network capacity appears to be free), transmission costs are hidden from both the originator and end-user (as flat-rate billing distorts the costs of sending live audio and video), regulation is nonexistent, receiving devices are becoming more affordable and more sophisticated, and, as in the 1920s, the net is being flooded with new programming.

In fact, the word "webcasting" itself already has several meanings. It is being used to describe push "broadcasting" and also to describe video and audio streaming that delivers clips and events on-demand. The essential point here with all due respect to WebTV and other new, low-cost client technologies, is that the Internet is NOT television, nor should it aspire to become TV. We are looking for new concepts and models of Internet webcasting. The Virtual Reality Webcasting (VRW) model is being announced as a new web communication paradigm.

4. Multi-User Technology

For streaming to take place, there must be at least a real time relationship between the delivery rate of the signal form and the presentation of the signal form. The characteristics of streaming include some specific criteria. One characteristic is the pushing of data across large area networks. At the early time of development (about the year 1997), the only partially public multi-user technology was the Sony Pictures Image Works (SPIW) Virtual Society (VS) Community Place System [5], [6]. It is based on the extended VRML 2.0 standard and it has several worlds already created that can be accessed by users from anywhere in the Internet. The problem is that the server system and browser for Community Place have not been made public (throughout the development of the VRML, many companies have placed parts of their work in the "public domain", thus encouraging further development in the area). This currently slows down the development and creation of multi-user worlds (Although SPIW has released Community Place Conductor VRML Authoring Software). At present, the only purpose for which Sony is using and developing Community Place is advertising this new technology as a sort of IRC (Internet Relay Chat) system for chatting and virtual meetings, where it has some advantages over the usual video-conferencing since it requires a narrow bandwidth (a 28.8 kbps modem is enough if an audio channel is not used). In a presentation at the SFeracon conference [6], this technology was demonstrated with real-time VR conferencing (chatting).

The SPIW server system supports audio streaming (so that users, represented by their "avatar", can also talk to each other), Java scripting and movement interpolators (that give these worlds a lot of embedded movement and dynamics).

Our general idea was to create a multi-user world where users would be actually computer-controlled by a program that would translate raw data received from an observing or positioning system (radar, laser rangefinding, differential Global Positioning System) to the real-time movements of the users that would be "avatarized" by ship models. Event position sensing is forming the cinematic data base and, with the object database, integrating the VR frame on the GIS background, Figure 12.1. Of course, we would have certain latency in data processing. After receiving the new position of an object (every few seconds), we would then extrapolate a path between the old and new positions.

5. The Real-Time VR Webcasting System

Virtual reality content streaming, including connections and links, will function differently from file transfers. In a stream, there is usually no return path to request a retransmission of a missed signal or data packet, and the receiver must therefore make the best of any corrupted data. Once a stream has started, receivers may join that stream without having to start at the beginning. The de-

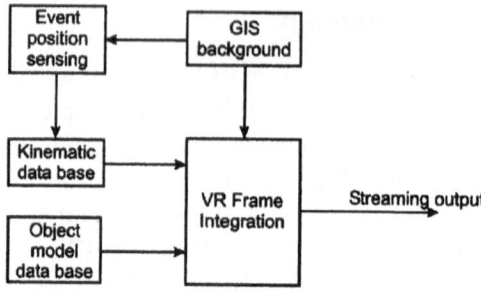

Figure 12.1.   Real-time frame integration.

livery rate is generally related to the presentation rate of the content. Streaming at other than the presentation rate is allowed. With real-time data streaming, we created a series of VRML models that differ only in the positions of the sailing boats processed from the raw data supplied to us by radar and laser rangefinding system on the vessels, Figure 12.2. These models were created with a constant time difference of several minutes. Browser programs reloaded the scene (that was static) every time it changed (after a pre-defined time interval). This was achieved by the following HTML code:

```
<META HTTP-EQUIV="Refresh" Content="300;URL=filename.html">
```

This code is resulting in the browser automatically reloading the HTML file every 5 minutes (300 seconds).

The interface that processed the raw data to an exact boat position in the VRML world was a rather simple console program for triangulation and simple conversion to the Cartesian square coordinate system used in the VRML. We obtained data refreshing in several minutes, since every other VRML file was created by extrapolation from the previous two positions. More frequent extrapolation would be counter-productive since we always lose some 20-30 seconds for reloading and initial rendering and we wanted to give users enough time to navigate a bit around the regatta field. The exact times between "reissuing" were empirically tested to determine the optimum for pleasant navigating and experiencing the scene.

6.   Modeling the Field

The model of the regatta field is based on maps and models created by the VRML Croatian Project of the Image Processing Group [7], Figure 12.3. Boat models are totally independent of the regatta field (and are actually made using a completely different modeling tool and even another operating system (Mac OS)), to be later incorporated in a single VRML file using the WWWInline VRML node. Again, the biggest role in differentiating each boat from the

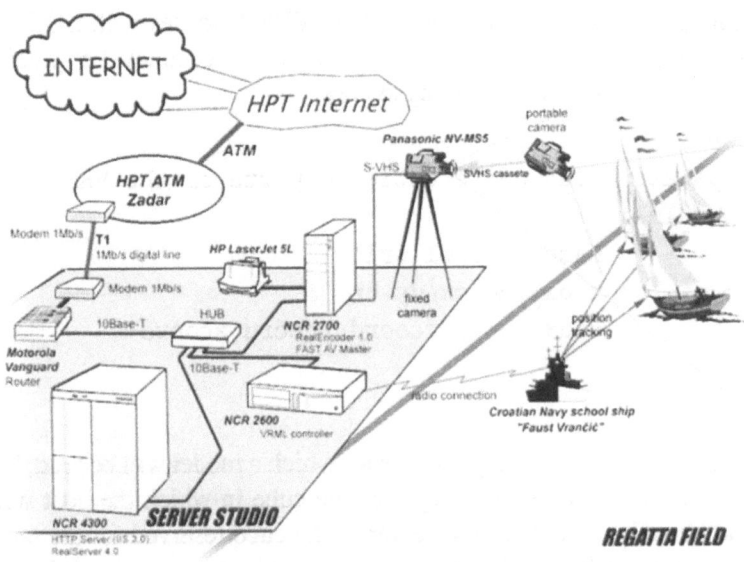

Figure 12.2. The webcasting infrastructure.

others is played by the lines on texture maps (university and college logos) included in the model.

Figure 12.3. VRML model of the regatta field.

WWWInline is a method by which we can put one VRML model in another (putting a ship on the regatta field). This is extremely useful in multi-user worlds where every user can create his/her own "avatar" (graphical representation) and then just report the URL of that model to the server, which will then handle it. Also, in-line processing is useful for creating VRML scenes with LOD (level of detail) nodes, where we define a different model for each level of precision. As a VRML node, it has the following syntax:

```
Separator {
    # separator node differentiates parts of the VRML source
        WWWInline {
                name "ship01.wrl"
                bboxSize dx dy dz
                bboxCenter xcoord ycoord zcoord
                }
        }
```

The name defines the arbitrary URL from which a model will be read, bbox-Size defines the width x height x depth of the cube in which the boat will be placed, and bboxCenter defines the center of the cube reserved for the model.

In-line processing has a big role in the currently proposed and discussed system for creating so-called "seamless worlds" that would actually be a single data space. The user would traverse through the scene and actually move between different worlds without noticing it [8], [9]. As it looks today, seamless worlds may not be included in the next VRML standard since they represent a complete and radical change. Up to now, worlds have always been constant and defined in size. The introduction of this new concept would change client-server relationships in the VRML considerably, that currently function well using the ordinary HTTP protocol. With the introduction of multi-user capabilities, new protocols were introduced (since for real-time communication we cannot use stateless protocols such as the HTTP). Moreover, seamless worlds will create the need for VRML servers and a server system, since up to now the VRML has been based on and used ordinary HTTP servers or a slightly enhanced HTTP server system.

7.   Hardware Configuration and Software Used

Hardware configuration:

```
Server NCR 4300
            (real-time video digitalization and management)
        4 x PentiumPro 200 MHz / 512 KB cache
        RAM: 512 MB
        HDD: 2 x 4 GB (SCSI-3)
        Fast Ethernet 100 Mbit
        Video capture adapter SVHS _ MPEG
```

```
(PCI Bus Master For Microsoft Windows NT)
```

```
Workstation NCR 2700
        (WWW server)
        Pentium 200 MHz / 512 KB cache
        RAM: 64 MB
        HDD: 2 GB (SCSI-2)
        Fast Ethernet 100 Mbit
```

```
2 x Workstation NCR 2600
        (Net management and administrative tasks)
        Pentium 166 MHz / 256 KB cache
        RAM: 16 MB
        HDD: 1 GB (EIDE)
        Ethernet 10 Mbit
```

```
Fast Ethernet Switch
```

Software used for webcasting:

```
        Windows NT Server 4.0
        Windows 95
        Adobe Premier 4.5
        Adobe Photoshop 4.0
        Digitalization software for Microsoft Windows NT
            (included with video capture adapter)
        MS NetShow
        MS Index Server 1.1
        MS FrontPage97
        MS J++ 1.1
        Liquid Reality
            (for developing dynamic Java controlled VRML worlds)
```

8. Hardware and Software Requirements on Client Side

The hardware platform shall at least rely on a Pentium III processor and a RAM memory size of 16 MB (tested), running with the operating system Microsoft Windows 95.

One of our primary goals was to make this VR Webcast as accessible as possible, which led us to orient ourselves to VRML 1.0. As a VRML browser, our recommendation is Live3D by the Netscape Corporation because it is a standard plug-in in all the versions of Navigator and Communicator since the Netscape Atlas PR2. Although newer versions of Live3D do support VRML 2.0, some users still have Netscape 3.0 (or 3.01) that comes with the Live3D version

1.0 (only VRML 1.0 support). For a UNIX workstation, we also recommend Live3D or TGS WebSpace for Solaris, HP-UX, AIX and SGI CosmoPlayer for Irix (SGI).

9.   Network Background

One of the most significant elements dictating the range and quality of information provided is the quality of the information flow through the network. The Internet infrastructure will not be discussed in this study because it depends on the network load and the quality of the infrastructure at all the nodes from the server to the client, i.e. spatially and temporally indeterminate, while the server is connected to the Internet via the most rapid available hook-up (100 Mbps). The average user is connected to the network by an analogue telephone line at a speed of 28,800 bits/sec with compression (over 90 % of the information is already compressed). For this type of sporting competition (regatta), we can reduce the time resolution to 5 minutes, with the transmission still remaining within the domain of real time.

The necessary target client bandwidth is M= 28,000/8* 1.1/1.024= 3.8672 KB/s, resulting in 1.13 MB for 5 min. under ideal transmission conditions.

The average speed of personal computers in that time and the quality of video presentation meet the requirements of the main video stream, with a resolution of 240 x 180 pixels, in a 16-bit palette, with a speed of 10 pictures per second and a duration of 30 seconds per 5 minutes, Table 12.1.

Table 12.1.   Video sizes for MPEG, 240 × 180 / 16 bit, audio 11kHz/16bit/mono.

Frames per second(*fps*)	1 s	30 s	1 min
5	14.65 KB	439.45 KB	878.91 KB
10	29.30 KB	878.91 KB	1,757.81 KB
15	43.95 KB	1,318.36 KB	2,636.72 KB
25	73.24 KB	2,197.27 KB	4,394.53 KB
30	87.89 KB	2,636.72 KB	5,273.44 KB

10.   Advantages of Virtual Reality Webcasting

Advantages of VRW over the usual real-audio and real-video webcasting:

- Much lower requirements for bandwidth (e.g. the University Regatta was webcast from the scene with the usual 33.6 kbps modem over a voice-grade line). Thus, it is easier to set up equipment; there is no need for high-speed networking and high speed Internet access. Also, on the user side, a 33.6 kbps modem is enough since VRML files are transported over the network in zipped state (size reduced 5-10 times - to some 50 kB) of original file (500-800kB), and texture images need to be downloaded only once since in later iterations they are called from the cache of the local computer.

- Lower hardware requirements. Our recommendation is, of course "as fast as possible", but we think that ordinary 486/66 would suffice (with Windows 95, since a similar configuration was used for the VRML presentation at the "Science in Croatia" Exhibit). Real-audio and especially real-video decoding are rather difficult processes for normal computers without hardware acceleration of such functions [with DSP (Digital Signal Processing) based accelerators].

- It is not necessary to be extremely near to the event. By using a forward deployed observer or observing system, the user can actually be on the other side of the globe while translating raw data from the observing system (laser range finding, radar, GPS) to the VRML environment. In the case of the University Regatta, the observer was onboard the ship in the middle of the regatta field, and data translation and model integration were performed on the shore.

- This system can provide a basis for position control and navigation systems which would allow the observer to enter the scene, providing a much better feeling for relations among the objects around him/her and a better sense of space, enabling him/her to react faster to new situations.

11. Shortcomings of the Virtual Reality Webcast

- There is presently no existing software available for creating multi-user real-time modified environments (a problem with the Sony SPIW Community Place Server System). Therefore, it is necessary to develop a quasi-real time, pseudo-real time or completely new real-time system from scratch (defining and applying new nonstateless protocols, extending the VRML standard and thus loosing the user base).

- VRML 2.0 browsers are still rather rare (since no new version of the Netscape Navigator has been recently shipped). Since Live3D in the version which supports VRML 1.0 was shipped with the Navigator 3.0 and 3.01 as a standard plug-in, this browser became the most common and de facto browser standard for the VRML. Thus, all our efforts were limited under the constraint that everything should be done using VRML 1.0 only.

- Of course, the quality of the image and video for videoconferencing is much better and more recognizable than the VRML model, which is always extremely limited by numerous factors (VRML is still not multimedia), and VRML is, of course, useless for webcasting lecture or speech.

12.   Epilogue about VRW

The project was intended to explore new simplified matched technologies for webcasting using VRML. Through the experiment, we wanted to gain certain experience and collect ideas about VR webcasting. We explored real time VR Webcasting possibilities. It is a globalized virtual reality-based extension in the real time of the real word (process) [10]. We proved the possibility to create a fully real-time VR webcasting system, to be based on the work by Virtual Reality Application and Development [11] and the VRML Project of Image Processing Group [7] that incorporate of a completely new browser, modification and extensive use of a multi-user system. Also, the process of data collection must be based on Global Positioning Systems on moving objects. Such a configuration of VR webcasting will represent a new Internet service.

13.   New Distributed Paradigm

Today, we have gone beyond simply predicting a convergence of television, computer and wide area networks. For video servers purists, big challenges lie ahead. We expect that the video and media servers will find their way into every genre of media delivery. Although the term media may have a completely different connotation, the delivery of moving media will anchor the concept of the video servers for years to come. The acceptance of compressed digital video, interactive multimedia systems and digital television are all contributors to a universal shift toward media and data servers in the new distributed environment. All earlier projects including the one described above accelerate the nascent stage of grid as a distributed emerging infrastructure with widespread use in multimedia and virtual reality technology. The grid is a new generation of information technologies that come from the web and enable us to compute and communicate in online spaces using cluster technology [12]. Clustering allows for common data sharing among various nodes. The nodes, in this application, can include both client workstations or media servers. Recalling that a node can be connected through a number of paths, the clients can access data through a number of different connected node paths. Clustering represents a system that provides high availability, and has real potential for media server architectures in grid environment. This approach is focusing the distributed media systems as the new teleimmersion application possibilities.

The distributed network oriented media technologies predict that applications will be driven by convergence of media. The advanced capabilities of the future Internet will give rise to true distributed applications, whose fundamental nature is the result of a congregation of resources. The Web is a database version of such an application: the Web is not located in any one place. In the future, application will be as hard to locate. In this way, the grid extends the notion of the Internet and supports advanced applications as teleimmersion. The systems accessed by teleimmersion include potentially large and distributed databases and simulations, all under some sort of real-time control. Increased access to high-performance networks and improved access to high-end virtual reality environments will cause advanced integral media communications, teleimmersion applications to come to the fore. Because grid technologies will be driven by teleimmersive applications, they will have to be closely coupled with, and designed a priori for, the types of interactive media applications that people prefer. This requirement distinguishes grids from the existing Internet and means, moreover, that grid services will need to track the latest developments in human-to-human integral media communications. For example, grids as a distributed media communication environment will have to support not only large number of video/voice grade connections, but also multichannel, high-resolution spatialized audio and video streams [12].

14.   Conclusion

The widely used applications come across repeatedly changing our life. We could not predict the details of the future development, but we are sure that the distributed computing in the new media communication is a pragmatic key technology. The very extensive development of computing and networking in global level has grown up to new distributed media communication. Media contents transformation and communication will be another challenge driven by teleimmersion application. Grid will need to support the delivery of HDTV streams to large number of users, ultra-high-resolution interactive teleimmersion environments, and high-resolution (in both space and time) media contents [13].

References

[1] J. Ramesh, "Merging virtual and real," *IEEE Transaction on Multimedia*, vol. 3, no. 4, pp. 254–258, 1996.

[2] D. Gracanin, "VRML 2.0," in *Broadband and Multimedia Workshop*, (Zagreb, Croatia), Nov. 11-12, 1996.

[3] L. Leemay, *3D graphics and VRML 2.0*. Indianapolis, USA: Sams Net Publishing, 1996.

[4] S. P. Mudur, D. S. Dixit, and S. H. Shanbag, "Delegation framework for walkthrough computations," in *Proceedings of the International Conference on Virtual Systems and Multimedia*, (Geneva, Switzerland), pp. 128–132, Sept. 10-12, 1997.

[5] "VRML at SGI." http://vrml.sgi.com/, April 1997.

[6] M. Juric, Z. Regvart, N. Biruski, and Z. Zelenika, "Chiba city blues," in *Proceedings of the SFeracon 97*, (Zagreb, Croatia), April 28, 1997.

[7] FER, "VRML project of image processing group." http://vrml.zesoi.fer.hr/, June 1997.

[8] K. Skala, "Evaluation of virtual board as a hypermedia visual interface," in *Proceedings of the Human Vision and Electronic Imaging*, (San Jose, California, USA), pp. 238–242, Jan. 29 to Feb. 1st, 1996.

[9] K. Skala, "Virtual world webcasting service," in *Proceedings on International Workshop on Intelligent Communications and Multimedia Terminals, COST #254*, (Ljubljana, Slovenia), pp. 186–190, 1998.

[10] K. Skala, "Real time virtual reality webcasting of sailing regatta," in *Internet Proceedings of Cost #254 Action Workshop*, (Budapest, Hungary), 1997.

[11] "Real-time virtual reality webcasting." http://www.regatta.sail.hr/en/97/project_en.htm, Aug. 1998.

[12] K. Skala, "GRID for scientific and economic development of Croatia," in *CARNet Users Conference*, (Zagreb, Croatia), Sep 25-27, 2002.

[13] L. Foster and C. Kesselman, *The Grid: Blueprint of a New Computing Infrastructure*. San Francisco, CA, USA: Morgan Kaufmann Publisher Inc., 1999.